Lecture Notes in Statistics

Edited by D. Brillinger, S. Fienberg, J. Gani, J. Hartigan, and K. Krickeberg

35

Linear Statistical Inference

Proceedings of the International Conference held at Poznań, Poland, June 4–8, 1984

Edited by T. Caliński and W. Klonecki

Springer-Verlag
Berlin Heidelberg New York Tokyo

Editors

T. Caliński
W. Klonecki
Mathematical Institute of the Polish Academy of Sciences
ul. Kopernika 18, 51-617 Wroclaw, Poland

AMS Subject Classification: 62-06, 62FXX, 62GXX, 62MXX

ISBN 3-540-96255-7 Springer-Verlag Berlin Heidelberg New York Tokyo
ISBN 0-387-96255-7 Springer-Verlag New York Heidelberg Berlin Tokyo

Library of Congress Cataloging-in-Publication Data. Main entry under title: Linear statistical inference. (Lecture notes in statistics; 35) 1. Linear models (Statistics)–Congresses. I. Caliński, T. II. Klonecki, W. (Witold) III. Series: Lecture notes in statistics (Springer-Verlag); QA276.L5484 1985 519.5 85-25096

Printed in Germany

Printing and binding: Beltz Offsetdruck, Hemsbach/Bergstr.
2147/3140-543210

FOREWORD

An International Statistical Conference on Linear Inference was held in Poznań, Poland, on June 4-8, 1984. The conference was organized under the auspices of the Polish Section of the Bernoulli Society, the Committee of Mathematical Sciences and the Mathematical Institute of the Polish Academy of Sciences.

The purpose of the meeting was to bring together scientists from various countries working in the diverse areas of statistical sciences but showing great interest in the advances of research on linear inference taken in its broad sense. Thus, the conference programme included sessions on Gauss-Markov models, robustness, variance components, experimental design, multiple comparisons, multivariate models, computational aspects and on some special topics. 38 papers were read within the various sessions and 5 were presented as posters. At the end of the conference a lively general discussion session was held.

The conference gathered more than ninety participants from 16 countries, representing both parts of Europe, North America and Asia. Judging from opinions expressed by many participants, the conference was quite successful, well contributing to the dissemination of knowledge and the stimulation of research in different areas linked with statistical linear inference. If the conference was really a success, it was due to all its participants who in various ways were devoting their time and efforts to make the conference fruitful and enjoyable. First who deserve thanks are all the speakers and authors, the chairmen of sessions, and the discussants. They made a good job, highly appreciated by the majority of the conference attendants. A smooth run of the conference is to be attributed to the efforts of the local organizing committee skillfully headed by Dr. R. Kala from the Department of Mathematical and Statistical Methods of the Poznań Academy of Agriculture and Dr. R. Zmyślony from the Wrocław branch of the Mathematical Institute of the Polish Academy of Sciences.

This volume contains about a half out of the number of 43 papers presented at the conference, and represents its main features and results. We would like to express thanks to all authors who decided to undertake the task of preparing their papers for publishing in the proceedings.

The standard of the papers owes very much to the work of a number of referees (a list of their names being included at the end of the volume) to whom not only the editors, we think, but also the authors are very greatful. We apologize, if some of the authors of submitted papers have possibly found the referee demands too restrictive.

In the final editorial work we were very much helped by many of our co-workers in Wrocław and in Poznań, and also by the administrative staff of the Mathematical Institute of the Polish Academy of Sciences in Warsaw. It is impossible to mention the names of all of them, to whom we owe so much, but we would like to thank in particular Dr. S. Zontek. Also we would like to thank Ms. A. Goździk and Ms. T. Rejniewicz for their excellent typing work. Last, but not least, let us express sincere thanks to the Publisher for their efficient and friendly co-operation.

T. Caliński

W. Klonecki

CONTENTS

SOME GEOMETRIC TOOLS FOR THE GAUSSIAN LINEAR MODEL WITH APPLICATIONS TO THE ANALYSIS OF RESIDUALS

H.Caussinus and J.Vaillant
Laboratoire de Statistique et Probabilités
Université Paul Sabatier
Toulouse, France

Summary. We consider the gaussian linear model (1) $Y = X\beta + \varepsilon$ where $\mathscr{L}(Y) = N_n(\mu, \sigma^2 I_n)$, $\sigma > 0$, $\mu \in Q$ (a linear subspace of $\mathbb{R}^n$). This model is invariant under the group of transformations $Y \to aY + \xi$ $(a > 0,\ \xi \in Q)$ and a maximal invariant is the vector T of normed residuals. Thus, if (1) is considered as a null hypothesis to be challenged, the restriction to invariant procedures leads to perform the analysis via T. The matrix approach is not very convenient to deal with T because the mapping $Y \to T$ is not linear. In fact, if model (1) is true, T is uniformly distributed on the unit sphere S of $Q^{\perp}$. Under some alternatives it is easy to compute the density of T with respect to the uniform probability on S, for example when $\mathscr{L}(Y) = N_n(\nu, W)$, $\nu \notin Q$.

The formulation which we advocate leads to straightforward results concerning the optimality of some procedures and enables us to give a clear account of the assumptions which are used. The problem of k outliers detection is more specially discussed, including the case where k is not fixed.

1. Introduction and notation

Throughout this paper $N_E(\nu, W)$ denotes the normal (gaussian) distribution on the Euclidean space E with mean ν and variance operator W. The scalar product on E is denoted by $\langle .,. \rangle$ and the corresponding

KEY WORDS: Gauss-Markov model; coordinate-free approach; analysis of residuals; outliers; Bayes procedures; distributions on spheres.

norm by $\|\cdot\|$. The dual space of E is identified to E by means of the scalar product in the usual way, so that W is a symmetric linear operator from E into E. The distribution $N_E(0,\sigma^2 I_E)$ ($\sigma > 0$, I_E is the identity operator from E into E) is called spherical normal; it is preserved by any orthogonal transformation and the projections onto orthogonal subspaces are independently spherical normal on each of these spaces.

The unit sphere of E is denoted by S_E, the uniform probability on S_E is U_E, the orthogonal projector onto a linear subspace Q is denoted by π_Q and $Q^\perp$ is the subspace of E orthogonal to Q. When the context is clear subspaces as indices may be dropped or replaced by simpler ones: for example S_n instead of $S_{\mathbb{R}^n}$, etc... .

The probability law of a random variable (generally vector) is denoted by $\mathscr{L}(\cdot)$.

In the classical linear model, a sequence of real random variables $Y_1, Y_2, \ldots, Y_n$ is given. These are considered as the coordinates of a vector Y with respect to a canonical basis $e = (e_1, \ldots, e_n)$ of $\mathbb{R}^n$. If the matrix of variances-covariances of the Y_i's is $\sigma^2\Gamma$ (known up to the factor σ^2 and nonsingular) the metric on $\mathbb{R}^n$ is chosen such that the $n \times n$ matrix of the scalar products $\langle e_i, e_j \rangle$, $i,j = 1,\ldots,n$, is Γ^{-1}, which implies that the variance operator of Y is $\sigma^2 I_n$.

Now the gaussian linear model is

$$\mathscr{L}(Y) = N_n(\mu, \sigma^2 I_n), \quad \sigma > 0, \quad \mu \in Q, \tag{1}$$

where Q is a given subspace of $\mathbb{R}^n$, $\dim(Q) = q$, $0 < q < n$.

This model is invariant under the group of transformations $Y \to aY + \xi$ ($a>0$, $\xi \in Q$) and a maximal invariant is the vector of normed residuals

$$T = \frac{1}{\|\pi_{Q^\perp}(Y)\|} \pi_{Q^\perp}(Y).$$

In the analysis of residuals model (1) is considered as a null hypothesis to be challenged. The restriction to invariant procedures leads to performing the analysis through T. In the framework of model

(1) the normed residuals, i.e. the coordinates of T in the basis e, do not have an easily handled distribution. However the distribution of the vector T is extremely simple since it is nothing more than the uniform probability on the unit sphere of $Q^{\perp}$: $\mathscr{L}(T) = U_{Q^{\perp}}$.

Under various alternatives to model (1) it is tempting to express the distribution of T with respect to $U_{Q^{\perp}}$ in a coordinate free approach. This is done in Section 2 for the case where $\mathscr{L}(Y) = N_n(\nu, W)$, $\nu \notin Q$. Section 3 gives applications to residual analysis emphasizing the case of outlier detection. Other applications are indicated in Section 4. Also in this section we discuss briefly a kind of robustness arising from the fact that $\mathscr{L}(T)$ could be $U_{Q^{\perp}}$ without $\mathscr{L}(Y)$ being normal. Finally Section 5 is devoted to the slightly different problem of "modified residuals". This section has been included for historical interest, still stressing the usefulness of the geometrical framework.

For earlier papers using the coordinate - free approach to Gauss-Markov estimation the reader could refer to Kruskal (1960) and Drygas (1970). However the geometrical tools developed herein are somewhat different and turn out to be also closely related to papers on spherical symmetry, e.g. King (1980).

2. Technical results

This section gives some useful probabilistic results. Lemmas 1 and 2 are well known properties of the normal distribution: see e.g. Dempster (1969), Chapter 12. Lemma 3 can be found in Degerine (1979) or Watson (1983), Chapter 2 (2.2.6). Lemma 4 is an obvious general result which leads to a straightforward derivation of Proposition 1.

Lemma 1: If Z is a random vector with values in the Euclidean space E, $\dim(E) = m$, then

$$\mathscr{L}(Z) = N_E(0, I_E) \Leftrightarrow \begin{cases} \text{(i)} & \mathscr{L}\left(\frac{Z}{\|Z\|}\right) = U_E \\ \text{(ii)} & \|Z\| \text{ and } \frac{Z}{\|Z\|} \text{ are independent} \\ \text{(iii)} & \mathscr{L}(\|Z\|^2) = \chi^2_m \text{ .} \end{cases}$$

Remark: Z is spherically distributed if (i) and (ii) hold.

Lemma 2: Let Q be a linear subspace of E. If $\mathcal{L}(Z) = N_E(0,I_n)$, then

(i) $\mathcal{L}(\pi_Q(Z)) = N_E(0,\pi_Q)$ considered as a distribution in E

(ii) $\mathcal{L}(\pi_Q(Z)) = N_Q(0,I_Q)$ considered as a distribution in Q.

Comment: Actually the variance operator is $\pi_Q {}^t\pi_Q$ and ${}^t\pi_Q$ is either π_Q or the canonical injection from Q into E according to whether π_Q is defined as an operator from E into itself or from E into Q. Using (ii) instead of (i) often results in simplifications.

Lemma 3: The moment-generating function of the uniform distribution U_E on S_E, $t \to \varphi_m(t)$, $t \in E$, is an increasing function of $\|t\|$ such that

$$\varphi_m(t) = \int e^{\langle t,x\rangle} dU_E(x) = \Gamma(\tfrac{m}{2}) \sum_{h=o}^{\infty} \frac{(\frac{1}{2}\|t\|)^{2h}}{h!\Gamma(h+\frac{m}{2})}, \qquad m = \dim(E).$$

Lemma 4: Let $(\Omega,\mathcal{A})$ and $(\Omega',\mathcal{A}')$ be two measurable spaces and T a measurable function from the former into the latter. Let P and P_o be two probability laws on $(\Omega,\mathcal{A})$, P_T and P_{oT} the probability laws induced by T on $(\Omega',\mathcal{A}')$. If P admits a density $\frac{dP}{dP_o}$ with respect to P_o, then P_T admits a density w.r.t. P_{oT} such that

$$\frac{dP_T}{dP_{oT}} \circ T = \mathbb{E}^T_{P_o}\left(\frac{dP}{dP_o}\right),$$

where $\mathbb{E}^T_{P_o}$ denotes the conditional expectation given T, the basic distribution being P_o.

Proof: The existence of $\frac{dP_T}{dP_{oT}}$ is obvious and for each $A' \in \mathcal{A}'$

$$P_T(A') = \int_{A'} \frac{dP_T}{dP_{oT}}\, dP_{oT} = P(T^{-1}(A')) = \int_{T^{-1}(A')} \frac{dP}{dP_o}\, dP_o,$$

while on the other hand

$$\int_{A'} \frac{dP_T}{dP_{oT}}\, dP_{oT} = \int_{T^{-1}(A')} \frac{dP_T}{dP_{oT}} \circ T \; dP_o .$$

The result follows from the definition of conditional expectation in $L^1(\Omega, \mathscr{A}, P_o)$.

Proposition 1: Let Z be a random vector with values in the Euclidean space E, $\dim(E) = m$, and $T = \frac{1}{\|Z\|} Z$.

(i) If $\mathscr{L}(Z) = N_E(\delta, V)$, V non singular, the distribution of T has a density $g(S_E \to \mathbb{R}_+)$ w.r.t. U_E given by

$$g(t;\delta,V) = \frac{1}{2^{\frac{m}{2}-1}\,\Gamma(\frac{m}{2})} |V|^{-\frac{1}{2}} e^{-\frac{1}{2}<\delta, V^{-1}\delta>} (<t,V^{-1}t>)^{-\frac{m}{2}} h_m\left(\frac{<t,V^{-1}\delta>}{<t,V^{-1}t>^{1/2}}\right),$$

where

$$h_m(u) = \int_0^\infty e^{uv} e^{-\frac{v^2}{2}} v^{m-1} dv.$$

Special cases of interest are the following.

(ii) If $\mathscr{L}(Z) = N_E(0,V)$,

$$g(t;0,V) = |V|^{-\frac{1}{2}} (<t,V^{-1}t>)^{-\frac{m}{2}}, \quad t \in S_E.$$

(iii) If $\mathscr{L}(Z) = N_E(\delta, \sigma^2 I_E)$, the density g becomes

$$g(t,\delta,\sigma) = \frac{1}{2^{\frac{m}{2}-1}\,\Gamma(\frac{m}{2})} e^{-\frac{\|\delta\|^2}{2\sigma^2}} h_m\left(\frac{<t,\delta>}{\sigma}\right), \quad t \in S_E.$$

Proof: We shall use Lemma 4 with $P_o = N_E(0, I_E)$ and $P = N_E(\delta, V)$. Now

$$\frac{dP}{dP_o}(Z) = |V|^{-\frac{1}{2}} \exp \tfrac{1}{2}(\|Z\|^2 - <Z-\delta, V^{-1}(Z-\delta)>).$$

Hence

$$g(T,\delta,V) = |V|^{-\frac{1}{2}} e^{-\frac{1}{2}<\delta,V^{-1}\delta>} \mathbb{E}^T_{P_o}[\exp \tfrac{1}{2}(\|Z\|^2 - \|Z\|^2 <T,V^{-1}T> + 2\|Z\| <T,V^{-1}\delta>)].$$

Since $P_o = N_E(0, I_E)$, $\|Z\|$ is independent of T and $\mathscr{L}(\|Z\|^2) = \chi^2_m$ by Lemma 1. Moreover, T behaves like a constant when computing $\mathbb{E}^T_{P_o}$. Thus, writing $X = \|Z\|^2$, $b = \langle T, V^{-1}\delta \rangle$, $c = \langle T, V^{-1}T \rangle$, we must compute

$$\mathbb{E}[\exp \tfrac{1}{2}[X(1-c)+2b\sqrt{X}] = \frac{1}{2^{\frac{m}{2}}\Gamma(\frac{m}{2})} \int_0^\infty e^{\frac{x}{2}-\frac{c}{2}x+b\sqrt{x}}\, e^{-\frac{x}{2}}\, x^{\frac{m}{2}-1}\, dx$$

$$= \frac{c^{-\frac{m}{2}}}{2^{\frac{m}{2}-1}\,\Gamma(\frac{m}{2})}\, h_m\left(\frac{b}{\sqrt{c}}\right)$$

from which Proposition 1 (i) follows.

The special case (ii) is obtained for $b = 0$. The integral $h_m(0)$ can be computed in closed form to reach the given explicit expression.

The probability law of T in the special case (iii) is called angular Gaussian distribution in Watson (1983): see Appendix C by Ch.Bingham.

Corollary 1: Let $\mathscr{L}(Y) = N_n(\nu, W)$, $\dim(Q^\perp) = m$, and $T = \frac{1}{\|\pi_{Q^\perp}(Y)\|}\pi_{Q^\perp}(Y)$.

Then the density of T with respect to $U_{Q^\perp}$ is $g(t,\delta,V)$ given in Proposition 1 (i), where $E = Q^\perp$, $\delta = \pi_{Q^\perp}(\nu)$ and $V = \pi_{Q^\perp} W_{|Q^\perp}$ is assumed non-singular. $W_{|Q^\perp}$ denotes the restriction of W to the subspace $Q^\perp$.

Proof: It is enough to notice that

$$\mathscr{L}(\pi_{Q^\perp}(Y)) = N_{Q^\perp}(\pi_{Q^\perp}(\nu), \pi_{Q^\perp} W_{|Q^\perp}).$$

For V to be non singular it suffices that W be non singular.

3. Application to the analysis of residuals

Since Anscombe (1960) and Anscombe and Tukey (1963) many articles have been devoted to the analysis of residuals, and many of them to the detection of outliers in the linear model. However, among these latter papers, few study theoretical optimality properties of the proposed rules of identification (Barnett and Lewis (1978)) and Hawkins (1980) who each include a chapter on outliers in the linear model, assign little or no space to this question). A notable exception is Ferguson (1960) who proved the Bayes optimal character of the procedure based on studentized residuals for detecting one outlier. Abbady (1980) and Butler (1981, 1983) extended this study to the detection of k outliers. Actually Butler dealt with a more general problem of subset selection in normal multivariate regression models.

In this section, we use Proposition 1 to prove the optimality of a procedure in a straightforward and natural way. This procedure is simple when the number of variables which can enter the model is fixed, say k (a special case is the detection of k outliers). The result in this case generalizes Abbady's work (in which the choice of the a priori distribution is more restrictive) and is very close to Butler's. Then we consider a more general case including the detection of at most k outliers.

The Bayesian framework is used throughout to prove an optimal character of a procedure, but we try also to discuss the realism of the assumed prior distributions.

3.1 General framework

Y is still a random vector, Q is a linear subspace of $\mathbb{R}^n$ ($\dim(Q) = q = n-m$) and $T = \frac{1}{\|\pi_{Q^\perp}(Y)\|}\pi_{Q^\perp}(Y)$. We consider J different subspaces Q_j ($Q_j \subset Q^\perp$, $\dim(Q_j) = k_j$) and the $J+1$ hypotheses

$$H_o: \mathscr{L}(Y) = N_n(\mu, \sigma^2 I_n), \quad \mu \in Q \qquad \text{(model (1))}$$

$$H_j: \mathscr{L}(Y) = N_n(\nu, \sigma^2 I_n), \quad \nu \in Q \oplus Q_j.$$

If H_o is true, $\mathscr{L}(T) = U_{Q^\perp}$. Let $\delta = \pi_{Q^\perp}(\nu) = \pi_{Q_j}(\nu)$ so that $\langle t,\delta\rangle = \langle\pi_{Q_j}(t),\delta\rangle$. We shall denote π_{Q_j} by π_j.

According to Proposition 1 and Corollary 1, if H_j is true the probability density function of T w.r.t. $U_{Q^\perp}$ is

$$t \to g_j(t;\delta,\sigma) = \frac{1}{2^{\frac{m}{2}-1}\Gamma(\frac{m}{2})} e^{-\frac{\|\delta\|^2}{2\sigma^2}} h_m\left(\frac{\langle \pi_j(t),\delta \rangle}{\sigma}\right) \qquad (2)$$

which depends only on $\frac{\delta}{\sigma}$.

Examples

(i) A problem of variable selection can be framed as follows. Vectors of $\mathbb{R}^n$ not belonging to Q, say $z_1, z_2, \ldots$, are given and we want to chose between entering none (H_o) or some (H_j) of these vectors in the model. If H_j means the inclusion of say, z_1 and z_2, then Q_j is spanned by $\pi_{Q^\perp}(z_1)$, $\pi_{Q^\perp}(z_2)$.

(ii) In model A for k outliers (at most k outliers), $z_1, z_2, \ldots$ are the vectors of the canonical basis $e = (e_1, \ldots, e_n)$ of $\mathbb{R}^n$, Q_j is spanned by a subset of $\{\pi_{Q^\perp}(e_1), \ldots, \pi_{Q^\perp}(e_n)\}$ including k (at most k) elements. For instance Y_1 and Y_2 are outliers means that θ_1 is added to $\mathbb{E}(Y_1)$ and θ_2 to $\mathbb{E}(Y_2)$ $(\theta_1, \theta_2 \neq 0)$; $\mathbb{E}(Y)$ becomes $\nu = \mu + \theta_1 e_1 + \theta_2 e_2$ and $\nu \in Q + [e_1, e_2] = Q \oplus [\pi_{Q^\perp}(e_1), \pi_{Q^\perp}(e_2)]$, where a sequence of vectors between brackets denotes the subspace spanned by these vectors (note that in the present case $Q_j = [\pi_{Q^\perp}(e_1), \pi_{Q^\perp}(e_2)]$ is generally but not necessarily of dimension 2).

(iii) Some change-point problems in the linear model enter also the above framework. Let Q be spanned by $x_1, \ldots, x_m$ so that $\mathbb{E}(Y) = \sum_{i=1}^{m} \beta_i x_i$. Let z_{ij}, $i = 1, \ldots, m$, $j = 1, \ldots, n$, be a vector whose ℓ^{th} coordinate in the canonical basis e is the same as ℓ^{th} coordinate of x_i if $\ell \geq j$ and 0 otherwise. Let Q_j be spanned by the m vectors $\pi_{Q^\perp}(z_{ij})$, $i=1,\ldots,m$. Then H_j accounts for a change in regression coefficients from the observation number j.

Prior distributions

A1: The prior probability of H_j is p_j $(\sum_{j=0}^{J} p_j = 1)$.

A2 : The conditional prior probability of $\frac{\delta}{\sigma}$ given $H_j, j\neq 0$, is spherical in Q_j. This implies that, given H_j:

- $\mathscr{L}(\frac{\delta}{\|\delta\|}) = U_{Q_j}$ (we shall write merely U_j)
- $\|\frac{\delta}{\sigma}\|$ is independent of $\frac{\delta}{\|\delta\|}$
- the prior probability law P_j^* of δ is characterized by the probability law P_j of $\frac{\|\delta\|}{\sigma}$.

Loss function

We introduce the following loss function for selecting H_i when the true hypothesis is H_j with the value δ of the parameter

$$\text{A3:}\quad \ell(i,j,\delta) = \begin{cases} 0 & \text{if } i=j \\ \ell & \text{if } j\neq i,\ j=0 \\ \ell(\|\delta\|) & \text{if } j\neq i,\ j\neq 0. \end{cases}$$

Bayes optimal procedure

A slight adaptation of Wald's results (Wald (1950), Chapter 5; see also Abbady (1980)) leads to the Bayes optimal invariant procedure

$$\text{select } H_j \text{ if:} \qquad (3)$$

$$p_j \int \ell(\|\delta\|) g_j(T,\delta,\sigma)\, dP_j^*(\delta) > \max_{i\neq j} p_i \int \ell(\|\delta\|) g_i(T,\delta,\sigma)\, dP_i^*(\delta).$$

For $i=0$, P_0^* is the Dirac measure on 0, $g_0(t,0,\sigma) = 1$ and

$$p_i \int \ell(\|\delta\|) g_i(t,\delta,\sigma)\, dP_i^*(\delta) = p_0 \ell.$$

For $j\neq 0$, we definc

$$f_j(T) = \int \ell(\|\delta\|) g_j(T,\delta,\sigma)\, dP_j^*(\delta).$$

Since in general $p_0 \ell$ is not know it is usual to adopt the following procedure

$$\text{select } H_o \text{ if } \max_{i \neq 0} p_i f_i(T) < c$$

$$\text{select } H_j,\ j \neq 0, \text{ if } p_j f_j(T) > \max[c, \max_{i \neq j} p_i f_i(T)] \tag{4}$$

c is chosen so that the level be α.

Computing c given α is beyond the scope of this paper.

Our purpose is now to transform the integrals $f_j(T)$ in order to give a simpler form to the selection procedure.

Taking into account A2, A3, expression (2) of g_j and using Fubini's theorem (recall $P_j^* = P_j \otimes U_j$) we obtain easily that

$$f_j(T) = \frac{1}{2^{\frac{m}{2}-1}\,\Gamma(\frac{m}{2})} \tag{5}$$

$$\int_{v=0}^{+\infty} e^{-\frac{v^2}{2}} v^{m-1} \int_{\|\delta\|=0}^{+\infty} \ell(\|\delta\|)\, e^{-\frac{\|\delta\|^2}{2\sigma^2}} \left[\int_{S_j} e^{<\pi_j(T), \frac{\delta}{\|\sigma\|}> \frac{\|\delta\| v}{\sigma}}\, dU_j\left(\frac{\delta}{\|\delta\|}\right)\right] dP_j\left(\frac{\|\delta\|}{\sigma}\right) dv.$$

3.2. <u>All hypotheses</u> H_j, <u>$j \neq 0$, "have the same dimension"</u>

Here we assume $k_j = k$ for all $j \neq 0$ and

A1' : the probabilities p_j are equal for $j \neq 0$,

A2' : $P_j^* = P^*$, $P_j = P$ for all $j \neq 0$.

The integral between brackets in (5) is the moment-generating function of the uniform distribution U_j on the unit sphere S_j of Q_j, taken at the point $\frac{\|\delta\| v}{\sigma}\|\pi_j(T)\|$. By Lemma 3 it is an increasing function of $\frac{\|\delta\| v}{\sigma}\|\pi_j(T)\|$. Integrating over v and $\frac{\|\delta\|}{\sigma}$ we find that $f_j(T)$ is an increasing function of $\|\pi_j(T)\|$. Finally A1' enables us to change (4) into

Proposition 2: If A1, A2, A1', A2', A3 hold, the following rule is Bayes optimal:

$$\text{select } H_o \text{ if } \max_{i\neq o} \|\pi_i(T)\| < c,$$

$$\text{select } H_j \text{ if } \|\pi_j(T)\| > \max\,[0, \max_{i\neq j} \|\pi_i(T)\|], \quad j\neq 0.$$

Remark: It is worth noticing that

$$\|\pi_j(T)\| = \frac{\|\pi_{Q_j}(Y)\|}{\|\pi_{Q^\perp}(Y)\|}.$$

If $Q_j = [\pi_{Q^\perp}(e_j)]$ is the linear subspace spanned by $\pi_{Q^\perp}(e_j)$ (this is the case when $\nu = \mu + ae_j$, $\mu \in Q$, $a \in \mathbb{R}$), then

$$\|\pi_j(T)\| = \frac{|\langle \pi_{Q^\perp}(e_j), Y\rangle|}{\|\pi_{Q^\perp}(Y)\| \, \|\pi_{Q^\perp}(e_j)\|} = \frac{|\langle \pi_{Q^\perp}(Y), e_j\rangle|}{\|\pi_{Q^\perp}(Y)\| \, \|\pi_{Q^\perp}(e_j)\|}.$$

is the absolute value of the jth studentized residual when the scalar product on $\mathbb{R}^n$ is canonical.

Comments: The present comments concern mainly the case of the detection of k outliers.

1. We may claim that Assumptions A1' and A2' are natural for reason of symmetry.

2. Concerning A2, let us consider $k=2$ and the subspace $[\pi_{Q^\perp}(e_1), \pi_Q(e_2)]$ (a sequence of vectors between brackets denotes the subspace spanned by these vectors). This corresponds to two outliers producing a shift of the mean $\mu (\mu \in Q)$ which becomes $\nu = \mu + \theta$, $\theta \in [e_1, e_2]$. It is clear that an assumption like A2 would be more realistic about θ than it is about $\delta = \pi_{Q^\perp}(\theta)$. And the sphericity of the distribution of θ does not imply in general the sphericity of the distribution of δ. However this is approximatly the case in

most problems when n is large compared to q. On the other hand, it can be argued that two outliers should be independent: for the canonical scalar product, since e_1 and e_2 are orthogonal, independence of the two corresponding outliers and sphericity of the distribution in $[e_1,e_2]$ is equivalent to the assumption of a spherical gaussian prior in $[e_1,e_2]$.

3. Concerning the loss function, the use of A3 results in a serious limitation. It might seem realistic that the loss be a function of $\|\delta\|$ which measures to what extent model (1) is falsified.
But often the problem is to find the outliers that matter (according to Andrews and Pregibon (1978)). If we are interested in estimating μ, it is clear that the loss function ought to take into account not only $\|\pi_{Q^\perp}(\theta)\|$ but also $\pi_Q(\theta)$, as Cook's residuals do (Cook (1977,1979); Cook and Weisberg (1980)).

Moreover a loos function which does not depend on the pair of hypotheses H_i and H_j is unsatisfactory because all pairs do not overlap in the same manner when $k_i, k_j > 1$.

3.3. All hypotheses H_j, $j\neq 0$, do not have the same dimension

Here the k_j's can differ and the assumption A2' has no meaning.We need a more precise knowledge of P_j to compute $f_j(T)$. We shall assume

A4: $P_j^* = N_{Q_j}(0,\rho\sigma^2 I_{Q_j})\quad \rho > 0,$

i.e. $P_j = \rho\sigma^2 . \chi^2_{k_i}$ (cf. Lemma 1, P_j^* being spherical),

$$\text{A3}' : \ell(i,j,\delta) = \begin{cases} 0 & \text{if } i=j \\ 1 & \text{if } i\neq j. \end{cases}$$

We use the expression of φ_{k_i} (Lemma 3) in (5) to obtain

$$f_j(T) = \frac{1}{2^{\frac{m}{2}-1}\Gamma(\frac{m}{2})} \int_{v=0}^{+\infty}\int_{x=0}^{+\infty} e^{-\frac{x}{2\sigma^2}} e^{-\frac{v^2}{2}} v^{m-1} \sum_{h=0}^{\infty} \frac{\|\pi_j(T)\|^{2h} x^h v^{2h}}{(2\sigma)^{2h} h!\,\Gamma(h+\frac{k_j}{2})} \frac{e^{-\frac{x}{2\rho\sigma^2}} x^{\frac{k_j}{2}-1}}{(2\rho\sigma^2)^{k_j/2}} dx\,dv.$$

The integration term by term leads to

$$f_j(T) = (1+\rho)^{-k_j/2} \quad A_m(\frac{\rho}{1+\rho} \quad || \pi_j(T)||^2),$$

where $A_m(t) = \frac{1}{\Gamma(\frac{m}{2})} \sum_{h=0}^{\infty} \frac{\Gamma(h+\frac{m}{2})}{h!} t^h = (1-t)^{-\frac{m}{2}}$.

Thus

$$f_j(T) = (1+\rho)^{-\frac{k_j}{2}} (1-\frac{\rho}{1+\rho}||\pi_j(T)||^2)^{-\frac{m}{2}} . \qquad (6)$$

Proposition 3: If A1, A3', A4 hold, the following rule is Bayes optimal with $f_j(T)$ given by (6):

select H_0 if $\max_{i \neq 0} p_i \; f_i(T) < c$

select H_j if $p_j \; f_j(T) > \max[c, \max_{i \neq j} p_i f_i(T)]$ for $j \neq 0$.

Comments: Proposition 3 cannot be used easily. We try now to transform this rule into a more useful one, at the cost of some approximations.

We consider the detection of k outliers, so that $k_j \leq k$, and we assume that n (hence m) is large with respect to q an k.

The first step is to choose the prior probabilities p_j.
In many practical situation, it is reasonable to suppose that each observation can be spurious with the same probability p independently of the others. Then $p_j = p^{k_j}(1-p)^{n-k_j}$. We assume that these quantities can be neglected for k_j greater than some k and we write

$$p_j = \gamma\beta^{k_j/2} \qquad 0 \leq k_j \leq k.$$

We note that, for consistency, the expected number of outliers np should be less than k but of the same order so that $\beta = O(\frac{1}{n^2})$.

On the other hand, $||\pi_{Q_j}(T)||^2$ is of order $\frac{k_j}{m}$, then we get the equivalence, for large n

$$(1 - \frac{\rho}{1+\rho}||\pi_j(T)||^2)^{-m/2} \sim e^{\frac{\rho}{1+\rho}||\pi_j(T)||^2 \frac{m}{2}} .$$

Now the comparison of the quantities $p_j \; f_j(T)$ can be done through the comparison of the quantities

$$\operatorname{Log} \frac{p_j f_j(T)}{\gamma} \sim \frac{k_j}{2} \operatorname{Log} \frac{\beta}{1+\rho} + \frac{\rho}{1+\rho} \frac{m}{2} \|\pi_j(T)\|^2. \tag{7}$$

The selection rule in Proposition 3 becomes:

select H_o if $\max_{i \neq 0} (\|\pi_i(T)\|^2 - ak_i) < b$

select H_j, $j \neq 0$, if $\|\pi_j(T)\|^2 - ak_j > \max[b, \max_{i \neq j} (\|\pi_i(T)\|^2 - ak_i]$.

Here a and b depend on the prior assumptions. In practice however they will be chosen directly rather than as a function of the quantities in these prior assumptions. b can be chosen such that the level be α. From (7) we have $a = \frac{1+\rho}{m\rho} \operatorname{Log} \frac{1+\rho}{\beta}$ from which we deduce that a should be positive and small: $a = O(\frac{1}{n} \operatorname{Log} n)$. Actually we used a Bayesian framework as a way of arriving at sensible procedures without believing very firmly in prior distributions. Adaptation of constants a and b will proceed from numerical trials (in progress) including comparisons with other procedures.

4. Further applications and comments

In the previous section, we gave an application of Proposition 1 (iii) to the analysis of residuals with a strong emphasis on model A for outliers. Proposition 1 (ii) could be used with model B (change of variances) and (i) with mixed causes of spuriousness.

On the other hand, Proposition 1 (ii) is a common starting point for deriving UMP (locally UMP...) invariant tests in testing problems involving an alternative to model (1) defined by $\mathscr{L}(Y) = N_n(\mu, W)$. In these cases the Neyman-Pearson lemma leads to rejection of model (1) for large values of $|V|^{-1/2}(\langle T, V^{-1}T \rangle)^{-m/2}$, i.e. small values of $\langle T, V^{-1}T \rangle$, where $V = \pi_{Q^\perp} W|_{Q^\perp}$. The only difficulty lies in inverting V.

It is clear that the properties of the spectral decomposition of V are central in all these problems which indlude, according to the specification of W, the Durbin-Watson problem (testing for disturbances autocorrelation), mixed models in ANOVA, random coefficients linear models (see for example Nyblom and Mäkeläinen (1983)), etc...

Finally, we note a robustness property of all statistical techniques involving T. Actually $\mathscr{L}(T) = U_{Q^\perp}$ does not require $\mathscr{L}(\pi_{Q^\perp}(Y)) = N_{Q^\perp}(0,\sigma^2 I_{Q^\perp})$. A look at Lemma 1 shows that this is only one aspect of normality. King (1980) (see also the included references) treats spherical symmetry (properties (i) and (ii) in Lemma 1). Philoche (1977) noticed that isotropy of $\mathscr{L}(\pi_{Q^\perp}(Y))$ or $\mathscr{L}(Y)$ is enough to ensure the validity of testing procedures involving T ($\mathscr{L}(Y)$ is isotropic if $P(\|Y-\mu\| \neq 0)=1$, $\mathscr{L}(\frac{Y-\mu}{\|Y-\mu\|})=U_R$; isotropy is preserved by projection). However the importance of this robustness property should not be overemphasized for the following reason. In the linear model, it is often felt that errors can be assumed independent. Now, if Y has independent coordinates w.r.t. the canonical basis and $\mathscr{L}(Y)$ is spherical w.r.t. μ in $\mathbb{R}^n$ provided with the canonical scalar product, then it is well known that $\mathscr{L}(Y)$ is $N_n(\mu,\sigma^2 I_n)$. Moreover, Letac (1981) proved a similar result for isotropy when $n > 2$.

5. Modified residuals

For many practical purposes, it is difficult to deal with the residuals $\pi_{Q^\perp}(Y)$ because their distribution depends on Q which differs from one problem to an other. To cope with this difficulty several modifications of residuals have been proposed (Theil (1965), Abrahamse and Koerts (1971),...) to build uncorrelated equally distributed normal "modified" or "ajusted" residuals.

Following the idea of Durbin (1961) one can also put forward residuals modified by introducing some random perturbation of T or $\pi_{Q^\perp}(Y)$. Tiao and Guttman (1967) did so in the cases of known or independently estimated variance, Hildreth (1971) considered unknown variance but his adjusted residuals are not gaussian. Geometric arguments give a straightforward solution.

Proposition 4: We still consider the linear model (1). Let w and X be independent random variables such that

$$\mathscr{L}(X) = N_Q(0,I_Q) \quad (\text{or} \quad N_n(0,\pi_Q)),$$

$$\mathscr{L}(w^2) = \chi^2_{n-q} \quad (w > 0),$$

(w,X) independent of T.

Then $\mathscr{L}(wT+X) = N_n(0,I_n)$.

<u>Proof</u>: Using the assumptions about w, $\mathscr{L}(T) = U_{Q^\perp}$ and Lemma 1 with $E = Q^\perp$, we have $\mathscr{L}(wT) = N_{Q^\perp}(0,I_{Q^\perp})$ which is the same as $N_n(0,\pi_{Q^\perp})$ (cf. Lemma 2). The result follows from the additivity of independent gaussian variables.

<u>Remarks</u>: 1. (w,X) can be obtained from a random variable Z whose law is $N_n(0,I_n)$ by putting $X=\pi_Q(Z)$ and $w=\|\pi_{Q^\perp}(Z)\|$.

2. Some properties concerning the "optimal closeness" of $(Y-\mu)$ and $R=wT+X$ are proven in Caussinus (1980). We recall them briefly.

Let $\mathscr{S}$ denote the set of statistics S valued in $\mathbb{R}^n$ such that (i) S is a function of the outcome Y and of any random variable stochastically independent of Y, (ii) $\mathscr{L}(S) = N_n(0,I_n)$. Then

a) $$\mathbb{E}(\|R-\tfrac{1}{\sigma}(Y-\mu)\|^2) = \inf_{S\in\mathscr{S}} \mathbb{E}(\|S-\tfrac{1}{\sigma}(Y-\mu)\|^2) = 2(n-\lambda^2_{n-q})$$

b) $$\mathbb{E}(\|\tfrac{1}{\|R\|}R - \tfrac{1}{\|Y-\mu\|}(Y-\mu)\|^2) = \inf_{S\in\mathscr{S}} \mathbb{E}(\|\tfrac{1}{\|S\|}S - \tfrac{1}{\|Y-\mu\|}(Y-\mu)\|^2) = 2(1-\tau^2_{n,q}),$$

where $\lambda_{n-q} = \sqrt{2\pi}/\beta(\frac{n-q}{2}, \frac{1}{2})$ $\quad \tau_{n,q} = \prod_{i=0}^{q-1} \frac{n-q+i}{n-q+\frac{1}{2}+i}$.

3. Although the derivation is not coordinate-free, it is clear that the paper by Durbin (1961) contains the main idea underlying Proposition 4. A careful reading of this paper (see p.52 and the subsequent editorial note) is of great interest for an historical view about the use of geometric arguments in statistics.

4. Several papers discuss the usefulness of various modified residuals and compare their efficiencies: see for example Ward (1973), Dubbelman (1978), Benhelli and Lambert (1981). Concerning the vector of modified

residuals wT+X, we have made some simulations for the main problem which is dealt with in this paper: the detection of outliers. The loss of efficiency with respect to ordinary residuals is very important so that these (and we guess, any other) modified residuals must not be used in this case (Caussinus (1980)). The perform much better when testing for autocorrelation of errors (Benhelli and Lambert (1981)). However, because of the increasing power of computing facilities, it may be thought that Proposition 4 is mainly interesting from a "theoretical" point of view.

Acknowledgment. The authors are grateful to Professor O.Bunke for valuable comments on a preliminary version of the paper.

References

[1] F.Abbady,(1980). Règles bayesiennes optimales pour la détection de valeurs aberrantes dans le modèle linéaire. Thèse, Université Paul Sabatier, Toulouse.

[2] A.P.J.Abrahamse, J.Koerts, (1971). New estimators of disturbances in regression analysis, J.A.S.A. 66 71-74.

[3] D.F.Andrews, (1971). Significance tests based on residuals. Biometrika 58 139-148.

[4] D.F.Andrews and D.Pregibon (1978): Finding the outliers that matter. J.R.S.S. B 40 85-93.

[5] F.J.Anscombe (1960). Examination of residuals. Proc. 4th Berkeley Symposium, vol. I 1-36.

[6] F.J.Anscombe and J.W.Tukey (1963). The examination and analysis of residuals. Technometrics 5 141-160.

[7] V.Barnett and T.Lewis (1978). Outliers in Statistical Data. Wiley, New-York.

[8] M.Benhelli et E.Lambert (1981). Comparaison de quelques tests d'indépendance des erreurs dans le modèle linéaire. Statistique et Analyse des Données 1 52-69.

[9] R.W.Butler (1981). The admissible Bayes character of subset selection techniques involed in variable selection, outlier detection and slippage problems. Ann.Stat. 9 960-973.

[10] R.W.Butler (1983). Optimal properties of one-step variable selection in regression. Ann. Stat. 11 219-224.

[11] H.Caussinus (1980). Sur l'analyse des résidus dans le modèle linéaire. Statistique et Analyse des Données 3 29-39.

[12] R.D.Cook (1971). Detection of influential observation in linear regression. Technometrics 19 15-18.

[13] R.D.Cook (1979). Influential observation in linear regression. J.A.S.A. 74 169-174.

[14] R.D.Cook and S.Weisberg (1980). Characterization of an empirical influence function for detecting influential cases in regression. Technometrics 22 495-508.

[15] S.Degerine (1979). Lois de Von Mises et lois liées. Ann. Inst. H.Poincaré, Section B XV 63-77.

[16] A.P.Dempster (1969). Elements of continuous multivariate analysis. Addison-Wesley Publishing Co.

[17] H.Drygas (1970). The coordinate-free approach to Gauss-Markov estimation. Lectures Notes in Op. Res. and Math.Syst., Springer--Verlag, Berlin.

[18] C.Dubbelman (1978). Disturbances in the linear model, estimation and hypothesis testing. Martinus Nijhoff Social Sciences Division, Leiden/Boston.

[19] J.Durbin (1961). Some methods for constructing exact tests. Biometrika 48 41-55.

[20] T.S.Ferguson (1960). On the rejection of outliers. Proc. 4th Berkeley Symposium,vol.I 253-287.

[21] D.M.Hawkins (1980). Identification of outliers. Chapman and Hall. London.

[22] C.Hilderth (1971). A note on approximate regression disturbances. Discussion Paper No2, Center for Economic Research,Department of Economics, University of Minnesota, Minneapolis, Minnesota.

[23] J.Karlin and D.Truax (1960). Slippage problems. Ann.Math.Stat. 31 296-324.

[24] M.L.King (1980). Robust tests for spherical symmetry and their applications to least squares regression. Ann.Stat. 8 1265-1271.

[25] W.Kruskal (1960). The coordinate-free approach to Gauss-Markov estimation, and its application to missing and extra observations. Proc. 4th Berkeley Symposium, vol. I 435-451.

[26] G.Letac (1981). Isotropy and sphericity: some characterizations of the normal distribution. Ann.Stat. 9 405-417.

[27] J.Nyblom and T.Mäkeläinen (1983). Comparisons of tests for the presence of random walk coefficients in a simple linear model. J.A.S.A. 78 856-864.

[28] J.L.Philoche, (1977). Une condition de validité pour le test F. Statistique et Analyse des Données, 1-1977 37-59.

[29] H.Theil (1965). The analysis of disturbances in regression analysis, J.A.S.A. 60 1067-79.

[30] G.C.Tiao and I.Guttman (1967). Analysis of outliers with adjusted residuals. Technometrics 9 541-559.

[31] A.Wald (1950). Statistical decision functions. Wiley, New-York.

[32] G.S.Watson (1983). Statistics on spheres. Wiley, New-York.

[33] L.L.Ward (1973). Is uncorrelating the residuals worth it ? Thèse, McGill University, Montréal.

APPROXIMATE DESIGN THEORY FOR A SIMPLE BLOCK DESIGN WITH RANDOM BLOCK EFFECTS

K.Christof and F.Pukelsheim
Institute of Mathematics
University of Augsburg
Federal Republic of Germany

Abstract. A simple block design with treatment effects fixed and block effects random is considered. It is shown that the sole design class which is amenable to an analysis is formed by all designs which are equi-blocksized. In this setting the block totals induce a new model, the interblock model, in which the mean vector depends linearly on the design matrix, and in which the covariance matrix is proportional to the identity matrix. The information matrix for the set of all treatment contrasts then is a convex function of the design matrix, quite distinct from the usual case where this relation is concave. The extreme points in the design space turn out to be the one-treatment-per-block designs, their information matrices depend on the treatment replication vector, only, and are called special C-matrices. We establish a disjunction which the matrix-ordering of two special C-matrices enforces on the associated replication vectors. This entails, in particular, that one-treatment-per-block designs are admissible provided no treatment gets more than one half of all observations.

1. Introduction and summary

The simple block design model with block effects random leads to a peculiar design problem since in the associated interblock model the dependence of the information matrix on the design matrix is, not concave, but convex. This is a marked difference to the usual case where this dependence is indeed concave. Here we shall discuss this

---

AMS 1980 Mathematics subject classification: Primary 62K05; 62K10.
Key words and phrases: Two-way-classification, mixed model; interblock model; one-treatment-per-block designs; convex information functions; special C-matrix; admissibility; Schur-concavity.

design problem from the approximate theory point of view, thus complementing the exact theory results obtained by Gaffke & Krafft (1980) in their 1978 Wisła Conference paper.

In Section 2 we outline the transition from the mixed effects model to the interblock model. It turns out that a manageable design problem can be formulated only among those designs which are equi-blocksized. Section 3 presents the approximate analogue of the exact results of Gaffke & Krafft (1980). As the optimization problem is one of maximizing a convex function, the *one-treatment-per-block designs*, i.e. the extreme points among all equi-blocksized designs, are crucial.

The information matrix for the set of all treatment contrasts of a one-treatment-per-block design depends on the treatment replication vector r, only, and takes the form $\Delta_r - rr'$ where Δ_r denotes the diagonal matrix with vector r on the diagonal. We shall call such matrices *special C-matrices*, they also play an important role in fixed effects models.

In Section 4 we study some implications of admissibility in the class of all special C-matrices; our results will be stated in terms of the replication vectors r which enter the definition, whereas the similar study of Giovagnoli & Wynn (1981) focused on eigenvalues. In the interblock model this yields admissibility of one-treatment-per-block designs in the class of all equi-blocksized designs, provided no treatment gets more than $1/2$ of all observations.

2. The interblock model

Consider a two-way classification $X_{ijk} = \mu + \alpha_i + \beta_j + \varepsilon_{ijk}$, with a fixed over-all mean $\mu \in R$, and a fixed treatment effects vector $\alpha = (\alpha_1, \ldots, \alpha_v)' \in R^v$. The block effects β_j, $j = 1, \ldots, b$, and the observational errors ε_{ijk} are assumed to be independent random variables with expectation 0, and variance τ^2 and σ^2, respectively. Denoting by 1_n the n-dimensional vector with all entries unity, the model has vector representation

$$X = \mu 1_n + A\alpha + B\beta + \varepsilon,$$

with known 0-1 matrices A of order $n \times v$ and B of order $n \times b$.

A design matrix N for this model is a $v \times b$ matrix with integer entries $n_{ij} \geq 0$. The treatment-block combination (i,j) does not appear in the model if $n_{ij} = 0$, and otherwise is replicated n_{ij} times, as indicated by the subscript k. The total number of observations then is $n = \Sigma\Sigma n_{ij}$. In order to discriminate between the influence of the sample size n and the distribution of observations between the various treatment-block combinations (i,j), as well as for purposes of the approximate design theory it is preferable to work with the <u>weight matrix</u>

$$W = N/n,$$

which is a normalized, per observation, version of the design matrix. Its marginals, i.e. row sums and column sums, equal

$$r = W1_b, \quad s = W'1_v,$$

and are the <u>treatment replication vector</u> and the <u>blocksize vector</u> in their normalized versions, respectively.

As in Gaffke & Krafft (1980, page 135) we shall study the treatment effects through the interblock estimators based on the b block totals $Y = B'X$. Its mean vector and dispersion matrix are found to be

$$E(Y) = n\mu s + nW'\alpha = n(s:W')\binom{\mu}{\alpha},$$

$$D(Y) = n^2\tau^2\Delta_s^2 + n\sigma^2\Delta_s = \Delta_S, \text{ say.}$$

Thus Y gives rise to a linear model which we shall call the <u>interblock model</u>.

The components of Y are uncorrelated, but have possibly unequal variances $S_j = n^2\tau^2 s_j^2 + n\sigma^2 s_j$. The information matrix for $\binom{\mu}{\alpha}$, i.e. the inverse of the dispersion matrix of the weighted least squares estimator, turns out to be $n^2\bar{M}(W)$, with

$$\bar{M}(W) = \begin{pmatrix} s'\Delta_S^+ s & s'\Delta_S^+ W' \\ W\Delta_S^+ s & W\Delta_S^+ W' \end{pmatrix},$$

a superscript + denoting Moore-Penrose inversion. Hence any weight matrix, and also any optimal weight matrix, has information matrix depending, through S, on the sample size n, the block variance τ^2, and the error variance σ^2. This is slightly different when the designs are assumed <u>equi-blocksized</u>, i.e. $s = 1_b/b$. Then plainly $D(Y) = (n^2\tau^2/b^2 + n\sigma^2/b)I_b$, and simple least squares estimation leads to the information matrix $M(W)/(\tau^2/b+\sigma^2/n)$, with

$$M(W) = \begin{pmatrix} 1 & r' \\ r & bWW' \end{pmatrix}.$$

However, there seems to be no convincing reason why an optimal weight matrix should necessarily be equi-blocksized.

Namely, $\bar{M}$ is a matrix-convex function on the set of all weight matrices with block marginals given to be s, as M is convex on the set of all designs which are equi-blocksized, see Marshall & Olkin (1979, page 468). This convexity breaks down when block marginals are allowed to vary, and we shall demonstrate this failure by example. Suppose $v = 2$, and $n^2\tau^2 = 1 = n\sigma^2$. Choose $W = \begin{pmatrix} 1/4 & 1/4 \\ 1/2 & 0 \end{pmatrix}$, and $V = \begin{pmatrix} 1/2 & 0 \\ 0 & 1/2 \end{pmatrix}$. Then

$$M(\tfrac{1}{2}W + \tfrac{1}{2}V) = \frac{1}{2145}\begin{pmatrix} 1410 & 690 & 720 \\ 690 & 362 & 328 \\ 720 & 328 & 392 \end{pmatrix},$$

$$\tfrac{1}{2}M(W) + \tfrac{1}{2}M(V) = \frac{1}{2100}\begin{pmatrix} 1360 & 710 & 650 \\ 710 & 610 & 100 \\ 650 & 100 & 550 \end{pmatrix},$$

and matrix convexity is violated because of $1410/2145 > 1360/2100$.

It then appears difficult to compare information matrices across varying block marginals, whence we shall follow Gaffke & Krafft (1980) and consider equi-blocksized designs, only.

3. Equi-blocksized designs for the interblock model

In the sequel we shall identify designs N and their weight matrices W. In the approximate theory the entries of W are not restricted to be rational numbers, i.e. we only have $w_{ij} \geq 0$, $\Sigma\Sigma w_{ij} = 1$. As we only consider equi-blocksized designs W, their information matrices for $\binom{\mu}{\alpha}$ are taken to be $M(W)$ from Section 2, thereby neglecting the common constant of proportionality $1/(\tau^2/b+\sigma^2/n)$.

Interest concentrates on the set of all treatment contrasts $(\alpha_1 - \overline{\alpha .}, \ldots, \alpha_v - \overline{\alpha .})' = K_v \alpha$, with

$$K_v = I_v - 1_v 1_v'/v.$$

The following lemma characterizes identifiability (i.e. estimability, testability) of $K_v\alpha$, with a proof slightly different from that given by Gaffke & Krafft (1980, page 136).

Lemma 1. In the interblock model, for every equi-blocksized weight matrix W the following assertions are equivalent:

(a) $K_v\alpha$ is identifiable under W.
(b) rank $(bWW' - rr') = v-1$.
(c) rank $(M(W)) = v$.

Proof. Set $K = \binom{0}{K_v}$. (a) means that the range (column space) of K is contained in the range of $M(W)$, see Pukelsheim (1983, page 194). It is easy to see that this is equivalent to

$$\text{range } (K_v) \subseteq \text{range } (bWW' - rr'). \qquad (*)$$

But the nullspace of K_v, being spanned by 1_v, is always contained in the nullspace of $bWW' - rr'$. Thus equality holds in (*), and (a) is equivalent to (b), and also to (c), since rank is additive on the Schur complement, see Ouellette (1981, page 199).

Lemma 1(c) leaves a one-dimensional subspace of non-identifiable linear forms, a spanning vector of which obviously is $\binom{-1}{1_v}$.

The information matrix for a parameter set $K'\binom{\mu}{\alpha}$ is given by $J(M(W)) = (K'M(W)^- K)^+$. In the case of treatment contrasts K equals $\binom{0}{K_v}$, and an easy calculation yields

$$J(M(W)) = (K_v(bWW' - rr')^+ K_v)^+ = bWW' - rr'.$$

Now M, as a function of W, is convex while J, as a function of M, is concave, see Marshall & Olkin (1979, page 468), Pukelsheim & Styan (1983, page 148). Hence no conclusion is possible for the composition $J \circ M$.

However, the explicit representation just obtained can be written as

$J(M(W)) = bWK_bW'$. Hence, after all, the information matrix $J \circ M$ for the treatment contrasts is a convex function of the design W. This distinct behaviour is due to the peculiar way by which in the interblock model the mean vector depends on the underlying design W.

Therefore, when the intention is to maximize information, we are led to maximize a convex functional of the design W, and the extreme designs become essential. The extreme points among all equi-blocksized designs were determined by Gaffke & Krafft (1980, page 141) to be what we shall call one-treatment-per-block designs T. By definition, such a design T has just one treatment i_j appearing in block j, for $j = 1,\ldots,b$, and the constraint of uniform block marginals forces $t_{i_j j} = 1/b$. An alternative way of introducing such designs is as follows, cf. Gaffke & Krafft (1980, page 142).

Lemma 2. In the interblock model, every equi-blocksized design W with rank $(bWW' - rr') = v - 1$, where $r = W1_b$, satisfies

$$J(M(W)) = bWW' - rr' \leq \Delta_r - rr',$$

and here equality holds if and only if W is a one-treatment-per-block design.

Proof. We have $bWW' - rr' = \Delta_r - rr' - (\Delta_r - bWW')$. The matrix $\Delta_r - bWW'$ is nonnegative definite since it is an information matrix in a fixed effects model, see Pukelsheim (1983, page 201). Hence the inequality. Equality occurs if and only if $bWW' = \Delta_r$, and this is a characteristic property of one-treatment-per-block designs.

We shall call $\Delta_r - rr'$ a special C-matrix. Thus we observe the intriguing feature that the special C-matrices are uniformly optimal information matrices among all equi-blocksized designs with given treatment marginals in both, the interblock model and the fixed effects model, see Pukelsheim (1983, page 202). It follows from the fixed effects model theory that there is no uniformly optimal member in the class of all special C-matrices. However, we may then pose the question of admissibility, and to this we turn next.

4. Admissibility of special C-matrices

It is easy to verify that $\Delta_r - rr'$ has rank $v-1$ if and only if the stochastic vector r is positive, i.e. each component of r is positive. Giovagnoli & Wynn (1981, page 412) explicitly construct the eigenvalue set Λ^* of all special C-matrices, for $v=3$ treatments. It follows from their results that $\Delta_t - tt' \leq \Delta_r - rr'$ need not imply $t=r$.

To see what can go wrong, in terms of the replication vectors, consider $t(\varepsilon) = (1-\varepsilon)e_i + \varepsilon 1_v/v$, where e_i is the i-th Euclidian unit vector. For small ε then $\Delta_{t(\varepsilon)} - t(\varepsilon)t(\varepsilon)'$ tends to vanish and thus falls below any matrix $\Delta_r - rr'$, whereas at the same time the treatment marginals $t(\varepsilon)$ converge to e_i and become extreme. In fact, this example qualitatively exhausts the possibilities, according to the following.

<u>Theorem</u>. Let t and r be two positive stochastic vectors in R^v. Then $\Delta_t - tt' \leq \Delta_r - rr'$ forces the alternative either (I) $t=r$, or (II) there exists some i such that $t_i > r_i$ and $t_j < r_j$, for all $j \neq i$.

<u>Proof</u>. Either $t_i \leq r_i$, for all i, and in view of $\Sigma t_i = \Sigma r_i$ then $t=r$. Or there exists some i such that $t_i > r_i$.
Now $\Delta_t - tt' \leq \Delta_r - rr'$ implies the converse ordering among the Moore-Penrose inverses $K_v \Delta_t^{-1} K_v$ and $K_v \Delta_r^{-1} K_v$, see Milliken & Akdeniz (1977, page 75). If $i=1$, premultiplication by $(-1_{v-1} : I_{v-1})$ and postmultiplication by its transpose yield

$$\Delta^{-1}_{(t_2,\ldots,t_v)'} + t_1^{-1} J_{v-1} \geq \Delta^{-1}_{(r_2,\ldots,r_v)'} + r_1^{-1} J_{v-1}.$$

Its diagonal elements then satisfy $1/t_j + 1/t_1 \geq 1/r_j + 1/r_1$ and therefore $1/t_j > 1/r_j$, for all $j \geq 2$. A similar argument obviously applies when $i \neq 1$, and so the proof is complete.

<u>Corollary 1</u>. (a) If $\Delta_t - tt' \leq \Delta_r - rr'$ and $t_i = r_i$, for some i, then actually $t=r$.
(b) If $\Delta_t - tt' \leq \Delta_r - rr'$ and $t \neq r$ then there exists some i such that $t_i > 1/2$, $r_i \in (1-t_i; t_i)$, and $t_j < 1/2$, $r_j \in (t_j; 1-t_j)$, for all $j \neq i$.
(c) If $\Delta_t - tt' \leq \Delta_r - rr'$ and $t_i \leq 1/2$, for all i, then actually $t=r$.

Proof. (a) rules out alternative (II). In (b) alternative (II) yields $t_i > r_i$, while the assumption implies $t_i - t_i^2 \leq r_i - r_i^2$. On the parabolic arc $y(1-y)$ this is possible under the stated conditions, only. A similar argument applies when $t_j < r_j$. Finally (c) is a consequence of (b).

Uniform treatment marginals $t_i = 1/v$ lead to the matrix K_v/v which, according to Corollary 1(c), is always maximal among special C-matrices.

Returning to the interblock model we can now prove admissibility of one-treatment-per-block designs whenever their treatment marginals are not too "unbalanced".

Corollary 2. In the interblock model, every one-treatment-per-block design T with treatment replications $t_i \in (0;1/2]$, for all i, and every other equi-blocksized design W satisfy

$$J(M(T)) \leq J(M(W)) \Rightarrow J(M(T)) = J(M(W)).$$

Proof. Positivity of the treatment marginals yields $J(M(T)) = \Delta_t - tt'$. Hence rank $J(M(W)) \geq$ rank $(\Delta_t - tt') = v-1$, and $J(M(W)) = bWW' - rr'$, where $r = W1_b$. By Lemma 2 then

$$\Delta_t - tt' \leq bWW' - rr' \leq \Delta_r - rr'. \qquad (**)$$

Corollary 1(c) forces $t = r$, as well as equality in (**). Lemma 2 provides the additional information that W must be another one-treatment-per--block design.

Finally we remark that, as in the fixed effects models, the question of how many different one-treatment-per-block designs there are leads into the combinatorial part of the theory. Recall that permuting blocks does not change the information matrix. Hence we may assume each treatment appearing once among the first v blocks, in order to ensure positive treatment marginals. The remaining $b-v$ (indistinguishable) blocks then must be distributed between the v treatments, and this can be done in $\binom{v+(b-v)-1}{b-v} = \binom{b-1}{b-v}$ ways. A further reduction is possible when real--valued criteria $k(J)$ are considered which are permutationally invariant. In the special case of the trace criterion we can even observe a simple ordering of the values trace $(\Delta_r - rr')$. For since the latter is a

Schur-concave function in r, see Marshall & Olkin (1979, page 67), its values will grow as r becomes more uniform. This provides an alternative derivation of Lemma 2 in Gaffke & Krafft (1980, page 138).

References

[1] N.Gaffke and O.Krafft (1980). D-optimum designs for the interblock--model. Lecture Notes in Statistics 2 134-143.

[2] A.Giovagnoli and H.P.Wynn (1981). Optimum continuous block designs. Proc.Roy.Soc.London Ser.A 377 405-416.

[3] A.W.Marshall and I.Olkin (1979). Inequalities: Theory of Majorization and Its Applications. New York: Academic Press.

[4] G.A.Milliken and F.Akdeniz (1977). A theorem on the difference of the generalized inverses of two nonnegative matrices. Comm.Statist. A - Theory Methods 6 73-79.

[5] D.V.Ouellette (1981). Schur complements and statistics. Linear Algebra Appl. 36 187-295.

[6] F.Pukelsheim (1983). On optimality properties of simple block designs in the approximate design theory. J.Statist. Plann. Inference 8 193-208.

[7] F.Pukelsheim and G.P.H. Styan (1983). Convexity and monotonicity properties of dispersion matrices of estimators in linear models. Scand. J.Statist. 10 145-149.

RECTANGULAR LATTICES REVISITED

L.C.A.Corsten
Department of Mathematics
Agricultural University
Wageningen, The Netherlands

Summary. The intrablock analysis of rectangular lattices can be clarified by knowledge of the canonical structure of block and treatment subspaces. In this case there are natural canonical subspaces in block space having factorial structure, in contrast to the situation e.g. of balanced incomplete block designs, lattice desings and group divisible or Latin square type PBIB designs, where canonical factorial subspaces arise in treatment space. This factorial structure for the dual of a rectangular lattice and the associated canonical efficiency factors lead to a simple explicit general expression for the treatment estimator and the efficiency factor at each type of treatment comparison. Recovery of interblock information requires only slight modifications.

1. Rectangular lattice designs

Before noting special properties of rectangular lattices it is required to recall what rectangular lattices, as introduced by Harshbarger (1947), are. These are in fact incomplete blocks designs for $k(k+1)$ treatments obtained from r (with $2 \le r \le k+1$) pairwise orthogonal classifications (so called replicates) of $(k+1)^2$ elements or treatments in $k+1$ classes or blocks of size $k+1$.

For example with $k = 3$ and $r \le 4$ we may start from the 5 replicates:

1 2 3 4	1 5 9 13	4 7 10 13	2 8 11 13	3 6 12 13
5 6 7 8	2 6 10 14	3 8 9 14	1 7 12 14	4 5 11 14
9 10 11 12	3 7 11 15	2 5 12 15	4 6 9 15	1 8 10 15
13 14 15 16	4 8 12 16	1 6 11 16	3 5 10 16	2 7 9 16

A rectangular lattice results by deleting one subset of (k+1) elements (treatments), in our example {13, 14, 15, 16} , which is represented exactly once in each block of each replicate, so that the remaining block size will be k. The elements to be deleted may be indicated by one class of a classification in k+1 blocks orthogonal to the r classifications retained: in our example with $r \leq 4$ the last block of the first classification (which will not be retained at all) is the set to be deleted, and the replicates of the rectangular lattice are obtained by striking out {13, 14, 15, 16} from at most four of the other replicates leading to the design:

Rep.1	Rep.2	Rep.3	Rep.4
1 5 9	4 7 10	2 8 11	3 6 12
2 6 10	3 8 9	1 7 12	4 5 11
3 7 11	2 5 12	4 6 9	1 8 10
4 8 12	1 6 11	3 5 10	2 7 9

of which at least two replicates will constitute a rectangular lattice.

If a classification orthogonal to r pairwise orthogonal classifications does not exist (e.g. the case of $r = 3$ and $k = 5$ where a Graeco-Latin square of order six does not exist) it may be possible to delete a set of k+1 treatments, represented exactly once in each block of each replicate, by trial and error. For example in the particular Latin square of order six:

```
 A   B   C   D   E  (F)
(B)  A   F   E   D   C
 C  (D)  A   B   F   E
 D   F  (E)  A   C   B
 E   C   B   F  (A)  D
 F   E   D  (C)  B   A
```

one may delete the set of six different encircled letters belonging to six different rows and to six different columns; now assigning treatment numbers from left to right in successive rows from top to bottom one will obtain the blocks of replication 1 according to rows, and using columns or latin letters in the order determined by the deletion of symbols from the original square, one will obtain the blocks of replications 2 and 3, respectively. At least two of the resulting replicates will constitute a rectangular lattice:

Rep.1					Rep.2					Rep.3				
1	2	3	4	5	10	15	20	25	30	7	14	17	24	26
6	7	8	9	10	1	11	16	21	26	2	13	20	23	29
11	12	13	14	15	2	6	17	22	27	4	9	16	25	28
16	17	18	19	20	3	7	12	23	28	5	8	15	21	27
21	22	23	24	25	5	9	14	19	29	1	6	12	18	30
26	27	28	29	30	4	8	13	18	24	3	10	11	19	22

It is very useful to note that the r blocks, arising in the r different classifications in connection with the deletion of the same element, and constituting the so called partner classes (in our examples arranged in the same row), have no treatment in common; the k treatments comprising any block will appear within each of the other replicates in each of the k partner classes (i.e.rows) different from the one to which that particular block belongs.

2. Intrablock analysis of incomplete blocks

For intra-block analysis we consider a block design as a two-way classification of the n elements of a random n-vector y in the Euclidean space R^n of n-tuples, one according to v classes (treatments) and denoted by A, and another according to w classes (blocks) and denoted by B. The number n_{ij} of elements belonging to treatment class i $(=1,\dots,v)$ as well as to block class $j(=1,\dots,w)$ is given by the incidence matrix $N=(n_{ij})$. The vectors $x_i(i=1,\dots,v)$ consisting of elements one merely in class i and elements zero elsewhere form an orthogonal basis of the subspace A of treatment effects; each vector in A consists of elements which are equal within each class i. Similarly, the vectors $z_j(j=1,\dots,w)$ with elements one merely in block class j and elements zero elsewhere form an orthogonal basis of the subspace B of block effects; each vector in B has all elements within class j equal.

The usual additivity assumption of treatment and block effects is equivalent to the assumption that Ey belongs to the subspace $\langle A,B\rangle$ of R^n spanned by A and B.

It should be noted that the operator P_A, representing orthogonal projection of any vector y in R^n on A, is equivalent to replacing each element of y by the average of all elements within that class of A to which it belongs, as $P_A y = \sum_{i=1}^{v} \frac{(x_i, y)}{(x_i, x_i)} x_i$, and the coefficient $\frac{(x_i, y)}{(x_i, x_i)}$ is the class i average of elements in the vector y concerned.

Similarly we have the operator P_B, representing orthogonal projection on B.

Least squares estimation of Ey amounts to finding regression coefficients for x_i denoted by the coordinate v-vector α of elements α_i, and similarly those for z_j denoted by the coordinate w-vector β of elements β_j, such that

$$\sum_i \alpha_i x_i + \sum_j \beta_j z_j$$

is the orthogonal projection of y on $\langle A,B \rangle$, or equivalently to finding $a \in A$ and $b \in B$ such that

$$y - (a+b) \perp \langle A,B \rangle.$$

Orthogonal projection of the left hand side on A and B, respectively, gives

$$P_A y - a - P_A b = 0; \quad P_B y - P_B a - b = 0. \tag{1}$$

Orthogonal projection of the first equation in (1) on B gives

$$P_B P_A y - P_B a - P_B P_A b = 0 \tag{2}$$

and subtraction of (2) from the second equation in (1) yields an equation for the block effects vector b only (I being the identity operator):

$$(I - P_B P_A) b = P_B y - P_B P_A y \tag{3}$$

which is equivalent to the reduced normal equations for block effects. The first equation gives subsequently

$$a = P_A(y-b). \tag{4}$$

Giving full attention to equation (3) in the unknown b, of great interest in rectangular lattices, we note that the right hand side

$$d = P_B(y-P_A y) \tag{5}$$

is orthogonal to $G = A \cap B$ in general, which can be proved as follows. Let A^* be the orthogonal complement of G in A, and similarly B^* that of G in B, then $y-P_A y$ is orthogonal to A and in particular to G, so that $P_B = P_G + P_{B^*}$ applied to $y - P_A y$ will not give a component belonging to G, and hence d will belong to B^*.

The solution of (3) for $I-P_B P_A$ restricted to B^* can in principle be obtained by the knowledge of the orthonormal canonical basis $u_1, \ldots, u_v$ for A and similarly $v_1, \ldots, v_w$ for B such that the inner product (u_i, v_j) is zero for all $i \neq j$, and ρ_i for $i = j$, while

$1 \geq \rho_1 \geq \rho_2 \geq \cdots \geq \rho_{\min(v,w)} \geq 0$. [See Corsten (1976)].

The vectors v_i belonging to the same canonical angle $\arccos \rho_i$ span a canonical subspace B_i of B. Obviously $\rho_i = 1$ corresponds to $G = A \cap B$. It follows that

$$P_B P_A v_i = P_B \rho_i u_i = \rho_i^2 v_i = \lambda_i v_i \quad \text{with} \quad \lambda_i = \rho_i^2$$

and similarly $P_A P_B u_i = \lambda_i u_i$; hence the equation

$$(I-P_B P_A)h = v_i$$

for h has the solution $\frac{1}{1-\lambda_i} v_i = \frac{1}{e_i} v_i$ in B^* (for which $\rho_i < 1$), where $e_i = 1-\lambda_i$ is the canonical efficiency factor. As all canonical correlation coefficients corresponding to B^* are smaller than one, the solution of (3) belonging to B^* is

$$b = \Sigma \frac{1}{1-\lambda_i} P_{B_i} d = \Sigma \frac{1}{e_i} P_{B_i} d, \tag{6}$$

summation extending to all canonical subspaces B_i of B^*. For connected designs (as are rectangular lattices) G is the one-dimensional space spanned by a vector consisting merely of elements one.

3. Canonical subspaces in block space of rectangular lattices

It can be proved quite simply that for rectangular lattices the block space B has four natural orthogonal canonical subspaces B_0, B_1, B_2 and B_3, i.e. with the property $Qv_i = \lambda_i v_i$ for all v_i in B_i where $Q = P_B P_A$:

B_0 is the trivial $G = A \cap B$ of dimension 1 and $\lambda_0 = 1$;

B_1 is the orthogonal complement of B_0 in the space F of replicate effects; since replicates and treatments are orthogonal, $\lambda_1 = 0$;

B_2 is the orthogonal complement of B_0 in the space W of partner effects (in our examples row effects) orthogonal to replication effects, with dimension k and $\lambda_2 = \frac{k+1-r}{rk}$ (equal to zero if $r = k+1$);

B_3 is concerned with the product classification $F \times W$ of F and W, i.e. the space B; the orthogonal complement of F and W in $F \times W$, i.e. the space of interaction contrasts $(F \times W)^*$ between replicates and partners form B_3 with dimension $k(r-1)$ and $\lambda_3 = \frac{k+1}{rk}$.

4. Intra-block analysis of rectangular lattices

The solution of (3) follows immediately from (6) as

$$b = \sum_{i=1}^{3} \frac{1}{e_i} P_{B_i} d, \qquad (7)$$

with $e_i = 1-\lambda_i$.

With the orthogonality of d, B_1, and B_2 to $B_0 = G$ and the well-known identity $P_{B_3} = P_B - P_F - P_W + P_{B_0}$ one can reduce (7) to

$$b = [\frac{1}{e_3} I + (\frac{1}{e_1} - \frac{1}{e_3}) P_F + (\frac{1}{e_2} - \frac{1}{e_3}) P_W] d \tag{8}$$

involving only simple averaging procedures, with $e_1 = 1$ and, possibly, replacement of $P_F d$ by $P_{F*} y$, as $P_F P_B = P_F$, $P_F P_A y = P_{B_O} y$ and $P_{F*} = P_F - P_{B_O}$.

Now the least squares estimator of the vector of treatment effects will be according to (4)

$$a = P_A(y-b) = P_A y - \frac{1}{e_3} P_A d - (\frac{1}{e_2} - \frac{1}{e_3}) P_A P_W d, \tag{9}$$

not containing P_F nor e_1.

This equation (9) exhibits, in addition to other terms, a linear mapping of $P_A y$ consisting of treatment averages by the operator

$$I + \frac{1}{e_3} P_A P_B + (\frac{1}{e_2} - \frac{1}{e_3}) P_A P_W . \tag{10}$$

From (10) we can deduce the submatrix V, related to α, of the effective (weak) inverse of the matrix of unreduced normal equations for α and β affecting the column of treatment totals only. Let N and $\tilde{N}$ be the binary incidence matrices of A and B, or A and W, respectively. Since

$\frac{1}{k} N'$ represents P_B, averaging relevant treatment elements over blocks,

$\frac{1}{r} N$ represents P_A, averaging relevant block elements over treatments classes, and similarly,

$\frac{1}{kr} \tilde{N}'$ represents P_W, averaging treatment elements over partner classes,

and $\frac{1}{r} \tilde{N}$ represents P_A, averaging partner class elements over treatment classes,

we find that

$$V = \frac{1}{r}[I_{k(k+1)} + \frac{1}{e_3 rk} NN' + \frac{1}{r^2 k} (\frac{1}{e_2} - \frac{1}{e_3}) \widetilde{NN'}]. \tag{11}$$

Hence for any pair of treatments i and j the difference in effects will be estimated as if both were uncorrelated each with variance, except for the factor σ^2, equal to

$$v_{ii} - v_{ij} = \frac{1}{r}[1 + \frac{r-\lambda_{ij}}{rk-k-1} - \frac{r-\tilde{\lambda}_{ij}}{(r-1)(k+1)(rk-k-1)}], \tag{12}$$

where λ_{ij} is the number of blocks containing both treatments i and j, and $\tilde{\lambda}_{ij}$ is the number of partners (rows in our examples) containing both treatments i and j. Note that λ_{ij} is binary, while $\lambda_{ij} \leq \tilde{\lambda}_{ij} \leq r$.

The factor in square brackets is the reciprocal of the efficiency factor involved in the comparison of treatments i and j. The values of this factor can vary widely, in particular for $r = 2$ and $k = 3$ up to the ratio 1.46.

5. Statistical consequences

In contrast to balanced incomplete blocks, lattice designs, group divisible, latin square type, or triangular partially balanced desings where the treatment space has factorial structure, rectangular lattices have natural canonical subspaces in the block space. In other words, the dual of a rectangular lattice has factorial structure. Spaces F and W in block space may be identified with main effects in the block space, If blocks are formed by litters of animals in an experiment where treatments are diets, F may be a classification according to stables or environments, W a classification according to common fathers and $(F \times W)^* = B_3$ the interaction contrasts of environmental and paternal effects. Moreover these three effects can be tested in a quasi-independent manner, as the sum of squares (b,d) for testing block effects can be partitioned in three orthogonal and (under normality) independent sums of squares according to (6) and (7) as

$$\sum_{i=1}^{3} \frac{1}{e_i} |P_{B_i} d|^2 . \tag{13}$$

6. Using inter-block information

Recovery of inter-block information with uniform block size under the assumption of random uncorrelated block effects with expectation zero and variance σ_b^2, while σ_e^2 is the residual variance, requires in general the unique solution for a from the equation

$$(I - gP_AP_B)a = P_Ay - gP_AP_By \tag{14}$$

with $g = \dfrac{k\sigma_b^2}{k\sigma_b^2 + \sigma_e^2}$; $I - gP_AP_B$ is non-singular unless $g = 1$ (i.e. $\sigma_b^2/\sigma_e^2 \to \infty$), which is the intrablock situation.

The solution of (14) for $a \in A$ and $g = 1$ is for incomplete blocks in general equal to the sum of P_Gy and the intrablock solution belonging to A*; this sum is given in this particular case by equation (9). The required solution of (14) involves simple replacement in (9) of λ_i by $g\lambda_i$, since $gP_AP_Bu = g\lambda_iu$ for any canonical vector u in the canonical space A_i, and of P_B by gP_B everywhere it occurs in (9) in the form (5), i.e. $d = P_B(y-P_Ay)$.

The inverse operator of $I - gP_AP_B$ affecting P_Ay in this solution appears to be

$$I - \frac{g}{1-g\lambda_3} P_AP_B + \left(\frac{1}{1-g\lambda_2} - \frac{1}{1-g\lambda_3}\right) gP_AP_W. \tag{15}$$

Hence the covariance matrix of the treatment effects estimator equals

$$\frac{\sigma_e^2}{r}\left[I_{k(k+1)} + \frac{1}{rk(\frac{1}{g} - \lambda_3)} NN' + \frac{1}{r^2k}\left(\frac{1}{\frac{1}{g} - \lambda_2} - \frac{1}{\frac{1}{g} - \lambda_3}\right)\tilde{N}\tilde{N}'\right], \tag{16}$$

a result essentially the same as that by Williams (1977), but not as directly obtained as here. Moreover, using p repeats of rectangular lattices requires only $\frac{1}{r}$ replaced by $\frac{1}{pr}$ in equations (11), (12) and (16), instead of new formulae as proposed by Williams and Ratcliff (1980).

References

[1] Corsten, L.C.A.(1976). Canonical correlation in incomplete blocks, In: Essays in Probability and Statistics, Shinko Tsusho, Tokyo, 125-154.

[2] Harshbarger, B.(1947). Rectangular lattices. Virginia Agricultural Experimental Station Memoirs 1.

[3] Williams, E.R. (1977). A note on rectangular lattice designs, Biometrics 33, 410-414.

[4] Williams, E.R. and Ratcliff, D.(1980). A note on the analysis of lattice designs with repeats, Biometrika 67,706-708.

MULTIPLE COMPARISONS BETWEEN SEVERAL TREATMENTS AND A SPECIFIED TREATMENT

C.W. Dunnett
McMaster University
Hamilton, Canada

1. Introduction

Suppose that there are several treatments to be compared in an experiment. The experimental design may be any one to which an analysis of variance model is applied; here, for simplicity, we assume the design is a one-way classification (i.e., the only factor in the design is the treatment factor). However, usually the analysis of variance F-tests do not satisfy the needs of the experimenter, who often wishes to make specific comparisons between certain treatment means. This leads to multiple tests if there are more than two treatments and the problem is how to make valid inferences concerning the treatment comparisons of interest in the experiment when there is more than one to be considered. The early work by Duncan (1951), Scheffé (1953) and Tukey (1953) laid the foundations for the subject of multiple comparisons or, as it is sometimes called, simultaneous inference. For a review, see Miller (1966 or 1981).

The most common situation is one where there is no structure between the treatments. Suppose we label the treatments A, B, C, etc. and the experimenter is interested in comparing treatment A with treatment B, treatment A with treatment C, and so on. In other words, if m denotes the number of treatments, he is interested in the $m(m-1)/2$ comparisons between all pairs of treatments. Tukey's (1953) method based on the Studentized range statistic is now a standard procedure described in many introductory textbooks. It provides simultaneous confidence limits (i.e., confidence limits with a specified confidence coefficient that all the pertinent parameter differences are simultaneously included within the stated confidence limits) for the $m(m-1)/2$ differences between pairs of treatments. The corresponding procedure for significance testing is the step-down method known as the Newman-Keuls method. Of course, Tukey's and the Newman-Keuls methods are not universally accepted. They are used when it is desired to control the experimentwise error rate rather than the per comparison error rate, but there may be situations in which the latter is more relevant. Other methods have been proposed, such as Duncan's "k-ratio" method (Duncan and Brant 1983). However, in this paper we will be concerned with methods of the Tukey and Scheffé type in which simultaneous confidence interval estimates are required.

Scheffé's (1953) method can be used to test a set of contrasts among the treatment means (including the pairwise differences between the means). However, if the contrasts of interest to the experimenter can be specified in advance, it may be possible to employ a method that is more efficient for that particular set of contrasts. For example, if the contrasts are mutually orthogonal, Bechhofer and Dunnett (1982) have proposed the use of the Studentized maximum modulus statistic and have provided tables.

Similarly, the experimenter may not be interested in the set of all pairwise comparisons between treatments, but only in a subset of these comparisons. If so, it may be possible to improve upon Tukey's method. A particular case to be considered here is the situation where one of the treatments has a special role, in that the only comparisons of interest are between this specified treatment and each of the others. This can happen if one of the treatments is a control or standard treatment and the remaining treatments are experimental treatments which are to be compared with the standard, or if the experiment consists of a new treatment under investigation which is to be compared with several reference standards.

A number of authors have been concerned with the methodology of multiple comparisons between several treatments and a single specified treatment, which were referred to by Miller (1966 or 1981) as many-to-one comparisons. The earliest known reference is Roessler (1946) who proposed that each comparison be made at significance level $1-(1-\alpha)^{1/m}$, where m is the number of treatments vs. control comparisons. This same adjustment of the level of significance was proposed later by Šidák (1967) for a more general multiple comparisons situation and shown to provide a conservative approximation. The exact solution for treatments vs. control multiple comparisons was obtained by Dunnett

(1955) and is given in Section 2 below. Other authors who have written on the treatments vs. control problem are Łukaszewicz and Sadowski (1960), Konijn (1960) and Šidák (1962). Shaffer (1977) has considered an extension to all pairwise intergroup comparisons when the treatments are divided into two groups.

Tukey's (1953) method for making all pairwise treatment comparisons assumes that the treatment groups have equal variances and equal sample sizes. Various alternatives have been proposed over the years to modify the method to cover unequal sample size and/or unequal variance situations: for a comprehensive review, see Stoline (1981). A recent result is the paper by Hayter (1984) which has finally provided the analytical proof of the conjecture that the so-called Tukey-Kramer modification is conservative in unequal sample size, homogeneous variance situations.

The main purpose of the present paper is to consider the treatments vs. control multiple comparisons problem in unequal sample size and/or unequal variance situations. If variance homogeneity can be assumed, but sample sizes are unequal, there is no problem in principle, since Dunnett (1955) provided the exact method for this situation. The problem is a computational one: we consider this problem in Section 3. When variance homogeneity cannot be assumed, then a single pooled variance estimate is not appropriate: this problem is treated in Section 4.

2. Multiple comparisons with a specified treatment when variance homogeneity can be assumed.

For the remainder of this paper we assume there are $k+1$ treatments in the experiment, indicated by subscripts $o, 1, \ldots, k$, with the subscript o referring to the specified treatment. We assume the data y_{ij} satisfy the one-way fixed effects ANOVA model:

$$y_{ij} = \mu_i + e_{ij}\,;\; i = o,1,\ldots,k \quad \text{and} \quad j = 1,\ldots,n_i, \tag{2.1}$$

where the e_{ij} are independent $N(0,\sigma^2)$ random variables and the μ_i and σ^2 are unknown parameters. The μ_i are the treatment means, μ_o being the treatment mean for the specified treatment. We assume that the experimenter is interested in the k quantities $\mu_i-\mu_o (i=1,\ldots,k)$; in particular, we assume that confidence interval estimates of the $\mu_i - \mu_o$ with stated joint confidence coefficient are required.

Denote the confidence limits as follows:

$$\mu_i - \mu_o \in \bar{y}_i - \bar{y}_o \pm hs\sqrt{\frac{1}{n_i} + \frac{1}{n_o}}, \tag{2.2}$$

where $\bar{y}_i$ and $\bar{y}_o$ are the sample means for the i-th and the specified treatments respectively, s^2 is the usual unbiased estimate of σ^2 based on ν df, and h is a constant to be determined to satisfy the probability requirement that (2.2) be correct for all $i = 1,\ldots,k$ simultaneously with probability $1-\alpha$, which is the specified joint confidence coefficient. It is easy to see that this requirement can be written

$$\Pr\{|t_i| < h;\ i = 1,\ldots,k\} = 1 - \alpha, \tag{2.3}$$

where $t_i = \dfrac{(\bar{y}_i - \bar{y}_o) - (\mu_i - \mu_o)}{s\sqrt{\frac{1}{n_i} + \frac{1}{n_o}}}$ is an ordinary

Student t variate. However, to determine h to satisfy (2.3), the joint distribution of the t_i is required, and this is the multivariate Student t-distribution defined by Dunnett and Sobel (1954) and also by Cornish (1954). The correlation coefficient between the numerators of t_i and t_j is given by

$$\rho_{ij} = \frac{1}{\sqrt{(1+\frac{n_o}{n_i})(1+\frac{n_o}{n_j})}}. \tag{2.4}$$

Denoting $b_i = 1/\sqrt{1+n_o/n_i}$, we see that the left-hand side of (2.3) can be expressed in the following integral form:

$$\int_0^{\infty}\int_{-\infty}^{+\infty}\prod_{i=1}^{k}\left[\Phi\left\{\frac{hs - b_i y}{\sqrt{1-b_i^2}}\right\} - \Phi\left\{\frac{-hs - b_i y}{\sqrt{1-b_i^2}}\right\}\right]\phi(y)dy\,p(s)ds, \tag{2.5}$$

where $\Phi(\cdot)$ and $\phi(\cdot)$ are the c.d.f. and density function of the standard normal distribution and $p(s)$ is the density function of s. Thus, the multiple comparisons problem is solved by evaluating the integral in (2.5) and determining the value of h to achieve the specified value $1-\alpha$ for the joint confidence coefficient. One-sided confidence intervals may be obtained in an analogous manner, the integral expression being the same as (2.5) except for the deletion of the second $\Phi(\cdot)$ expression within the rectangular brackets. The author has a Fortran computer program to evaluate the integral in (2.5). However, it would be convenient to have tables of h available. It would be a hopeless task to compute general tables for completely unequal sample sizes due to the large number of combinations of values of the correlation coefficients given by (2.4) that would be required.

Some special cases are useful. If $n_1 = \ldots = n_k$ (but $\neq n_o$), then all the ρ_{ij} are equal to a common value $\rho = 1/(1+n_o/n_1)$. If, in addition, $n_o = n_1$ then $\rho = 0.5$. Tables for the case $\rho = 0.5$ were computed by Dunnett (1955, 1964). Hahn and Hendrickson (1971) gave tables for the two-sided case for $\rho = .0, .2, .4, .5$ and $\alpha = .10, .05, .01$. Gupta et al (1983) have tables for the one-sided case for $\rho = 0.0\ (.1)\ 0.9$ and $\alpha = .25, .1, .05, .01$. The present author has a set of tables for both one-sided and two-sided cases for $\rho = 0.0\ (.1)\ 0.9$ and $\alpha = .2, .1, .05$ and $.01$. In addition, values for sample sizes in the ratio $n_o/n_i = \sqrt{k}$, representing optimal allocation between control and treatments, are included.

3. Homogeneous variances and unequal sample sizes

As described in Section 2 above, the homogeneous variance, unequal sample size situation is solved in principle, but unless a computer program is available to the user to evaluate the integral in (2.5), he would be restricted to $n_1 = \ldots = n_k \neq n_o$ situations due to the absence of tables other than for the equal-ρ case. In this section, we provide an approximate solution, but a surprisingly accurate one, which enables existing tables to be utilized.

The method proposed is as follows. Compute the $k(k-1)/2$ correlation coefficients, ρ_{ij}, given by (2.4) above. Denote by $\bar{\rho}$ their arithmetic average value. Then, instead of determining h to satisfy the integral in (2.5), which requires a computer program to evaluate the integral, we use the value $\bar{\rho}$ as the common correlation coefficient and interpolate on ρ in available tables. Numerical studies by the author indicate that this method is quite accurate, certainly adequate for all practical purposes, and also differs from the exact result in the conservative direction.

4. Heterogeneous variances

Consider now the model given by (2.1) but with the e_{ij} being $N(0,\sigma_i^2)$ random variables, $\sigma_i^2 \neq \sigma^2$. Let s_i^2 be an unbiased estimate of $\sigma_i^2 (i = o,1,\ldots,k)$. This is the same model considered in Dunnett (1980) for the problem of making all pairwise comparisons between treatments.

Proceeding as in Dunnett (1980), we propose to use confidence interval estimates of the following form, instead of (2.2), for the heterogeneous variances situation:

$$\mu_i - \mu_o \in \bar{y}_i - \bar{y}_o \pm h\sqrt{\frac{s_i^2}{n_i} + \frac{s_o^2}{n_o}} \quad . \tag{4.1}$$

The problem then is to determine the value of h to approximately achieve the probability requirement.

The correlation coefficient between $\bar{y}_i - \bar{y}_o$ and $\bar{y}_j - \bar{y}_o$ is no longer given by (2.4), but is

$$\rho_{ij} = \frac{1}{\sqrt{\left(1 + \frac{\sigma_i^2/n_i}{\sigma_o^2/n_o}\right)\left(1 + \frac{\sigma_j^2/n_j}{\sigma_o^2/n_o}\right)}} \tag{4.2}$$

which is unknown. Thus, the multivariate Student t-distribution is no longer applicable.

Computer simulations have been done, analogous to those described in Dunnett (1980), to investigate alternative possibilities. One possibility is to replace the unknown variances in (4.2) by their estimates, and average the $k(k-1)/2$ resulting estimates $\hat{\rho}_{ij}$, and then determine h from tables of multivariate Student's t for the equal correlation case using the average $\hat{\rho}$ obtained. In many situations, this leads to conservative intervals, but sometimes they are slightly non-conservative (analogous to the GH procedure described in Dunnett (1980)). Other conservative procedures are available, however, analogous to the C and T3 procedures described in Dunnett (1980); these will be described more fully in a forthcoming publication (Dunnett (1984)).

5. Numerical example

Consider a hypothetical example comparing $k=4$ treatments with a control, in which the sample sizes are 10, 8, 12 and 6 on the four treatments and 10 on the control. Under the assumption of variance homogeneity, the pooled variance estimate from the five samples has $\nu = 41$ df. The values of the b_i used in calculating the correlation coefficients $\rho_{ij} = b_i b_j$ from (2.4) are .7071, .6667, .7385 and .6124, respectively. The values of the ρ_{ij} are accordingly .4714, .5222, .4330, .4924, .4082 and .4523, with an average value of $\bar{\rho} = .4633$.

Suppose we require simultaneous two-sided confidence limits on the four treatment differences from control. The exact value of h required to make the integral in (2.5) equal to the value .95, when using the four values of the b_i obtained above, turns out to be $h = 2.5489$. Thus, the exact confidence limits are obtained by multiplying each standard error of a treatment difference from the control value by this constant, and adding it to and subtracting it from the observed treatment difference $\bar{y}_i - \bar{y}_o$ in accordance with (2.2) above.

On the other hand, approximate limits can be obtained by interpolating for the value $\rho = .4633$ in tables of the multivariate Student t with $\nu = 41$ df. This is equivalent to computing the integral in (2.5) with each b_i replaced by $\sqrt{.4633} = .6806$. The approximate value of h obtained in this way is $h = 2.5495$, which would yield slightly wider limits. The actual confidence coefficient using this approximate value of h is 95.007 % differing from the specified value of 95 % in the conservative direction by being slightly larger. Thus, it is clear that the approximation proposed in Section 3 yields quite accurate results in this example, and experience has shown that this is generally the case.

References

[1] Bechhofer, R.E. and Dunnett, C.W. (1982), Multiple comparisons for orthogonal contrasts: Examples and Tables, Technometrics, 24: 213-222.

[2] Cornish, E.A. (1954), The multivariate t-distribution associated with a set of normal sample deviates, Australian Journal of Physics, 7: 531-542.

[3] Duncan, D.B.(1951), A significance test for differences between ranked treatments in an analysis of variance, Virginia Journal of Science, 2: 171-189.

[4] Duncan, D.B. and Brant, L.J.(1983), Adaptive t-tests for multiple comparisons, Biometrics,39: 790.

[5] Dunnett, C.W. (1955), A multiple comparison procedure for comparing several treatments with a control, Journal of American Statistical Association, 50: 1096-1121.

[6] Dunnett, C.W. (1964), New tables for multiple comparisons with a control, Biometrics, 20: 482-491.

[7] Dunnett, C.W.(1980), Pairwise multiple comparisons in the unequal variance case, Journal of American Statistical Association, 75: 796-800.

[8] Dunnett, C.W. and Sobel, M.(1954), A bivariate generalization of Student's t-distribution, with tables for certain special cases, Biometrika, 41: 153-169.

[9] Gupta, S.S., Panchapakesan, S. and Sohn, J.K.(1983), On the distribution of the studentized maximum of equally correlated normal random variables, Technical Report No.83-31, Department of Statistics, Purdue University.

[10] Hayter, A.J. (1984), A proof of the conjecture that the Tukey-Kramer multiple comparisons procedure is conservative, The Annals of Statistics, 12: 61-75.

[11] Konijn, H.S. (1960), Multiple comparisons with controls, The Australian Journal of Statistics, 2: 16-18.

[12] Łukaszewicz, J. and Sadowski, W. (1960), On comparing several populations with a control population, Zastosowania Matematyki, V: 309-320.

[13] Miller, R.G. (1966, 1981), Simultaneous statistical inference, McGraw-Hill Book Co.

[14] Roessler, E.B. (1946), Testing the signifficance of observations compared with a control, Proceedings of the American Society for Horticultural Science, 47: 249-251.

[15] Scheffé, H.(1953), A method for judging all contrasts in the analysis of variance, Biometrika, 40: 87 - 104.

[16] Shaffer, J.P. (1977), Multiple comparisons emphasizing selected contrasts: An extension and generalization of Dunnett´s procedure, Biometrics, 33: 293-304.

[17] Šidák, Z. (1962), Unequal numbers of observations in comparing several treatments with one control, (In Czech), Aplikace Matematiky, 7: 292-314.

[18] Šidák, Z.(1967), Rectangular confidence regions for the means of multivariate normal distributions, Journal of the American Statistical Association 62: 626-633.

[19] Stoline, M.R.(1981), The status of multiple comparisons: Simultaneous estimation of all pairwise comparisons in one-way ANOVA designs, The American Statistician, 35: 134-141.

[20] Tukey, J.W.(1953), The problem of multiple comparisons, Princeton University.

MINIMAX PREDICTION IN LINEAR MODELS

H. Drygas
Department of Mathematics
University of Hessen
Kassel, F.R.G.

1. Introduction

Let us consider the linear regression model

$$y = X\beta + \sigma\varepsilon, \tag{1.1}$$

where - as usual - y and ε are random $n \times 1$ - vectors, X is a known $n \times k$ - matrix (the regression matrix) and the $k \times 1$ - vector β and σ are unknown parameters. Moreover, let

$$E(\varepsilon) = 0, \; E(\varepsilon\varepsilon') = Q_{11}, \tag{1.2}$$

where Q_{11} is a given n.n.d. (non-negative definite) $n \times n$ - matrix. We want to assume that there is a restriction on β in the form of an ellipsoidal inequality, i.e., there exists a n.n.d. $k \times k$ - matrix T and an element $\beta_o \in \mathbb{R}^k$ such that

$$\beta \in B_1 = \{\beta : (\beta-\beta_o)'T(\beta-\beta_o) \leq k_o\}. \tag{1.3}$$

In contrary to the treatment in the literature (see, e.g. Rödel and Toutenburg (1978)) it is not assumed that T is regular. This case is often encountered in economics. Consider for example, the estimation of the consumption function $C = \alpha+\beta Y+\varepsilon$, where it is known that $.5 \leq \beta \leq .7$ and middle-point $\beta_o = .6$ may act as a reasonable choice for β_o. In this case

$$T = \begin{pmatrix} 0 & 0 \\ 0 & 1 \end{pmatrix}, \quad \beta_o = \begin{pmatrix} 0 \\ .6 \end{pmatrix}, \quad k_o = .01, \tag{1.4}$$

and T is singular. Rödel and Toutenburg overcome this problem by approximating T by

$$T_n = \begin{pmatrix} n & 0 \\ 0 & 1 \end{pmatrix}, \tag{1.5}$$

n a large number. In this paper we make a proposal for the use of the prior information (1.3) in the case of a singular T.

Since estimation is a special case of prediction we here will deal with the problem of prediction. Therefore let y_* be a random $m \times 1$ - vector to be predicted such that

$$E\begin{pmatrix} y \\ y_* \end{pmatrix} = \begin{pmatrix} X \\ X_* \end{pmatrix}\beta, \quad \mathrm{Cov}\begin{pmatrix} y \\ y_* \end{pmatrix} = \begin{pmatrix} \sigma^2 Q_{11} & \sigma Q_{12} \\ \sigma Q_{21} & Q_{22} \end{pmatrix}. \tag{1.6}$$

2. Minimax-prediction

Let us assume that we want to predict $p'y_*$, $p \in \mathbb{R}^m$, by the linear function $a'y + d$, $a \in \mathbb{R}^n$, $d \in \mathbb{R}$. Then the risk $R(a,d,p;\beta,\sigma)$ is given by

$$\begin{aligned} R(a,d,p;\beta,\sigma) &= E((a'y+d-p'y_*)^2) \\ &= (a'X\beta-p'X_*\beta+d)^2+\sigma^2 a'Q_{11}a-2\sigma a'Q_{12}p+p'Q_{22}p. \end{aligned} \tag{2.1}$$

We want to minimize

$$\sup_{\beta \in B_1} R(a,d,p;\beta,\sigma)$$

subject to a and d. It is convenient to rewrite the function $R(a,d,p;\beta,\sigma)$ as

$$\begin{aligned} R(a,d,p;\beta,\sigma) = ((\beta-\beta_o)'(X'a-(X_*)'p)+c)^2 + \\ \sigma^2 a'Q_{11}a-2\sigma a'Q_{12}p+p'Q_{22}p, \end{aligned} \tag{2.2}$$

where

$$c = d + \beta_o'(X'a - (X_*)'p). \tag{2.3}$$

Now let us at first see what happens if $Tx = 0$, $x \neq 0$. Consider $\beta_\lambda = \beta_o+\lambda x$, $\lambda \in \mathbb{R}$. Then it follows that $\beta_\lambda \in B_1$ and

$$R(a,d,p;\beta_\lambda,\sigma) = (\lambda x'(X'a-(X_*)'p)+c)^2+ \sigma^2 a'Q_{11}a-2\sigma a'Q_{12}p+p'Q_{22}p. \tag{2.4}$$

Clearly $\lim\limits_{\lambda\to\infty} \frac{1}{\lambda^2} R(a,d,p;\beta_\lambda,\sigma) = (x'(X'a-(X_*)'p)^2$. On the other hand $\sup\limits_{\beta\in B_1} R(a,d,p;\beta,\sigma) < \infty$ implies $\lim\limits_{\lambda\to\infty} \sup \frac{1}{\lambda^2} R(a,d,p;\beta_\lambda,\sigma) = 0$. Therefore $\sup\limits_{\beta\in B_1} R(a,d,p;\beta,\sigma) = \infty$ unless $x'(X'a-(X_*)'p) = 0$. This gives rise to the following

<u>2.1 Definition</u>: $p'y_*$, $p\in\mathbb{R}^m$, is called minimax-B_1-predictable under the model (1.6) with respect to the restrictions (1.3) if, for some $a\in\mathbb{R}^n$, $x'(X'a-(X_*)'p) = 0$ for each $x\in T^{-1}(0)$.

This is equivalent to $(X_*)'p\in \operatorname{im}(X'\vdots T)$.
The latter assertion follows from $(T^{-1}(0))^\perp = \operatorname{im}(T)$ and $\operatorname{im}(X'\vdots T) = \operatorname{im}(X')+\operatorname{im}(T)$. Because of $E(x'(a'y-p'y_*)) = 0$ if $Tx = 0$, $a'y$ could also be called a T-unbiased predictor of $p'y_*$. In order to compute $\sup\limits_{\beta\in B_1} R(a,d,p;\beta,\sigma)$ in the case $(X_*)'p\in \operatorname{im}(X'\vdots T)$ we represent T in the form $T = D'D$, where D is an $r\times k$-matrix. The case $r = \operatorname{rank}(T)$ will be the most convenient choice.

<u>2.2. Lemma</u>: If $p'y_*$ is minimax-B_1-predictable, then

$$\sup_{\beta\in B_1} R(a,d,p;\beta,\sigma) = (|c| + \sqrt{k_0}\,\|Db\|)^2+\sigma^2 a'Q_{11}a - 2\sigma a'Q_{12}p+p'Q_{22}p,$$

where $(X_*)'p - X'a = Tb = D'Db$ and c is defined by (2.3).

<u>Proof</u>: Before Definition 2.1 it has been shown that the supremum of $R(a,d,p;\beta,\sigma)$ over $\beta\in B_1$ is finite iff $a'y$ is T-unbiased for $p'y_*$. This is equivalent to $X'a-(X_*)'p\in \operatorname{im}(T)$ or $(X_*)'p = X'a+Tb = X'a+D'Db$, say. Therefore we have to find the supremum of

$$((\beta-\beta_0)'D'Db+c)^2 = ((\beta-\beta_0)'D'Db)^2+2c(\beta-\beta_0)'D'Db+c^2. \tag{2.5}$$

Let us at first assume $c\geq 0$. Then it follows from Cauchy-Schwarz inequality

$$((\beta-\beta_0)'D'Db+c)^2 \leq \|D(\beta-\beta_0)\|^2\|Db\|^2+ \tag{2.6}$$

$$+2c\,\|D(\beta-\beta_o)\|\,\|Db\| + c^2 \le k_o\|Db\|^2 + 2c\sqrt{k_o}\,\|Db\| + c^2 = (c+\sqrt{k_o}\,\|Db\|)^2,$$

since $\beta \in B_1$ is equivalent to $\|D(\beta-\beta_o)\| \le \sqrt{k_o}$. If $c < 0$, then $((\beta-\beta_o)'D'D(-b)+(-c))^2 = ((\beta-\beta_o)'D'Db+c)^2$ and in view of $\|D(-b)\| = \|Db\|$ it is obtained that

$$(c+(\beta-\beta_o)'D'Db)^2 \le (-c+\sqrt{k_o}\,\|Db\|)^2. \tag{2.7}$$

Now if $\|Db\| = 0$ nothing has to be proved. If $\|Db\| \neq 0$, then choose $\beta = \beta_o + \sqrt{k_o}\,\|Db\|^{-1}b$ and $|c| + (\beta-\beta_o)'D'Db = |c| + \sqrt{k_o}\,\|Db\|$ is obtained. This finishes the proof of the lemma. Q.E.D.

We have found that under minimax - B_1 - predictability $\sup_{\beta\in B_1} R(a,d,p;\beta,\sigma)$ equals $(|c| + \sqrt{k_o}\,\|Db\|)^2 + \sigma^2 a'Q_{11}a - 2\sigma a'Q_{12}p + p'Q_{22}p$.

This expression has to be minimized with respect to c and b. Obviously the optimal choice of c is $c = 0$, i.e. $d = -\beta_o'(X'a-(X_*)'p) = \beta_o'D'Db = (D_{\beta_o})'Db$. Finally $\|Db\|^2 = b'D'Db = b'Tb = b'TT^-Tb = ((X_*)'p-X'a)'T^-((X_*)'p-X'a)$ and the task of minimizing

$$\begin{aligned} &k_o(X'a-(X_*)'p)'T^-(X'a-(X_*)'p) + \sigma^2 a'Q_{11}a \\ &-2\sigma a'Q_{12}p + p'Q_{22}p \end{aligned} \tag{2.8}$$

subject to $X'a-(X_*)'p \in \mathrm{im}(T) = \mathrm{im}(D')$ arises in connection with minimax-prediction (T^- is a generalized inverse of T, i.e. $TT^-T = T$). This problem will be solved by the following theorem.

<u>2.3. Theorem</u>: $a'y+d$ is minimax-B_1-predictor of $p'y_*$ iff $a'y+d$ is BLUP (Best Linear Unbiased Predictor) of $p'y_*$ in one of the following two linear prediction models:

$$\text{(i)}\quad E\begin{pmatrix} y \\ T^-T\beta_o \\ y_* \end{pmatrix} = \begin{pmatrix} X \\ T^-T \\ X_* \end{pmatrix}\beta\ , \quad \mathrm{Cov}\begin{pmatrix} y \\ T^-T\beta_o \\ y_* \end{pmatrix} = \begin{pmatrix} \sigma^2Q_{11} & 0 & \sigma Q_{12} \\ 0 & k_oT^- & 0 \\ \sigma Q_{21} & 0 & Q_{22} \end{pmatrix},$$

where T^- is a symmetric and reflexive and thus n.n.d. g-inverse of T.

(ii) $E\begin{pmatrix} y \\ D\beta_o \\ y_* \end{pmatrix} = \begin{pmatrix} X \\ D \\ X_* \end{pmatrix}\beta$, $\mathrm{Cov}\begin{pmatrix} y \\ D\beta_o \\ y_* \end{pmatrix} = \begin{pmatrix} \sigma^2 Q_{11} & 0 & \sigma Q_{12} \\ 0 & kDT^-D' & 0 \\ \sigma Q_{21} & 0 & Q_{22} \end{pmatrix}$.

Here $T = D'D$ and DT^-D' is independent of the g-inverse T^- of T. If Rank $(D) = \mathrm{Rank}(T) = r$, $D \in \mathbb{R}_{r \times k}$, then $DT^-D' = I_r$.

This theorem needs some explanation because β_o is not a random vector, but a fixed vector. If T and Q_{11} are both regular, then it is known (see, e.g., Rödel and Toutenburg (1978)) that the minimax-B_1- estimator of β is given by $\tilde{\beta} = (\sigma^{-2}X'Q_{11}^{-1}X + k_o^{-1}T)^{-1}(\sigma^{-2}X'Q_{11}^{-1}y + k_o^{-1}T\beta_o)$. This formula allows a Bayesian interpretation, too. If y,β are both normally distributed $y|\beta \sim N(X\beta, \sigma^2 Q_{11})$, $\beta \sim N(\beta_o, k_o T^{-1})$, then the posterior distribution $\beta|y$ is given by $N(\tilde{\beta}, V)$, where $V = (\sigma^{-2}X'Q_{11}^{-1}X + k_o^{-1}T)^{-1}$. Now we change the interpretation of β_o and β by letting β be an unknown but fixed parameter and by letting β_o be an (observed) random vector uncorrelated with y and having expectation β and covariance-matrix k_oT^{-1}. If we adopt this interpretation of β_o, then $\tilde{\beta}$ is also BLUE (Best Linear Unbiased Estimator) of β in the model

$$E\begin{pmatrix} y \\ \beta_o \end{pmatrix} = \begin{pmatrix} X \\ I \end{pmatrix}\beta, \quad \mathrm{Cov}\begin{pmatrix} y \\ \beta_o \end{pmatrix} = \begin{pmatrix} \sigma Q_{11} & 0 \\ 0 & k_o T^{-1} \end{pmatrix}.$$

Theorem 2.3 extends this result to the case of prediction and the case of possibly rank-deficient matrices X, Q_{11}, T.

<u>Proof of the theorem</u>: a) $a'y+d$ is minimax-B_1-predictor of $p'y_*$ iff $d = -\beta_o'(X'a-(X_*)'p)$, $X'a-(X_*)'p \in \mathrm{im}(T) = \mathrm{im}(TT^-)$ and

$$k_o(X'a-(X_*)'p)'T^-(X'a-(X_*)'p)+\sigma^2 a'Q_{11}a-2\sigma a'Q_{12}p+p'Q_{22}p \tag{2.9}$$

is minimized subject to $X'a-(X_*)'p \in \mathrm{im}(T)$.

b) $a_o'y + b'T^-T\beta_o$ is BLUP of $p'y_*$ iff the expectation of $a_o'y+b'T^-T\beta_o$ coincides with $E(p'y_*)$ in model (i) and $E((a_o'y+b'T^-T\beta_o-p'y_*)^2)$ is minimized in this class of unbiased predictors of $p'y_*$.

The requirement of unbiasedness yields $a_o'X\beta + b'T^-T\beta = p'X_*\beta$, i.e., $-X'a_o+(X_*)'p = TT^-b \in \mathrm{im}(T)$ and $a_o'y+b'T^-T\beta_o = a_o'y+\beta_o'((X_*)'p-X'a_o)$. This shows that the class of unbiased linear predictors of $p'y_*$ in model

(i) coincides with the class of predictors $a'y+d$, where $d = -\beta_o'(X'a-(X_*)'p)$. The function that has to be minimized in model (i) is

$$\sigma^2 a_o' Q_{11} a_o + k_o b'T^- b - 2\sigma a_o' Q_{12} p + p' Q_{22} p. \tag{2.10}$$

As T^- is symmetric and reflexive, $T^- = T^- T T^- = T^- T (T^-)'$, which shows that T^- is also n.n.d. Since T^- is reflexive the term $k_o b'T^- b$ in (2.10) also equals $k_o b'T^- T T^- T T^- b = k_o((X_*)'p - X'a_o)'T^-((X_*)'p - X'a_o)$, where use was made of the symmetry of T^-. Thus it is proved that the expressions (2.9) and (2.10) are identical and have to be minimized under the same restriction $X'a-(X_*)'p \in \mathrm{im}(T)$. Consequently both minimization procedures lead to the same result.

c) $a_1'y + a_2' D\beta_o$ is BLUP of $p'y_*$ in model (ii) iff $E(a_1'y + a_2' D\beta_o) = a_1' X\beta + a_2' D\beta = E(p'y_*) = p' X_* \beta$ and $E((a_1'y + a_2' D\beta_o) - p'y_*)^2)$ is minimized in this class of predictors. The requirement of unbiasedness is thus equivalent to $(X_*)'p - X'a_1 = D'a_2 \in \mathrm{im}(D') = \mathrm{im}(T)$ and $a_1'y + a_2' D\beta_o = a_1'y + ((X_*)'p - X'a_1)'\beta_o = a_1'(y - X\beta_o) + p'X_*\beta_o$.
The function to be minimized is

$$\begin{aligned} &\sigma^2 a_1' Q_{11} a_1 - 2\sigma a_1' Q_{12} p + p' Q_{22} p + k_o a_2' D T^- D' a_2 \\ &= \sigma^2 a_1' Q_{11} a_1 - 2\sigma a_1' Q_{12} p + p' Q_{22} p + k_o ((X_*)'p - X'a_1)' T^- ((X_*)'p - X'a_1) \end{aligned} \tag{2.11}$$

and the restriction is again $((X_*)'p - X'a_1) \in \mathrm{im}(D') = \mathrm{im}(T)$. This minimization problem coincides with the minimization problem given in (2.10) and, therefore, also with the determination of the minimax-B_1-predictor of $p'y_*$.

d) Since

$$D'DT^-D'D = TT^-T = T = D'D = TT^+T = D'DT^+D'D \tag{2.12}$$

is equivalent to $DT^-D' = DT^+D'$, it follows that DT^-D' is independent of the choice of the g-inverse T^- of T. If rank (T) = rank $(D') = r$, then $D'^{-1}(0) = 0$ and from $D'DT^-D'D = T = D'D$ it follows that $DT^-D'D = D$, $D' = D'D(T^-)'D' = D'DT^-D'$, $DT^-D' = I_r$. Evidently, also DT^-D' is the orthogonal projection onto $\mathrm{im}(D)$ (Rao and Mitra (1971), p.111).

<u>2.4. Theorem</u>: $\tilde{y}_* = Cy + d$, $C \in \mathbb{R}_{m\times n}$, $d \in \mathbb{R}^m$ has the property that $p'\tilde{y}_* = p'(Cy+d)$ is minimax-B_1-predictor of any minimax-B_1-predictable $p'y_*$ iff

$$(X_*-CX)'X_*'^{-1}(\mathrm{im}(X' \vdots T)) \subseteq \mathrm{im}(T), \; p'd = p'(X_*-CX)\beta_o \tag{2.13}$$

for all $p \in X_*'^{-1}(\mathrm{im}(X' \vdots T))$ and

$$\sigma^2 p'CQ_{11}C'p - 2\sigma p'CQ_{12}p + p'Q_{22}p + k_o p'(X_*-CX)T^-(X_*-CX)'p \tag{2.14}$$

is minimized uniformly for all $p \in X_*'^{-1}(\mathrm{im}(X' \vdots T))$. This again is the case iff $p'\tilde{y}_*$ is BLUP of any predictable $p'y_*$ (i.e. $X_*'p \in \mathrm{im}(X' \vdots T)$) in one of the two linear prediction models of the preceding theorem.

<u>Proof</u>: The theorem follows simply from $p'(Cy+d) = (C'p)'y + p'd$ and application of Theorem 2.3.

3. Minimax-estimation

If we let $m = k$, $X_* = I_k$, $Q_{12} = 0$, $Q_{21} = 0$, $Q_{22} = 0$, then the prediction problem becomes an estimation problem. The minimax-B_1-estimator is thus given by

$$a'(y - X\beta_o) + p'\beta_o, \tag{3.1}$$

where $p - X'a \in \mathrm{im}(T) = \mathrm{im}(D')$ and

$$\sigma^2 a'Q_{11}a + k_o(p - X'a)'T^-(p - X'a) \tag{3.2}$$

is minimized subject to $p - X'a \in \mathrm{im}(T)$.

<u>3.1. Theorem</u>: a) $a'(y - X\beta_o) + p'\beta_o$ is minimax-B_1-estimator of $p'\beta$ iff $p - X'a \in \mathrm{im}(T) = \mathrm{im}(D')$ and a minimizes

$$\sigma^2 a'Q_{11}a + k_o(p - X'a)'T^-(p - X'a) \tag{3.3}$$

subject to $p - X'a \in \mathrm{im}(T)$. This is the case iff $a'(y - X\beta_o) + p'\beta_o$ is BLUE (Best Linear Unbiased Estimator) of $p'\beta$ in one of the following models:

(i) $E\begin{pmatrix} y \\ T^-T\beta_o \end{pmatrix} = \begin{pmatrix} X \\ T^-T \end{pmatrix}\beta$, $\mathrm{Cov}\begin{pmatrix} y \\ T^-T\beta_o \end{pmatrix} = \begin{pmatrix} \sigma^2 Q_{11} & O \\ O & k_o T^- \end{pmatrix}$,

(ii) $E\begin{pmatrix} y \\ D\beta_o \end{pmatrix} = \begin{pmatrix} X \\ D \end{pmatrix}\beta$, $\mathrm{Cov}\begin{pmatrix} y \\ D\beta_o \end{pmatrix} = \begin{pmatrix} \sigma^2 Q_{11} & O \\ O & k_o DT^-D' \end{pmatrix}$.

b) $Cy+d$, $C \in \mathbb{R}_{k\times n}$, $d \in \mathbb{R}^k$ has the property that $p'(Cy+d)$ is minimax-B_1-estimator of $p'\beta$ whenever $p'\beta$ is B_1-estimable (i.e., $p \in \mathrm{im}(X' \vdots T)$) iff

$$d-(I-CX)\beta_o \in T^{-1}(O) \cap X^{-1}(O), \quad (I-CX)'\mathrm{im}(X' \vdots T) \subseteq \mathrm{im}(T) \tag{3.4}$$

and

$$\sigma^2 p'CQ_{11}C'p + k_o p'(I-CX)T^-(I-CX)'p \tag{3.5}$$

is minimized uniformly for all $p \in \mathrm{im}(X' \vdots T)$.

This theorem follows easily from Theorem 3.1 a) by noting that $p'(Cy+d) = (C'p)'y+p'd$ or by specializing Theorem 2.4. Note that again $T^-T\beta_o$ $(D\beta_o)$ are treated as if they are random vectors with expectation $T^-T\beta$ $(D\beta)$ and covariance matrix T^- (DT^-D') and uncorrelated with y.

We have used the connection between estimation and prediction to solve the estimation problem. This relation can be conversed. If the estimation problem is solved then also the prediction problem can be solved by solving an additional system of linear equations. This method was described in Drygas (1975). This equation is in our case

$$A\begin{pmatrix} \sigma^2 Q_{11} & O \\ O & k_o DT^-D' \end{pmatrix} = (\sigma Q_{21} \vdots O) \tag{3.6}$$

and the optimal predictor is obtained as

$$\tilde{y}_* = X_*\hat{\beta} + A\begin{pmatrix} y - X\hat{\beta} \\ D\beta_o - D\hat{\beta} \end{pmatrix}, \tag{3.7}$$

where $\hat{\beta}$ is a BLUE of β in model (ii) in the sense that $p'\hat{\beta}$ is BLUE of $p'\beta$ whenever $p'\beta$ is estimable. Let us split up $A = (A_{11} \vdots A_{12})$, then (3.4) becomes

$$\sigma^2 A_{11}Q_{11} = \sigma Q_{21}, \quad A_{12}k_o DT^-D' = O. \tag{3.8}$$

An obvious solution of the second equation is $A_{12} = O$ and we thus get the following theorem.

3.2. Theorem: A minimax-B_1-predictor of y_* is given by

$$\tilde{y}_* = X_*\hat{\beta} + \sigma^{-1}A_{11}(y-X\hat{\beta}) \tag{3.9}$$

in the sense that $p'\tilde{y}_*$ is minimax-B_1-predictor of $p'y_*$ whenever $p'y_*$ is minimax-B_1-predictable. Here A_{11} is any solution of the equation $A_{11}Q_{11} = Q_{21}$ and $\hat{\beta}$ is a BLUE of β in one of the models (i) or (ii) in the sense that $p'\hat{\beta}$ is BLUE of $p'\beta$ whenever $p'\beta$ is estimable.

The solution of the equation $A_{11}Q_{11} = Q_{12}$ is also needed when the ordinary best linear predictors (T=O) have to be determined. Therefore if the estimation part of the problem is solved the same problem has to be solved for all prediction problems.

4. Computation of the minimax-estimator

For computation of the minimax-predictor or the minimax-estimator the BLUE's of estimable functions of β in the model

$$E\begin{pmatrix} y \\ D\beta_o \end{pmatrix} = \begin{pmatrix} X \\ D \end{pmatrix}\beta, \quad \mathrm{Cov}\begin{pmatrix} y \\ D\beta_o \end{pmatrix} = \begin{pmatrix} \sigma^2 Q_{11} & O \\ O & k_o DT^- D' \end{pmatrix} \tag{4.1}$$

have to be determined. Consider therefore an estimable function $A\beta$ of β, i.e., $A = B_1X+B_2D$. A BLUE is determined by $G_1y+G_2D\beta_o$, where

$$G_1X + G_2D = A \text{ (unbiasedness)}, \tag{4.2}$$

$$\sigma^2 G_1Q_{11}a_1+k_oG_2DT^-D'a_2 = O \tag{4.3}$$

if $X'a_1+D'a_2 = O$ (optimality). Inserting $G_2D = A-G_1X$ and $D'a_2 = -X'a_1$ yields the equation

$$G_1(\sigma^2 Q_{11}+k_oXT^-X')a_1 = k_oAT^-X'a_1 \tag{4.4}$$

whenever $X'a_1 \in \mathrm{im}(D')$. Fortunately the equation $G_1(\sigma^2 Q_{11}+k_oXT^-X') = k_oAT^-X'$ has a solution. From $\mathrm{im}(\sigma^2 Q_{11}+k_oXT^-X') = \mathrm{im}(\sigma^2 Q_{11})+\mathrm{im}(k_oXT^-)$ (T^- is n.n.d.!) we get $(\sigma^2 Q_{11} + k_oXT^-X')(\sigma^2 Q_{11}+k_oXT^-X')^-k_oXT^- = k_oXT^-$.

By transposing we obtain that

$$G_1 = k_o AT^- X' (\sigma^2 Q_{11} + k_o XT^- X')^- \tag{4.5}$$

and the estimator obtained in this way is

$$A\hat{\beta} = k_o AT^- X' (\sigma^2 Q_{11} + k_o XT^- X')^- (y - X\beta_o) + A\beta_o . \tag{4.6}$$

This result, however, holds only under the provision that the equation $A - G_1 X = G_2 D$ possesses a solution G_2. A counter-example is $T = O$ in which case the estimation formula (4.6) is only valid for the trivial case $A = O$.

If for example $\mathrm{im}(X') \subseteq \mathrm{im}(D')$ then (4.4) holds for all $a_1 \in \mathbb{R}^n$. Then, however, $X' = D'B$, $X = B'D$ and $A = (B_1 B' + B_2) D$ and $A - G_1 X = (B_1 D' + B_2 + G_1 B') D$ whatever G_1 may be. This shows that the estimation formula (4.6) is applicable in this case.

An alternative formula to (4.6) can be obtained in the case $\mathrm{im}(X) \subseteq \mathrm{im}(Q_{11})$. We prepare it by

<u>4.1. Lemma</u>: Let $\mathrm{im}(X) \subseteq \mathrm{im}(Q_{11})$, $\mathrm{im}(X') \subseteq \mathrm{im}(T)$ and Q^- be a n.n.d. g-inverse of Q_{11}. Then

$$(\sigma^2 Q_{11} + k_o XT^- X')^- = \sigma^{-2} [Q^- - Q^- X (k_o^{-1} \sigma^2 T + X' Q^- X)^- X' Q^-] . \tag{4.7}$$

<u>Proof</u>: We abbreviate $k_o^{-1} \sigma^2 T + X' Q^- X$ by G. Then since Q^- is n.n.d., $GG^- X' Q^- = X' Q^-$. From this we get, in view of $\mathrm{im}(X) \subseteq \mathrm{im}(Q)$, $\mathrm{im}(X') \subseteq \mathrm{im}(T)$, that $Q_{11} Q^- X = X$, $XT^- T = X$ and

$$(\sigma^2 Q_{11} + k_o XT^- X') \sigma^{-2} (Q^- - Q^- X G^- X' Q^-) = Q_{11} Q^- + k_o \sigma^{-2} XT^- X' Q^- - X G^- X' Q^-$$

$$- k_o \sigma^{-2} XT^- X' Q^- XT^- G^- X' Q^- = Q_{11} Q^- + k_o \sigma^{-2} XT^- X' Q^- - XG^- X' Q^-$$

$$- k_o \sigma^{-2} XT^- (G - k_o^{-1} \sigma^2 T) G^- X' Q^- = Q_{11} Q^- , \tag{4.8}$$

$$Q_{11} Q^- (\sigma^2 Q_{11} + k_o XT^- X') = \sigma^2 Q_{11} + k_o XT^- X' . \tag{4.9}$$

From this formula we get that (4.6) can be written as follows

$$A\hat{\beta} = k_o \sigma^{-2} AT^- (X' Q^- - X' Q^- X (k_o^{-1} \sigma^2 T + X' Q^- X)^- X' Q^-) (y - X\beta_o) + A\beta_o . \tag{4.10}$$

Using that $X'Q^-X = G-k_o^{-1}\sigma^2T$, where again $G = k_o^{-1}\sigma^2T+X'Q^-X$, we get

$$A\hat{\beta} = k_o\sigma^{-2}AT^-(k_o^{-1}\sigma^2TG^-X'Q^-)(y-X\beta_o)+A\beta_o$$

$$=A(k_o^{-1}\sigma^2T+X'Q^-X)^-X'Q^-(y-X\beta_o)+A\beta_o, \tag{4.11}$$

where we have used $\operatorname{im}(A')\subseteq\operatorname{im}(T)$. This estimator is the well-known ridge-estimator.

The ridge-type structure of minimax-estimator can also be obtained via the Unified Least Squares approach (see Drygas (1984)). Since $\operatorname{im}(D)=\operatorname{im}(DT^-T)\subseteq\operatorname{im}(DT^-)=\operatorname{im}(DT^-D')$ this approach is considerably simplified. Let W be any symmetric n.n.d. matrix such that $\operatorname{im}(\sigma^2Q_{11}W^-X)\subseteq\operatorname{im}(X)\subseteq\operatorname{im}(W)$ (e.g. $W=\sigma^2Q_{11}+XX'$). Then

$$W_1 = \begin{pmatrix} W & O \\ O & k_oDT^-D' \end{pmatrix} \tag{4.12}$$

is ULS for the whole model and

$$A(X'W^-X+k_o^{-1}D'(DT^-D')^-D)^-X'W^-y+k_o^{-1}D'(DT^-D')^-D\beta_o \tag{4.13}$$

is BLUE of $A\beta$, whenever $A\beta$ is estimable.
Since $DT^-T=D$ in view of $D'DT^-T=D'D$, it follows that $(D^-)'T(D^-)$ is a g-inverse of DT^-D'. Indeed, $DT^-D'(D^-)'TD^-DT^-D'DT^-D'(D^-)'D'DD^-DT^-D' = DT^-D'DT^-D' = DT^-TT^-D' = DT^-D'$. Therefore $D'(DT^-D')^-D$ can be chosen equal to $D'(D')^-TD^-D = D'(D^-)'D'DD^-D = D'D = T$ and (4.13) simplifies to

$$A(X'W^-X+k_o^{-1}T)^-(X'W^-y+k_o^{-1}T\beta_o). \tag{4.14}$$

5. General quadratic loss functions

Another approach to estimation and prediction is the use of a general quadratic loss function. Let us consider this approach for the problem of estimation. Let

$$R(\hat{\beta},\beta) = E((\hat{\beta}-\beta)'A(\hat{\beta}-\beta)) \tag{5.1}$$

be the quadratic risk, A being a symmetric n.n.d. $k\times k$-matrix. For simplicity we consider only estimators of the form $\hat{\beta}=\beta_o+C(y-X\beta_o)$.

Then again $\sup_{\beta \in B_1} R(\hat{\beta},\beta) < \infty$ iff $\operatorname{im}(I-CX)' \subseteq \operatorname{im}(T)$ (T-unbiasedness).

It now follows that

$$R(\hat{\beta},\beta) = (\beta-\beta_o)'(CX-I)'A(CX-I)(\beta-\beta_o)+\sigma^2 \operatorname{tr}(ACQ_{11}C') \tag{5.2}$$

and with $(I-CX)' = TB = D'DB$

$$\begin{aligned} \sup_{\beta \in B_1} R(\hat{\beta},\beta) &= \sup_{\|D(\beta-\beta_o)\| \leq \sqrt{k_o}} R(\hat{\beta},\beta) = k_o \lambda_{max}(DBAB'D') \\ &+ \sigma^2 \operatorname{tr}(ACQ_{11}C'), \end{aligned} \tag{5.3}$$

where $\lambda_{max}(Z)$ denotes the largest eigenvalue of the matrix Z. Now if $D = T^{1/2}$, then $(I-CX)' = T^{1/2}T^{1/2}B$, $T^{1/2}B = (T^-)^{1/2}(I-CX)'$ and (5.3) transforms into

$$\begin{aligned} \sup_{\beta \in B_1} R(\hat{\beta},\beta) &= k_o \lambda_{max}((T^-)^{1/2}\,(CX-I)'A(CX-I)(T^-)^{1/2}) \\ &+ \sigma^2 \operatorname{tr}(ACQ_{11}C') = \lambda_{max}(A(CX-I)T^-(CX-I)') \\ &+ \sigma^2 \operatorname{tr}(ACQ_{11}C'), \end{aligned} \tag{5.4}$$

since $\lambda_{max}(DBAB'D') = \lambda_{max}(AB'TB) = \lambda_{max}(AB'TT^-TB)$, and optimization problem is to minimize (5.4) subject to $\operatorname{im}(CX-I)' \subseteq \operatorname{im}(T)$. This problem has been dealt with in the literature for the regular case (Q_{11} and T regular, $\operatorname{rank}(X) = k$) by Kuks (1972), Kuks and Olman (1972) and Läuter (1975). (See also Bunke (1975)). Läuter obtained an explicit formula for general A. It is to be expected that his approach will carry over to the singular case by a slight modification of some formulae (e.g., (4.14) might be a useful expression for such a modification). Kuks, Kuks and Olman and Läuter showed that if $A = aa'$, $a \in \mathbb{R}^k$, then a minimax estimator exists, which is independent of a. This loss-function corresponds to the approach used in this paper.

Acknowledgement. The author is gratefully indebted to the two referees for their stimulating suggestions.

References

[1] Bunke, O.(1975): Minimax, linear, ridge and shrunken estimators for linear parameters, Math. Operationsforsch. Statist. 6, 697-701.

[2] Drygas, H.(1975): Estimation and prediction for linear models in general spaces, Mathematische Operationsforsch. Statist. 6, 301-304.

[3] Drygas, H.(1984): On the unified theory of least squares. To appear in "Probability and Mathematical Statistics".

[4] Kuks, J. (1972): Minimax estimator of the regression coefficient (Russian), Izv. Acad. Nauk Eston. SSR 21, 73-78.

[5] Kuks, J. and Olman, W.(1972): Minimax linear estimator of the coefficient of regression, Izv. Acad. Nauk Est. SSR 21, 66-72.

[6] Läuter, H.(1975): Minimax linear estimators for linear parameters, Math. Operationsforsch. Statist. 6, 689-695.

[7] Rao, C.R. and Mitra, S.K. (1971): Generalized Inverse of Matrices and its Applications, Wiley, New York.

[8] Rödel, E. and Toutenburg, H.(1978): Mathematisch-statistische Methoden in der Ökonomie, Akademie-Verlag, Berlin.

SINGULAR INFORMATION MATRICES, DIRECTIONAL DERIVATIVES, AND SUBGRADIENTS IN OPTIMAL DESIGN THEORY

N. Gaffke
Technical University Aachen
Federal Republic of Germany

Summary. A linear regression setup is considered, where the parameters of interest are only part of or a linear function of the full set of unknown parameters. Under the standard statistical assumptions the problem of finding an optimal approximate design leads to a convex minimization problem of the following type:

> Minimize $\Phi(M) = \varphi(J(M))$ over the set of all information matrices M of those approximate designs which allow a linear unbiased estimator for a given $\mathbb{R}^s$ - valued linear function of the unknown parameters.

The optimality criterion φ is a convex and decreasing real-valued function on the set of all positive definite $(s\times s)$-matrices, and $J(M)$ is the reduced information matrix of M. The possibility of an optimal design with a singular M has caused a great many difficulties when trying to obtain characterizations for optimality via directional derivatives as in the "regular case". Recently the author has given a description of directional derivatives by means of shorted matrices and minimum seminorm generalized inverses. The present paper gives some further results. The set of all subgradients of the objective function Φ is found. As a result, a general equivalence theorem for optimality can be derived directly using a major theorem of convex analysis, as pointed out by Pukelsheim and Titterington (1983). A saddle point characterization for optimal designs is proved which has a nice interpretation in terms of a least favourable transformation of the linear regression setup.

1. Preliminaries

We consider a linear regression setup

$$y(x) = \sum_{j=1}^{k} a_j f_j(x) = a'f(x), \; x \in \mathscr{X} \;,$$

where x is the controlled variable, which can be chosen from the experimental region $\mathscr{X}$, the vector $a = (a_1, \ldots, a_k)' \in \mathbb{R}^k$ is an unknown parameter vector, and $f = (f_1, \ldots, f_k)'$ is a known $\mathbb{R}^k$-valued function on $\mathscr{X}$ with compact range $\{f(x): x \in \mathscr{X}\}$. Let K be a given $(s \times k)$-matrix of rank s, and let Ka represent the parameters of interest. An approximate design ξ is a probability measure on $\mathscr{X}$ with finite support. Under the standard statistical assumptions, the information matrix of ξ is given by

$$M(\xi) = \int_{\mathscr{X}} f(x) f'(x) d\xi(x),$$

and if

$$\text{range } (K') \subset \text{range } (M(\xi)), \tag{1}$$

then the covariance matrix of the Gauss-Markov estimator for Ka under ξ is proportional to $KM^-(\xi)K'$. Here M^- denotes a generalized inverse of M, and by (1) the matrix $KM^-(\xi)K'$ is positive definite and does not depend on the particular g-inverse $M^-(\xi)$. Optimality of a design refers to the minimization of $\psi(KM^-(\xi)K')$ over the set of all designs ξ which satisfy (1), where ψ is a given optimality criterion, a real valued function defined on $PD(s)$, the set of all positive definite $(s \times s)$-matrices. We will, however, use another formulation for ψ-optimality. Firstly, we make use of the associated function

$$\varphi(C) = \psi(C^{-1}), \quad C \in PD(s),$$

since for most optimality criteria the function φ is convex, whereas ψ may not be convex (cf. Kiefer (1975), Gaffke (1981)). Secondly, we consider the set of information matrices rather than the set of designs. Let $\mathscr{M}$ denote the set of all information matrices $M(\xi)$, where ξ ranges over the set of all approximate designs. Obviously, $\mathscr{M}$ is a compact and convex set and is actually the convex hull of $\{f(x)f'(x): x \in \mathscr{X}\}$. By $NND(k)$ we denote the convex cone of all nonnegative definite $(k \times k)$-matrices and by $\mathscr{A}(K)$ the convex cone

$$\mathscr{A}(K) = \{A \in NND(k): \text{range } (K') \subset \text{range } (A)\}.$$

The mapping $J: \mathscr{A}(K) \to PD(s)$ is defined by

$$J(A) = (KA^{-}K')^{-1}, \quad A \in \mathscr{A}(K).$$

For an information matrix $M \in \mathscr{M} \cap \mathscr{A}(K)$ the matrix $J(M)$ may be called the reduced information matrix for Ka. The problem of finding a ψ-optimal design (for Ka) now reads as

$$\text{minimize } \Phi(M) = \varphi(J(M)) \quad \text{over} \quad M \in \mathscr{M} \cap \mathscr{A}(K).$$

Of course, we assume that $\mathscr{M} \cap \mathscr{A}(K)$ is nonempty, which is satisfied, for instance, if the components $f_1, \dots, f_k$ of f are linearly independent. Also we assume throughout that φ is convex and decreasing, i.e.

$$\varphi(\alpha C + (1-\alpha)D) \leq \alpha\varphi(C) + (1-\alpha)\varphi(D)$$

$$\text{for all } C,D \in PD(s), \text{ and } \alpha \in [0,1],$$

and

$$C \leq D \text{ implies } \varphi(C) \geq \varphi(D) \text{ for all } C,D \in PD(s),$$

where, by definition, $C \leq D$ iff $D - C$ is nonnegative definite. This semiordering "$\leq$" on the set of all symmetric matrices of a given size is called the Löwner semiordering. These assumptions on φ are met by all commonly used optimality criteria. The broad class of information functionals as considered by Pukelsheim (1980) and Pukelsheim and Titterington (1983) is included by taking

$$\varphi(C) = -\log j(C), \quad C \in PD(s),$$

where j is an information functional. Well-known examples of information functionals are

$$j_{-p}(C) = \{\tfrac{1}{s}\mathrm{tr}(C^{-p})\}^{-1/p}, \quad C \in PD(s),$$

for $-1 \leq p \leq \infty$, which correspond to Kiefer's (1975) φ_p-criteria,

$$\varphi_p(C) = \log\{\tfrac{1}{s}\mathrm{tr}(C^{-p})\}^{1/p}, \quad C \in PD(s);$$

for $p = 0$ or $p = \infty$ these formulas are to be understood in the limiting sense.

To obtain characterizations for the solutions of the above minimization problem the simple approach via directional derivatives (Whittle (1973)) will be used in Section 2. This approach worked quite easily in the "regular case", when the information matrices in $\mathcal{M} \cap \mathcal{A}(K)$ are non-singular, (cf. Kiefer (1974)). In the general case considerable difficulties arise when computing directional derivatives of Φ at singular matrices, (cf. Kiefer (1961), Section 2, Silvey (1980), p. 44, Näther and Reinsch (1981)). In this context the notion of shorted matrices (or shorted operators, cf. Anderson (1971), Krein (1947)) turns out to be of great value. We shall make use of the following results on shorted matrices.

Theorem 1.1 (W.N. Anderson (1971), Krein (1947)).

Let $M, M_o \in NND(k)$. There exists

$$\max \{N \in NND(k): N \leq M, \text{ range } (N) \subset \text{range } (M_o)\},$$

where "$\leq$" and the maximum refer to the Löwner semiordering. This maximum is denoted by $S(M,M_o)$, the shorted matrix of M with respect to (the range of) M_o.

Theorem 1.2 (W.N. Anderson (1971)).

Let $M,M_o \in NND(k)$. There exists a unique decomposition $M = S + T$, such that $S,T \in NND(k)$, range $(S) \subset$ range (M_o), and range $(T) \cap$ $\cap$ range $(M_o) = \{0\}$. In this decomposition $S = S(M,M_o)$.

The following representation of the shorted matrix due to Mitra and Puri (1979) stands in duality to that of Theorem 1.1. For the definition of a minimum seminorm g-inverse we refer to Rao and Mitra (1971), Chapter 3.1.

Theorem 1.3 (Mitra and Puri(1979)).

Let $M,M_o \in NND(k)$. Then

$$S(M,M_o) = \min_{M_o^-} M_o(M_o^-)' \, M \, M_o^- \, M_o,$$

where the minimum refers to the Löwner semiordering and is taken over all g-inverses M_o^- of M_o. The minimum is attained at any minimum M-seminorm g-inverse of M_o.

In Section 2 the results on directional derivatives are summarized, which have been proved in Gaffke (1984). In Section 3 two complements are given: Firstly, the set of all subgradients of the objective function Φ is described. As a result, a general equivalence theorem can be derived directly by using a major theorem of convex analysis as pointed out by Pukelsheim and Titterington (1983). Secondly, a saddle point characterization of an optimal design is proved, which has a nice interpretation in terms of a least favourable transformation of the linear regression setup.

2. Directional derivatives

The matrix function

$$J(A) = (KA^-K')^{-1}, \quad A \in \mathscr{A}(K),$$

is concave and increasing, (cf. Pukelsheim and Styan (1983)). So the composition $\Phi = \varphi \circ J$ is a convex and decreasing real function on the convex cone $\mathscr{A}(K)$. For $M_o \in \mathscr{A}(K)$ and $M \in \mathrm{NND}(k)$ the limit

$$D_\Phi(M_o, M) = \lim_{\alpha \to 0+} \frac{1}{\alpha} \{\Phi((1-\alpha)M_o + \alpha M) - \Phi(M_o)\}$$

exists in $\mathbb{R} \cup \{-\infty\}$ and is called the directional derivative of Φ at M_o in the direction of $M - M_o$. For the problem of minimizing $\Phi(M)$ over $M \in \mathscr{M} \cap \mathscr{A}(K)$ the "steepest descent"

$$\inf_{M \in \mathscr{M}} D_\Phi(M_o, M) \quad \text{for} \quad M_o \in \mathscr{M} \cap \mathscr{A}(K)$$

is of particular interest. As pointed out by Whittle (1973) the following simple lemma serves as a starting point for equivalent characterizations of optimal designs.

Lemma 2.1. Let $M_o \in \mathscr{M} \cap \mathscr{A}(K)$. Then

$$\Phi(M_o) = \min_{M \in \mathcal{M} \cap \mathcal{A}(K)} \Phi(M), \text{ iff } D_\Phi(M_o, M) \geq 0 \text{ for all } M \in \mathcal{M},$$

or - equivalently - iff $\inf_{M \in \mathcal{M}} D_\Phi(M_o, M) = 0$.

Let us briefly discuss the classical situation, in which the information matrix M_o is nonsingular and the function φ is differentiable on $PD(s)$. Here $PD(s)$ is considered as an open subset of the space $Sym(s)$ of all symmetric $(s \times s)$-matrices, which is endowed with the scalar product $tr\{C \cdot D\}$, $C, D \in Sym(s)$. Let $\nabla\varphi[C]$ denote the gradient of φ at $C \in PD(s)$, $\nabla\varphi[C] \in Sym(s)$, and, since φ is decreasing, $-\nabla\varphi[C] \in NND(s)$. For $M_o \in PD(k)$ the composition $\Phi = \varphi \circ J$ is differentiable at M_o, and its gradient is given by

$$\nabla\Phi[M_o] = M_o^{-1} K' J(M_o) \nabla\varphi[J(M_o)] J(M_o) K M_o^{-1}. \tag{2}$$

Hence for $M \in NND(k)$

$$D_\Phi(M_o, M) = tr\{\nabla\Phi[M_o] \cdot (M - M_o)\}. \tag{3}$$

Since this directional derivative is affine-linear in its second argument and $\mathcal{M}$ is the convex hull of $\{f(x)f'(x) : x \in \mathcal{X}\}$, one obtains

$$\min_{M \in \mathcal{M}} D_\Phi(M_o, M) = \min_{x \in \mathcal{X}} f'(x) \nabla\Phi[M_o] f(x) - tr\{\nabla\varphi[J(M_o)] \cdot J(M_o)\}.$$

This together with Lemma 2.1 yields the classical equivalence theorems, (cf. Kiefer (1974)).

Let us next turn to the case where φ is differentiable on $PD(s)$, but $M_o \in \mathcal{A}(K)$ may be singular. By replacing the inverse M_o^{-1} in (2) by an arbitrary g-inverse M_o^- we obtain matrices

$$\nabla\Phi[M_o, M_o^-] = (M_o^-)' K' J(M_o) \nabla\varphi[J(M_o)] \cdot J(M_o) K M_o^-. \tag{4}$$

Of course, for a singular M_o none of the matrices defined by (4) is the gradient of Φ at M_o (which does not exist). Actually these matrices are subgradients of Φ at M_o, (see Section 3). Also, formula (3) no longer holds true for all $M \in NND(k)$, when the gradient of Φ at M_o is replaced by some $\nabla\Phi[M_o, M_o^-]$. However, the following lemma can be proved.

Lemma 2.2. If $M_o \in \mathscr{A}(K)$, $M \in NND(k)$, and range $(M) \subset$ range (M_o), then

$$D_\Phi(M_o,M) = tr\{\nabla\Phi[M_o,M_o^-] \cdot (M-M_o)\},$$

which does not depend on the particular g-inverse M_o^- of M_o.

The case of an arbitrary $M \in NND(k)$ can be led back to the situation of Lemma 2.2 by using the shorted matrix $S(M,M_o)$. I owe the proof of the following lemma to Professor S. K. Mitra.

Lemma 2.3. If $M_o \in \mathscr{A}(K)$ and $M \in NND(k)$, then

$$J((1-\alpha)M_o+\alpha M) = J((1-\alpha)M_o + \alpha S(M,M_o)), \; 0 \leq \alpha < 1,$$

and hence

$$D_\Phi(M_o,M) = D_\Phi(M_o,S(M,M_o)).$$

Proof. Write $(1-\alpha)M_o + \alpha M = A + B$, where

$$A = (1-\alpha)M_o + \alpha S(M,M_o), \quad B = \alpha(M - S(M,M_o)).$$

By Theorems 1.1 and 1.2

$$A,B \in NND(k), \text{ and } \text{range}(A) \cap \text{range}(B) = \{0\}.$$

It can easily be seen that each g-inverse of $A+B$ is a g-inverse of A, and thus

$$K(A+B)^-K' = KA^-K',$$

which proves the first identity of the lemma. The second is an immediate consequence.

From Lemma 2.2, Lemma 2.3 and Theorem 1.3 one obtains the following quasilinear representation of the directional derivative $D_\Phi(M_o,M)$.

Lemma 2.4. If $M_o \in \mathscr{A}(K)$ and $M \in NND(k)$, then

$$D_\Phi(M_o,M) = \max_{M_o^-} tr\{\nabla\Phi[M_o,M_o^-] \cdot (M-M_o)\},$$

where the maximum is taken over all g-inverses M_o^- of M_o. As for the "steepest descent" $\inf_{M\in\mathcal{M}} D_\Phi(M_o,M)$, if $M_o \in \mathcal{M} \cap \mathcal{A}(K)$, a minimax result can be proved.

Theorem 2.1. If $M_o \in \mathcal{M} \cap \mathcal{A}(K)$, then

$$\min_{M\in\mathcal{M}} \max_{M_o^-} \operatorname{tr}\{\nabla\Phi[M_o,M_o^-]\cdot M\} = \max_{M_o^-} \min_{M\in\mathcal{M}} \operatorname{tr}\{\nabla\Phi[M_o,M_o^-]\cdot M\},$$

and hence

$$\min_{M\in\mathcal{M}} D_\Phi(M_o,M) = \max_{M_o^-} \min_{x\in} f'(x)\nabla\Phi[M_o,M_o^-]f(x) - \operatorname{tr}\{\nabla\varphi[J(M_o)]\cdot J(M_o)\}.$$

Let us finally indicate the necessary alterations in Lemma 2.4 and Theorem 2.1, if the function φ is not assumed to be differentiable on $PD(s)$. The gradient of φ is then replaced by subgradients of φ, where by definition a matrix $E \in Sym(s)$ is a subgradient of φ at $C_o \in PD(s)$, iff for all $C \in PD(s)$

$$\varphi(C) - \varphi(C_o) \geq \operatorname{tr}\{E\cdot(C-C_o)\}.$$

Since the domain $PD(s)$ is an open set (in the space $Sym(s)$), the set $\partial\varphi[C_o]$ of all subgradients of φ at C_o is nonempty, convex and compact, (cf. Rockafellar (1972), § 23). For an information functional j and $\varphi(C) = -\log j(C)$, $C \in PD(s)$, the set of subgradients $\partial\varphi[C_o]$ was described in Gaffke (1984), Lemma 3. If $M_o \in \mathcal{A}(K)$, then instead of (4) we consider for any g-inverse M_o^- and any $E \in \partial\varphi[J(M_o)]$ the matrices

$$\nabla\Phi[M_o,M_o^-,E] = (M_o^-)'K'J(M_o)EJ(M_o)KM_o^-. \tag{5}$$

Again, these turn out to be subgradients of Φ at M_o, (see Section 3). Of course, if φ is differentiable, then (5) reduces to (4), since $\partial\varphi[J(M_o)]$ consists of the single element $\nabla\varphi[J(M_o)]$. Lemma 2.4 and Theorem 2.1 can be generalized.

Lemma 2.4 a. If $M_o \in \mathcal{A}(K)$ and $M \in NND(k)$, then

$$D_\Phi(M_o,M) = \max_{M_o^-,E} \operatorname{tr}\{\nabla\Phi[M_o,M_o^-,E]\cdot(M-M_o)\},$$

where the maximum is taken over all g-inverses M_o^- of M_o and all $E \in \partial\varphi[J(M_o)]$.

Theorem 2.1 a. If $M_o \in \mathcal{M} \cap \mathcal{A}(K)$, then

$$\min_{M \in \mathcal{M}} \max_{M_o^-, E} \operatorname{tr}\{\nabla\Phi[M_o, M_o^-, E] \cdot (M - M_o)\}$$

$$= \max_{M_o^-, E} \min_{M \in \mathcal{M}} \operatorname{tr}\{\nabla\Phi[M_o, M_o^-, E] \cdot (M - M_o)\},$$

and hence

$$\min_{M \in \mathcal{M}} D_\Phi(M_o, M) = \max_{M_o^-, E} \{\min_{x \in \mathcal{X}} f'(x)\nabla\Phi[M_o, M_o^-, E]f(x) - \operatorname{tr}[E \cdot J(M_o)]\}.$$

3. Complements

From Whittle's Lemma 2.1 and from Theorem 2.1a we obtain the following equivalence theorem for design optimality, which was proved (for information functionals) by Pukelsheim (1980) via duality theory (Theorem 7 of that paper).

Theorem 3.1. Let $M_o \in \mathcal{M} \cap \mathcal{A}(K)$. Then

$\Phi(M_o) = \min_{M \in \mathcal{M} \cap \mathcal{A}(K)} \Phi(M)$, iff there exist a g-inverse M_o^- of M_o and a subgradient E of φ at $J(M_o)$, such that

$$f'(x)\nabla\Phi[M_o, M_o^-, E]f(x) \geq \operatorname{tr}\{E \cdot J(M_o)\} \quad \text{for all} \quad x \in \mathcal{X}.$$

It was pointed out by Pukelsheim and Titterington (1983) that a major theorem of convex analysis can be used to characterize design optimality. Under the assumption that the components of f are linearly independent, one obtains from Theorem 27.4 in Rockafellar (1972) that

> $M_o \in \mathcal{M} \cap \mathcal{A}(K)$ minimizes $\Phi(M)$ over $M \in \mathcal{M} \cap \mathcal{A}(K)$,
> iff there exists a subgradient B of Φ at M_o,
> such that
> $f'(x)Bf(x) \geq \operatorname{tr}\{B \cdot M_o\}$ for all $x \in \mathcal{X}$.

This will be brought together with Theorem 3.1 by the following description of the set $\partial\Phi[M_o]$ of all subgradients of Φ at $M_o \in \mathcal{A}(K)$ (see Lemma 3.2 below). Recall that a matrix $B \in \mathrm{Sym}(k)$ is defined to be a subgradient of Φ at M_o, iff for all $M \in \mathcal{A}(K)$

$$\Phi(M) - \Phi(M_o) \geq \mathrm{tr}\,\{B\cdot(M-M_o)\}.$$

Lemma 3.1. Let $M_o \in \mathcal{A}(K)$ and $B \in \mathrm{Sym}(k)$. Then B is a subgradient of Φ at M_o iff

$$-B \in \mathrm{NND}(k), \quad BM_o = BK'J(M_o)K,$$
$$\text{and} \quad KBK' \in \partial\varphi[J(M_o)].$$

Proof.

1. Let $B \in \partial\Phi[M_o]$. Since Φ is decreasing, $-B \in \mathrm{NND}(k)$. For $C \in \mathrm{PD}(s)$ clearly $K'CK \in \mathcal{A}(K)$, and from Lemma 1 in Gaffke and Krafft (1977) $J(K'CK) = C$. Putting $C = J(M_o)$ and using the subgradient inequality with $M = K'J(M_o)K$, we obtain

$$0 \geq \mathrm{tr}\,\{B(K'J(M_o)K - M_o)\}.$$

Since $-B \in \mathrm{NND}(k)$ and $M_o - K'J(M_o)K \in \mathrm{NND}(k)$ (cf. Gaffke and Krafft (1982), Corollary 2.9), it follows $BM_o = BK'J(M_o)K$. For an arbitrary $C \in \mathrm{PD}(s)$ and with $M = K'CK$ the subgradient inequality yields

$$\begin{aligned}\Phi(K'CK) - \Phi(M_o) &= \varphi(C) - \varphi(J(M_o))\\ &\geq \mathrm{tr}\,\{B(K'CK - M_o)\} = \mathrm{tr}\,\{B(K'CK - K'J(M_o)K)\}\\ &= \mathrm{tr}\,\{KBK'(C - J(M_o))\},\end{aligned}$$

and hence $KBK' \in \partial\varphi[J(M_o)]$.

2. Let B satisfy the three conditions stated in the lemma. Using $K'J(M)K \leq M$ for $M \in \mathcal{A}(K)$ we obtain

$$\begin{aligned}&\mathrm{tr}\,\{B(M - M_o)\} \leq \mathrm{tr}\,\{B(K'J(M)K - M_o)\}\\ &= \mathrm{tr}\,\{B(K'J(M)K - K'J(M_o)K\} = \mathrm{tr}\,\{KBK'(J(M) - J(M_o))\}\end{aligned}$$

$$\leq \varphi(J(M)) - \varphi(J(M_o)) = \Phi(M) - \Phi(M_o),$$

and hence $B \in \partial\Phi[M_o]$.

Lemma 3.2. Let $M_o \in \mathcal{A}(K)$. Then $\partial\Phi[M_o]$ consists of all matrices

$$\nabla\Phi[M_o, M_o^-, E] - D,$$

where M_o^- is any g-inverse of M_o, E is any element of $\partial\varphi[J(M_o)]$, and D is any element of $NND(k)$ with $\text{range}(D) \subset \text{nullspace}(M_o)$. The same set $\partial\Phi[M_o]$ is obtained, when the set of all g-inverses M_o^- is restricted to all M_o^- of the form $M_o^- = (M_o + N)^{-1}$, where N is any element of $NND(k)$ such that $\text{range}(N)$ is complementary to $\text{range}(M_o)$.

Proof. One can easily check that any such matrix $\nabla\Phi[M_o, M_o^-, E] - D$ satisfies the conditions of Lemma 3.1 and is hence a subgradient of Φ at M_o. Now let B be any matrix in $Sym(k)$ which satisfies the conditions of Lemma 3.1. Choosing any matrix $N_o \in NND(k)$ with $\text{range}(N_o) = \text{nullspace}(M_o)$ we get from Theorem 1.2 that there exist matrices $D, A \in NND(k)$ such that $-B = D + A$, $\text{range}(D) \subset \text{nullspace}(M_o)$, and $\text{range}(A) \cap \text{nullspace}(M_o) = \{0\}$. The latter implies that

$$\text{nullspace}(A) + \text{range}(M_o) = \mathbb{R}^k.$$

Choosing a subspace $\mathcal{L} \subset \text{nullspace}(A)$ which is complementary to $\text{range}(M_o)$ and a matrix $N \in NND(k)$ with $\text{range}(N) = \mathcal{L}$, we obtain

$$(M_o + N)A(M_o + N) = M_o A M_o,$$

and hence

$$A = (M_o + N)^{-1} M_o A M_o (M_o + N)^{-1}.$$

From $-BM_o = AM_o$ and $-BK' = AK'$ we get $AM_o = AK'J(M_o)K$ and $E = -KAK' \in \partial\varphi[J(M_o)]$, and hence

$$\begin{aligned} B = -A - D &= -(M_o + N)^{-1} M_o A M_o (M_o + N)^{-1} - D \\ &= (M_o + N)^{-1} K' J(M_o) E J(M_o) K (M_o + N)^{-1} - D \\ &= \nabla\Phi[M_o, (M_o + N)^{-1}, E] - D. \end{aligned}$$

Remarks

1. If $\varphi(C) = -\log j(C)$, $C \in PD(s)$, where j is an information functional, then from Lemma 3.1 and from Lemma 3 in Gaffke (1984) the characterization of $\partial\Phi[M_o]$ of Pukelsheim and Titterington (1983), (Lemma 1), can easily be obtained: For $M_o \in \mathscr{A}(K)$, a matrix $B \in \mathrm{Sym}(k)$ is a subgradient of Φ at M_o, iff $-B \in NND(k)$ and

$$1 = \mathrm{tr}\,\{-BM_o\} = j(J(M_o)) \cdot j^o(-KBK'),$$

where j^o is the polar function of j.

2. The fact that for $\partial\Phi[M_o]$ we only have to take into account g-inverses M_o^- of the form $M_o^- = (M_o + N)^{-1}$ with $N \in NND(k)$ and range (N) complementary to range (M_o) is closely related to Silvey's condition for design optimality using complementary subspaces, [cf. Silvey (1978), (1980), Theorem 3.13.1]. By Lemma 3.2 and Theorem 3.1 $M_o \in \mathscr{M} \cap \mathscr{A}(K)$ minimizes $\Phi(M)$ over $M \in \mathscr{M} \cap \mathscr{A}(K)$, iff there exist an $N \in NND(k)$ with range (N) complementary to range (M_o) and an $E \in \partial\varphi[J(M_o)]$, such that for all $x \in \mathscr{X}$

$$f'(x)\nabla\Phi[M_o,(M_o + N)^{-1},E]f(x) \geq \mathrm{tr}\,\{E \cdot J(M_o)\}.$$

Since $J(M_o) = J(M_o + N)$, and

$$\{\nabla\Phi[M_o,(M_o + N)^{-1},E] : E \in \partial\varphi[J(M_o)]\} = \partial\Phi[M_o + N],$$

we obtain Theorem 2 in Pukelsheim and Titterington (1983). For a differentiable φ this was proved in Gaffke (1984), (Corollary 2).

Another idea for characterizing design optimality is the following. The problem of singular information matrices arises from the fact, that in the linear regression setup

$$y(x) = a'f(x), \quad x \in \mathscr{X}, \quad a \in \mathbb{R}^k,$$

the parameters of interest are given by a lower dimensional vector Ka, where K is a given $(s \times k)$-matrix. One may ask whether it is possible to reduce (or transform) the linear regression setup in such a way that design optimality for Ka in the original model corresponds to design optimality for the full parameter set in a reduced regression setup.

This can in fact be accomplished by considering transformations $f_L(x) = L\ f(x)$, $x \in \mathscr{X}$, and a reduced linear regression setup

$$y_L(x) = b'f_L(x), \quad x \in \mathscr{X}, \quad b \in \mathbb{R}^s,$$

where L is a g-inverse of K', i.e. L is an $(s \times k)$-matrix with $LK' = I_s$ (unit matrix of order s). Clearly, the set of information matrices for regression y_L is given by

$$\mathscr{M}_L = \{M_L = LML' : M \in \mathscr{M}\},$$

and the optimal design problem for the full parameter set in y_L reads as

$$\text{minimize } \varphi(LML') \quad \text{over} \quad M \in \mathscr{M},$$

where $\varphi(C)$ is defined to be ∞ for a singular $C \in NND(s)$. The connection with the original problem results from

$$J(M) = \min_L LML', \quad \text{for} \quad M \in \mathscr{A}(K), \tag{6}$$

where the minimum refers to the Löwner semiordering and is taken over all $(s \times k)$-matrices L with $LK' = I_s$, (cf. Gaffke and Krafft (1982), Theorem 2.4). So for $M \in \mathscr{M} \cap \mathscr{A}(K)$

$$\varphi(J(M)) = \max_L \varphi(LML'), \tag{6 a}$$

and the original problem reads as

$$\text{minimize } \max_L \varphi(LML') \text{ over } M \in \mathscr{M} \cap \mathscr{A}(K).$$

Here the intersection $\mathscr{M} \cap \mathscr{A}(K)$ can be replaced by $\mathscr{M}$ as a result of the following lemma, where, as above, $\varphi(C) = \infty$ for $C \in NND(s) \setminus PD(s)$.

Lemma 3.3. Let $M \in NND(k)$. Then $M \in \mathscr{A}(K)$ iff LML' is nonsingular for all $(s \times k)$-matrices L with $LK' = I_s$.

Proof. If $M \in \mathscr{A}(K)$ and $LK' = I_s$, then by (6) $J(M) \leq LML'$, and hence LML' is nonsingular. If $M \notin \mathscr{A}(K)$, then there exists a $z \in \mathbb{R}^k$ with $Mz = 0$ and $Kz = w \neq 0$. By Corollary 1, p. 27, in Rao and Mitra (1971) $z = K^{-}w$ for some g-inverse K^{-} of K, i.e. $z = L'w$ for some L with $LK' = I_s$. Since $ML'w = Mz = 0$, LML' is singular.

From Lemma 3.3,

$$\inf_{M \in \mathcal{M} \cap \mathcal{A}(K)} \varphi(J(M)) = \inf_{M \in \mathcal{M}} \max_{L} \varphi(LML'). \tag{7}$$

Thus $M_o \in \mathcal{M} \cap \mathcal{A}(K)$ attains the infimum on the left hand side of (7) iff M_o is a minimax strategy for player 1 in the game with kernel $\kappa(M,L) = \varphi(LML')$. Theorem 3.2 below implies that, if there exists such an M_o, then the game is definite and there exists a maximin strategy L_o for player 2, i.e.

$$\min_{M \in \mathcal{M}} \max_{L} \varphi(LML') = \max_{L} \inf_{M \in \mathcal{M}} \varphi(LML').$$

Thus, if M_o is optimal for Ka in regression $y(x)$, M_o is optimal for the full parameter set in a reduced regression $y_{L_o}(x)$, where L_o is any maximim strategy for player 2, and might be called a least favourable transformation.

<u>Theorem 3.2</u>. Let $M_o \in \mathcal{M}$. Then (i) and (ii) below are equivalent.

(i) $M_o \in \mathcal{M} \cap \mathcal{A}(K)$ and $\varphi(J(M_o)) = \min_{M \in \mathcal{M} \cap \mathcal{A}(K)} \varphi(J(M))$.

(ii) There exists an $(s \times k)$-matrix L_o with $L_oK' = I_s$, such that

$$\varphi(LM_oL') \leq \varphi(L_oM_oL_o') \leq \varphi(L_oML_o')$$

for all $M \in \mathcal{M}$ and all $(s \times k)$-matrices L with $LK' = I_s$.

<u>Proof</u>.

1. Let (i) be satisfied. By Theorem 3.1 there exist a g-inverse M_o^{-} and a subgradient $E \in \partial\varphi[J(M_o)]$ such that

$$\operatorname{tr}\{\nabla\Phi[M_o, M_o^{-}, E] \cdot M\} \geq \operatorname{tr}\{E \cdot J(M_o)\}$$

for all $M \in \mathcal{M}$. Putting $L_o = J(M_o)KM_o^{-}$ we have $L_oK' = I_s$, and using (5) we obtain

$$\operatorname{tr}\{EL_oML_o'\} \geq \operatorname{tr}\{EJ(M_o)\}$$

for all $M \in \mathcal{M}$. Hence for all $M \in \mathcal{M}$

$$\begin{aligned}\varphi(L_oML_o') &\geq \varphi(J(M_o)) + \operatorname{tr}\{E\cdot(L_oML_o' - J(M_o))\}\\ &\geq \varphi(J(M_o)) = \varphi(L_oM_oL_o').\end{aligned}$$

The inequality $\varphi(LM_oL') \leq \varphi(L_oM_oL_o')$ for $LK' = I_s$ follows from (6 a).

2. Let (ii) be satisfied. Since $\mathcal{M} \cap \mathcal{A}(K)$ was assumed to be non-empty, $M_o \in \mathcal{A}(K)$ by Lemma 3.3. From (6 a) and since

$$\varphi(L_oM_oL_o') = \max_L \varphi(LM_oL')$$

we get $\varphi(L_oM_oL_o') = \varphi(J(M_o))$. For $M \in \mathcal{M} \cap \mathcal{A}(K)$, using (6 a), we obtain

$$\varphi(J(M_o)) = \varphi(L_oM_oL_o') \leq \varphi(L_oML_o') \leq \varphi(J(M)),$$

and hence (i).

Acknowledgements. I wish to thank Professor S.K. Mitra for communicating the proof of Lemma 2.3 to me. Thanks also to a referee who helped to improve the presentation.

References

Anderson, W.N. (1971). Shorted operators. SIAM J. Appl. Math. 20, 520-525.

Gaffke, N. (1981). Some classes of optimality criteria and optimal designs for complete two-way layouts. Ann. Statist. 9, 893-898.

Gaffke, N. (1984). Directional derivatives of optimality criteria at singular matrices in convex design theory. Math. Operationsforsch. Statist., Ser Statistics. (To appear).

Gaffke, N., Krafft, O. (1977). Optimum properties of latin square designs and a matrix inequality. Math. Operationsforsch. Statist., Ser. Statistics 8, 345-350.

Gaffke, N., Krafft, O. (1982). Matrix inequalities in the Löwner ordering. In: Korte, B. (ed.). Modern Applied Mathematics, Optimization and Operations Research. North-Holland, Amsterdam, 595-622.

Kiefer, J. (1961). Optimum designs in regression problems, II. Ann. Math. Statist. 32, 298-325.

Kiefer, J. (1974). General equivalence theory for optimum designs (approximate theory). Ann. Statist. 2, 849-879.

Kiefer, J. (1975). Construction and optimality of generalized Youden designs. In: Srivastava, J.N. (ed.). A survey of statistical design and linear models. North-Holland, Amsterdam, 333-353.

Krein, M. (1947). The theory of self-adjoint extensions of semi-bounded Hermitian transformations and its applications I. Mat. Sbornik N.S. 20 (62), 431-495, and 21 (63), 365-404. (Russian, with English summary).

Mitra, S. K., Puri, M.L. (1979). Shorted operators and generalized inverses of matrices. Lin. Algebra Appl. 25, 45-56.

Näther, W., Reinsch, V. (1981). D_s-optimality and Whittle's equivalence theorem. Math. Operationsforsch. Statist., Ser. Statistics 12, 307--316.

Pukelsheim, F. (1980). On linear regression designs which maximize information. J. Statist. Plann. Infer. 4, 339-364.

Pukelsheim, F., Styan, G.P.H. (1983). Convexity and monotonicity properties of dispersion matrices of estimators in linear models. Scand. J. Statist. 10, 145-149.

Pukelsheim, F., Titterington, D.M. (1983). General differential and Lagrangian theory for optimal experimental design. Ann. Statist. 11, 1060-1068.

Rao, C.R., Mitra, S.K. (1971). Generalized Inverse of Matrices and its Applications, Wiley. New York.

Rockafellar, R.T. (1972). Convex Analysis. Second printing, Princeton.

Silvey, S.D. (1978). Optimal design measures with singular information matrices. Biometrika 65, 553-559.

Silvey, S.D. (1980). Optimal Design. Chapman-Hall, London.

Whittle, P. (1973). Some general points in the theory of optimal experimental design. J. Roy. Statist. Soc. B 35, 123-130.

A NOTE ON ADMISSIBILITY OF IMPROVED UNBIASED ESTIMATORS IN TWO VARIANCE COMPONENT MODELS

S.Gnot
Department of Mathematics
Agricultural Academy
Wrocław, Poland

J.Kleffe
Institute of Mathematics
Academy of Sciences of the GDR
Berlin, GDR

Introduction. The fact that multiples of unbiased quadratic estimators may have uniformly smaller mean squared error risks received some attention in statistics. Hodges and Lehmann (1951) improved the sample variance by multiplying it by $(n-1)/(n+1)$, where n is the sample size. Perlman (1972) did a more comprehensive study of this phenomenon and LaMotte (1979) proved that for a normal random model each invariant quadratic form, as an estimator of its expectation, is dominated by an appriopriate scalar multiple of itself. These modified estimators, however, may not be admissible in the class of invariant quadratic estimators without the unbiasedness condition. A discussion of the problem for the balanced one-way ANOVA model was given by LaMotte (1979). Generally it is not known how to improve quadratic unbiased estimators by admissible quadratic estimators. In this paper we make a step in this direction and examine admissibility of multiples of admissible invariant quadratic unbiased estimators for a mixed linear model with two variance components only. The results are based on an explicit description of the class of admissible invariant quadratic estimators given by Gnot and Kleffe (1983). The above paper also helps to follow the technicalities presented in this note. Special attention receives estimation of the individual variance components. A generalized Hodges-Lehmann modification for estimation of residual variance is derived.

2. Classes of admissible invariant estimators

Let y be a normally distributed random n-vector with

$$Ey = X\beta, \quad \text{Cov } y = \theta_1 I + \theta_2 V, \ \beta \in R^p, \ \theta_1 > 0, \ \theta_2 \geq 0, \tag{2.1}$$

where X, V (V is n.n.d) are known matrices, rank $X = p$ while β and $\theta = (\theta_1, \theta_2)$, are unknown parameter vectors. Under this model and the quadratic loss function there are two classes of admissible invariant quadratic estimators for a given function $f'\theta$. Both are formed by linear combinantions of just two quadratic forms

$$Q_1(u,v) = y'(M\Lambda_{u,v}M)^+y, \ Q_2(u,v) = y'(M\Lambda_{u,v}M)^+VMy, \tag{2.2}$$

where $\Lambda_{u,v} = I+2uV+(u^2+v)VMV$, $M=I-XX^+$, $u \geq 0$, $v \geq 0$, and are defined as follows

$$\begin{aligned} \mathscr{A}^f_{IU} &= \{\lambda_1 Q_1(u,v) + \lambda_2 Q_2(u,v), \quad H\lambda = f, \quad u \geq 0, \quad v \geq 0\} \\ \mathscr{A}^f_{I} &= \{\bar\lambda_1 Q_1(u,v) + \bar\lambda_2 Q_2(u,v), \quad H\bar\lambda = \bar f = (2I+HC)^{-1}HCf, \quad u \geq 0, \ v \geq 0\}, \end{aligned} \tag{2.3}$$

where

$$H = \begin{pmatrix} r, & s \\ s, & m \end{pmatrix}, \quad C = \begin{pmatrix} 1, & u \\ u, & u^2+v \end{pmatrix},$$

$r = \text{tr}(M\Lambda_{u,v}M)^+$, $s = \text{tr}(M\Lambda_{u,v}M)^+V$, $m = \text{tr}(M\Lambda_{u,v}M)^+VMV$ and $H\lambda = f$ is the unbiasedness condition. The class $\mathscr{A}^f_{IU}$ together with the limiting estimator obtained as v tends to infinity and given by

$$\gamma_0 = \begin{cases} \dfrac{f_1 - f_2 s_0/m_0}{n-p-m_0}\, y'[M-MV(MVM)^+]y + \dfrac{f_2}{m_0}\, y'(MVM)^+y & \text{if } m_0 < n-p \\ \lambda_{10} y'(MVM)^+(MVM)^+y + \lambda_{20} y'(MVM)^+y & \text{if } m_0 = n-p \end{cases} \tag{2.4}$$

forms the entire class of admissible estimators in the set of invariant quadratic unbiased estimators for $f'\theta$, also obtained by Olsen, Seely and Birkes (1976) in a less explicit form. Denote this class by $\bar{\mathscr{A}}^f_{IU}$. The notation in (2.4) is as follows $s_0 = \lim_{v\to\infty} vs = \text{tr}(MVM)^+$, $r_0 = \lim_{v\to\infty} vr = \text{tr}(MVM)^+(MVM)^+$ if $m_0 = n-p$, $m_0 = \lim_{v\to\infty} vm = \text{rank } MVM$ and $\lambda_0 = (\lambda_{10}, \lambda_{20})'$

satisfies the equation

$$\begin{pmatrix} r_0 , & s_0 \\ s_0 , & m_0 \end{pmatrix} \lambda_0 = f.$$

For a given $u \geq 0$ the limiting estimator in $\mathscr{A}_I^f$ is

$$\bar{\gamma}_0(u) = \bar{\lambda}_{10} y'[M-MV(MVM)^+]y + \bar{\lambda}_{20} y'(MVM)^+ y, \tag{2.5}$$

where

$$\bar{\lambda}_{10} = \lim_{v\to\infty} \bar{\lambda}_1 = \frac{(2+m_0)f_1 + (2u-s_0)f_2}{(2+n-p-m_0)(2+m_0)},$$

while

$$\bar{\lambda}_{20} = \lim_{v\to\infty} \bar{\lambda}_2/v = f_2/(m_0+2).$$

The estimator $\bar{\gamma}_0(u)$ depends on u if $m_0 < n-p$, (or equivalently $M \neq MV(MVM)^+$), and $f_2 \neq 0$. However, it has been shown by Gnot and Kleffe (1983) that $\bar{\gamma}_0(u)$ is admissible for each $u \geq 0$ so that the class $\mathscr{A}_I^f$ together with the limiting estimators (2.5) forms the entire class of admissible estimators in the set of invariant quadratic estimators for $f'\theta$. Denote this class by $\bar{\mathscr{A}}_I^f$. It does not contain any unbiased estimator.

The estimators in $\bar{\mathscr{A}}_{IU}^f$ and $\bar{\mathscr{A}}_I^f$ are Bayes invariant quadratic estimators introduced by Kleffe and Pincus (1974) and Kleffe (1975). They correspond to all prior distributions τ on the space of parameter vector θ with the second moment $E_\tau \theta\theta' = C$. Let us therefore call these estimators being best at u and v within their classes. An estimator might well be best at several pairs (u,v) simultaneously. The case $v = 0$ refers to locally best estimation. The limiting case $v\to\infty$ may be considered as corresponding to the prior distribution given by $C = \begin{pmatrix} 0,0 \\ 0,1 \end{pmatrix}$. It follows from the definitions of $\bar{\mathscr{A}}_{IU}^f$ and $\bar{\mathscr{A}}_I^f$ that for any given (u,v) the estimator best at (u,v) in $\bar{\mathscr{A}}_I^f$ coincides with the best at (u,v) in $\bar{\mathscr{A}}_{IU}^{\bar{f}}$, where

$$\bar{f} = (2I+HC)^{-1} HCf. \tag{2.6}$$

3. Does $\bar{\mathscr{A}}_I^f$ contain multiples of estimators from $\bar{\mathscr{A}}_{IU}^f$?

A multiple $\kappa\gamma$ of an admissible invariant quadratic unbiased estimator $\gamma \in \bar{\mathscr{A}}_{IU}^f$ is admissible in the larger class of invariant quadratic estimators iff $\kappa\gamma \in \bar{\mathscr{A}}_I^f$. Necessary and sufficient conditions for this to happen are given in the following lemma.

Lemma 3.1. Let $\gamma \in \bar{\mathscr{A}}_{IU}^f$ and let $\mathscr{N}$ be the set of all pairs (u,v) for which γ is best unbiased estimator for $f'\theta$, $f \neq 0$. Then there exists κ such that $\kappa\gamma \in \bar{\mathscr{A}}_I^f$ iff f is an eigenvector of HC for some $(u,v) \in \mathscr{N}$. In this case $\kappa = \eta(\eta+2)^{-1}$, where $\eta \geq 0$ is the eigenvalue of HC corresponding to f.

Proof. Assume that $\gamma \in \bar{\mathscr{A}}_{IU}^f$ is best at $(u,v) \in \mathscr{N}$. Then by (2.6) there exists a number κ such that $\kappa\gamma \in \bar{\mathscr{A}}_I^f$ iff there exists $(u,v) \in \mathscr{N}$ such that

$$\kappa f = (2I+HC)^{-1}HCf,$$

or equivalently iff $HCf = 2\kappa(1-\kappa)^{-1}f$. Thus $\eta = 2\kappa(1-\kappa)^{-1}$ and $\kappa = \eta(\eta+2)^{-1}$. To prove that for each $u \geq 0$, $v \geq 0$ both eigenvalues of HC are nonnegative note that $trHC = n-p > 0$ and $detHC = v(rm-s^2) \geq 0$.

Since

$$HC = \begin{pmatrix} r+us, & ur+(u^2+v)s \\ s+um, & us+(u^2+v)m \end{pmatrix}$$

we obtain

$$HCf = \begin{pmatrix} (f_1+uf_2)(r+us)+vsf_2 \\ (f_1+uf_2)(s+um)+vmf_2 \end{pmatrix}.$$

If $f_1 = 0$ and $f_2 \neq 0$ then f is an eigenvector of HC corresponding to the eigenvalue η iff

$$u(r+us)+vs = 0, \quad u(s+um)+vm = \eta.$$

Moreover, $u(r+us)+vs = 0$ and $s = 0$ imply $u = v = 0$ and $u(s+um)+vm=$

$\eta = 0$.

From the above we have the following corollary.

Corollary 3.1. The only multiple $\kappa\gamma$ of $\gamma \in \bar{\mathscr{A}}^f_{IU}$, for estimation of θ_2, which belongs to $\bar{\mathscr{A}}^f_I$ has $\kappa = 0$.

Let us now concentrate on the case $f_1 \neq 0$ and put $f = (1,\rho)$, where $\rho = f_2/f_1$. Then f is an eigenvector of HC iff

$$\rho = \frac{(1+u\rho)(s+um)+\rho vm}{(1+u\rho)(r+us)+\rho vs}, \tag{3.1}$$

and the eigenvalue of HC corresponding to f is $\eta = (1+u\rho)(r+us)+\rho vs$. It is easy to solve the consistent quadratic equation (3.1) for any given u and v in order to find ρ. The inverse problem, i.e. finding u and v given ρ is not easy and needs further investigation. If $v = 0$ then $\rho = (s+um)/(r+us)$ and $\eta = n-p$, so that multiples of certain locally best invariant quadratic unbiased estimates (C.R.Rao's MINQUE, (1971)) are in the class $\bar{\mathscr{A}}^f_I$. The multipler is

$$\kappa = (n-p)/(n-p+2). \tag{3.2}$$

The inverse problem of finding u given ρ is now somewhat simpler. Using the results given in the Appendix it is easy to show that $s+um$ (given $v = 0$) is a monotonic decreasing function of u and decreases from tr MV for $u = 0$ to zero as $u \to \infty$. Also $r+us$ is a monotonic decreasing function with values in the range from $n-p$ to $n-p-m_o$. Hence to solve the equation

$$\rho(r+us)-(s+um) = 0$$

it is necessary to find the roots of the difference of two monotonic decreasing functions.

The main interest centers around estimation of the individual variance component θ_1. Then $\rho = 0$ and it follows from (3.1) that $f = (1,0)'$ is eigenvector of HC iff $(s+um)/(r+us) = 0$. This condition is never satisfied for finite u and v but holds iff v tends to infinity and $m_o < n-p$. To prove it note that $vr \to \infty$ as $v \to \infty$ and $m_o < n-p$, (see Appendix). Further, it follows from (2.4) and (2.5) that $\bar{\gamma}_o = \kappa\gamma_o$, where $\kappa = (n-p-m_o)/(n-p-m_o+2) = \nu_K/(\nu_K+2)$ and $\nu_K = n-p-m_o$ (see Appendix).

Corollary 3.2. There exists an admissible estimator of the form $\kappa\gamma$, $\gamma \in \bar{\mathscr{A}}^f_{IU}$, for estimation of θ_1 iff $m_O < n-p$. This estimator is uniquely given by $\kappa\gamma_O$, where γ_O is the limiting estimator in $\bar{\mathscr{A}}^f_{IU}$ as $v \to \infty$ and

$$\kappa = \nu_K/(\nu_K+2) \tag{3.3}$$

The estimator $\kappa\gamma_O$ will be called here a generalized Hodges-Lehmann modification (see Klotz, Milton and Zacks (1969)). The next section gives the answer to the question whether $\kappa\gamma_O$ is uniformly better than γ_O.

It follows from Corollary 3.2 that if $m_O < n-p$ then the only estimator $\gamma \in \bar{\mathscr{A}}^f_{IU}$ for $f = (1,0)'$ $(\rho=0)$ such that $\kappa\gamma \in \bar{\mathscr{A}}^f_I$ is the uniquely given limiting estimate γ_O. It is clear that $\kappa\gamma_o$ is the limiting estimator for θ_1 in $\bar{\mathscr{A}}^f_I$. It is interesting to known for which f with $\rho \neq 0$ there exists κ such that $\kappa\gamma_o \in \bar{\mathscr{A}}^f_I$, where γ_o is the limiting estimator for $f'\theta$ in $\bar{\mathscr{A}}^f_{IU}$. To answer this question note that if v tends to infinity then (3.1) becomes $\rho = \rho m_o/[(1+u\rho)\nu_K+\rho s_o]$ and $\rho = (m_o-\nu_K)/(u\nu_K+s_o)$. From the above we obtain the following corollary.

Corollary 3.3. There exists an admissible estimator of the form $\kappa\gamma_o$ for $f'\theta$ iff

$$\rho = (m_o-\nu_K)/(u\nu_K+s_o). \tag{3.4}$$

The multipler $\kappa = m_o/(m_o+2)$. For $m_o = \nu_K$ formula (3.4) reduces to the result presented in Corollary 3.2. Note that $\kappa\gamma_o$ is the limiting estimator in $\bar{\mathscr{A}}^f_I$ and if $\nu_K > 0$ then it depends on u as well as the function $f'\theta$ with ρ given by (3.4).

4. Improving estimators in $\bar{\mathscr{A}}^f_{IU}$ by multiplication with a constant

A general result by Perlman (1972) is that for any unbiased estimator γ and for each θ

$$\operatorname{var}_\theta\gamma - \operatorname{MSE}_\theta(\kappa\gamma) \geq 0 \tag{4.1}$$

iff

$$\frac{1+\kappa}{1-\kappa} \geq \sup_\theta\, (E_\theta\gamma)^2/\operatorname{var}_\theta\gamma = c. \tag{4.2}$$

It follows that (4.1) holds for all κ with

$$\max(0, \frac{c-1}{c+1}) \leq \kappa \leq 1.$$

LaMotte (1979) derived an upper bound for c in case of quadratic invariant estimation. He proved that the inequality

$$(E_\theta \gamma)^2 \leq \frac{n-p}{2} \operatorname{var}_\theta \gamma$$

holds for all θ and all invariant quadratic unbiased estimators $\gamma = y'Ay$. This implies that $\kappa\gamma$ with $\kappa = (n-p-2)/(n-p+2)$ is uniformly better than γ under the mean squared error risk function. Of course, the same applies to the multipler (3.2) which is greater than the one considered here. There are in fact multiples of certain MINQUE's which are uniformly better than MINQUE's and admissible, i.e. they belong to $\bar{\mathscr{A}}^f_I$. But the above simple argument does not apply to the estimator considered in Lemma 3.2. We need to find a better bound for κ corresponding to special functions $f'\theta$. To do it, it is convenient to use a canonical form representation of the model (2.1) which is given in the Appendix. There it is shown that the variance of an invariant quadratic estimator can be expressed as

$$\operatorname{var}_\theta \gamma = 2\theta_1^2 \Sigma a_i^2 (1+\alpha_i q)^2/\nu_i, \quad q = \theta_2/\theta_1.$$

Hence (4.1) holds iff

$$\Sigma a_i^2 (1+\alpha_i q)^2/\nu_i - \frac{1-\kappa}{2(1+\kappa)} (f_1+f_2 q)^2 \geq 0. \tag{4.3}$$

This inequality can be rewritten in the form

$$(\xi_1 - \delta f_2^2) q^2 + 2(\xi_2 - \delta f_1 f_2) q + (\xi_3 - \delta f_1^2) \geq 0, \tag{4.4}$$

where $\delta = (1-\kappa)/2(1+\kappa)$, $\xi_1 = \Sigma a_i^2 \alpha_i^2/\nu_i$, $\xi_2 = \Sigma a_i^2 \alpha_i/\nu_i$, $\xi_3 = \Sigma a_i^2/\nu_i$ are all positive constants. If (4.4) is satisfied for all q, then necessarily

$$\xi_1 \geq \delta f_2^2 \quad \text{and} \quad \xi_3 \geq \delta f_1^2. \tag{4.5}$$

A sufficient condition for (4.4) to hold for all $q \geq 0$ is (4.5) and $\xi_2 \geq \delta f_1 f_2$. But in case $\xi_2 < \delta f_1 f_2$ a necessary and sufficient condition for (4.4) is (4.5) and the inequality

$$(\xi_2 - \delta f_1 f_2)^2 - (\xi_1 - \delta f_2^2)(\xi_3 - \delta f_1^2) \leq 0$$

which can be presented as

$$\delta(\xi_1 f_1^2 - 2\xi_2 f_1 f_2 + \xi_3 f_2^2) \leq \xi_1 \xi_3 - \xi_2^2. \tag{4.6}$$

It is shown in the Appendix that $\xi_1\xi_3 - \xi_2^2 > 0$ and $\xi_1 f_1^2 - 2\xi_2 f_1 f_2 + \xi_3 f_2^2 > 0$. Therefore (4.6) holds iff

$$\delta \leq \frac{\xi_1\xi_3 - \xi_2^2}{\xi_1 f_1^2 - 2\xi_2 f_1 f_2 + \xi_3 f_2^2}.$$

In consequence we have

<u>Lemma 4.1</u>. A necessary and sufficient condition for $\kappa\gamma$ to be uniformly better than $\gamma \in \bar{\mathscr{A}}^f_{IU}$ is

(i) $f_2^2\delta \leq \xi_1$

(ii) $f_1^2\delta \leq \xi_3$,

(iii) $f_1 f_2 \delta \leq \xi_2$ or $(\xi_1 f_1^2 - 2\xi_2 f_1 f_2 + \xi_3 f_2^2)\delta \leq \xi_1\xi_3 - \xi_2^2$,

where $\delta = (1-\kappa)/2(1+\kappa)$.

Explicit expressions for ξ_1, ξ_2 and ξ_3 related to the estimator $\gamma = y'Ay$ are given in the Appendix. We are mainly interested in application of these results to estimators considered in Corollaries 3.1 and 3.2. Note that for $f = (0,1)'$ and $\kappa = 0$ we have $\delta = 1/2$ and according to Lemma 4.1 and Corollary 3.1 0 is uniformly better than $\gamma(u=0, v=0)$ and admissible for θ_2 iff $\xi_1 \geq 1/2$. A more explicit inequality can be obtained by putting $a_i = (\lambda_1 + \lambda_2\alpha_i)\nu_i$ in ξ_1, where $H\lambda = (0,1)'$ and the elements of H are $r = n-p$, $s = \Sigma\nu_i\alpha_i$, $m = \Sigma\alpha_i^2\nu_i$.

If κ is given by (3.3), then $\delta = 1/2(n-p-m_o+1)$. Further using $f = (1,0)'$ in (2.4) and $\xi_3 = \text{tr}A^2$ we have $\xi_3 = 1/(n-p-m_o)$ and $\delta < \xi_3$, so that by Lemma 4.1 the estimator $\kappa\gamma_o$ is uniformly better than γ_o. Note also that the estimator obtained by multiplying γ_o with the universal constant $\kappa = (n-p-2)/(n-p+2)$ is inadmissible in the class of invariant quadratic estimators of θ_1.

References

[1] Gnot, S., Kleffe, J.(1983). Quadratic estimation in mixed linear models with two variance components. J.Statist. Planning and Inference 8, 267-279.

[2] Gnot, S., Kleffe, J., Zmyslony, R.(1984). Nonnegativity of admissible invariant quadratic estimates in mixed linear models with two variance components. Submitted to J.Statist.Planning and Inference.

[3] Hodges, J.L., Lehmann, E.L. (1951). Some applications of the Cramer-Rao inequality. Proc.Sec.Berkeley Symp.Math.Prob.1, 13-22.

[4] Kleffe, J.(1975). Bayes invariant quadratic estimators for variance components in linear models. Math. Operationsf. Statist. 6, 753-767.

[5] Kleffe, J., Pincus, R.(1974). Bayes invariant quadratic unbiased estimators for parameters of the covariance matrix in a normal linear model. Math. Operationsf. Statist. 5, 43-67.

[6] Klotz, J.H., Milton, R.C., Zacks, S.(1969). Mean square efficiency of estimators of variance components. J.Americ. Statist. Assoc.64 1383-1402.

[7] LaMotte, L.R.(1979). On admissible invariant quadratic estimators of variance components. Variance Components and Animal Breeding. L.Dale van Vleck and Shayle R. Searle ed. Cornell University, Ithaca, N.Y.181-188.

[8] Olsen A., Seely, J., Birkes, D. (1976). Invariant quadratic unbiased estimation for two variance components. Ann. Statist. 4, 878-890.

[9] Perlman, M.D. (1972). Reduced mean squared error estimation for several parameters. Sankhya Ser.B, 89-92.

[10] Rao, C.R.(1971). Estimation of variance and covariance components -MINQUE theory. J.Mult. Analysis 1, 257-275.

Appendix

Olsen, Seely and Birkes (1976) considered invariant quadratic estimators of $f'\theta$ as quadratic forms of the maximal invariant statistics $t=By$, where B is such that $B'B = M$, $BB' = I_{n-p}$. The linear model for t is

$$Et = 0, \quad Ett' = \theta_1 I + \theta_2 W, \quad W = BVB'.$$

Further, if $W = \alpha_1 E_1 + \alpha_2 E_2 + \ldots + \alpha_K E_K$ with $\alpha_1 > \alpha_2 > \ldots > \alpha_K \geq 0$ is the spectral decomposition of W, then quadratic forms (2.2) simplify to

$$Q_1(u,v) = \Sigma_{i=1}^K w_i t'E_i t, \quad Q_2(u,v) = \Sigma_{i=1}^K \alpha_i w_i t'E_i t,$$

where $w_i = [1+2u\alpha_i + (u^2+v)\alpha_i^2]^{-1} > 0$. The terms of matrix H are

$$r = \Sigma w_i \nu_i, \quad s = \Sigma \alpha_i w_i \nu_i, \quad m = \Sigma \alpha_i^2 w_i \nu_i,$$

where $\nu_i = \mathrm{tr}E_i$ is the multiplicity of α_i. It is easy to prove that $m_o < n-p$ implies $M \neq MVM(MVM)^+$ and $\alpha_K = 0$, so that in this case $\lim_{v\to\infty} vr = \infty$. Further each admissible invariant quadratic estimator takes the form $\gamma = t'A_* t = \Sigma_{i=1}^K a_i t'E_i t/\nu_i$ with $a_i = (\lambda_1 + \lambda_2 \alpha_i) w_i \nu_i$ and the statistics $t'E_i t/\nu_i$ are independent with expectations $\theta_1 + \theta_2 \alpha_i$ and variances $2(\theta_1+\theta_2\alpha_i)^2/\nu_i$. This makes it easy to derive the variance of a quadratic estimator as given in Section 4. To express a given estimator in a canonical form let us note that for $y'Ay$, $AX = 0$ holds

$$A = MA = MAM, \quad y'Ay = t'BABt = t'A_* t.$$

Routine computation yields the constants needed in Lemma 4.1 as

$$\xi_1 = \mathrm{tr}AVAV, \quad \xi_2 = \mathrm{tr}AVA, \quad \xi_3 = \mathrm{tr}A^2$$

and it follows that $\xi_1\xi_3 - \xi_2^2 > 0$ and $\xi_1 f_1^2 - 2\xi_2 f_1 f_2 + \xi_3 f_2^2 > 0$ by observing that

$$\begin{pmatrix} \xi_1, & \xi_2 \\ \xi_2, & \xi_3 \end{pmatrix} = \begin{pmatrix} \mathrm{tr}AVAV, & \mathrm{tr}AVA \\ \mathrm{tr}\,AVA, & \mathrm{tr}A^2 \end{pmatrix}$$

is positive definite if MVM is not a multiple of M, what we assume for the reason of estimability. But actual computations are generally easier to perform in terms of the spectral decomposition for we mostly need to known the eigenvalues α_i and their multiplicities ν_i only. Some hints how to derive them are given in Gnot, Kleffe and Zmyślony (1984).

Acknowledgement. The first author contributed to this work when he was staying at the Institute of Mathematics of GDR Academy of Sciences in Berlin in 1983. He wants to express his gratitude to Dr.J.Kleffe for making this visit possible.

LINEAR STATISTICAL INFERENCE BASED ON L-ESTIMATORS

J.Jurečková
Institute of Probability and
Mathematical Statistics
Charles University
Prague, Czechoslovakia

1. Introduction

Consider the linear model

$$\underset{\sim}{Y} = \underset{\sim}{X}\underset{\sim}{\beta} + \underset{\sim}{E}, \tag{1.1}$$

where $\underset{\sim}{Y} = (Y_1,\ldots,Y_n)'$ is a vector of independent observations, $\underset{\sim}{X}$ is a known (nxp)-matrix, $\underset{\sim}{\beta} = (\beta_1,\ldots,\beta_p)'$ is an unknown parameter and $\underset{\sim}{E} = (E_1,\ldots,E_n)'$ is a vector of independent errors identically distributed according to a distribution function (d.f.) F, which is either uknown or only partially known. We want to find an estimator for $\underset{\sim}{\beta}$ which has high efficiency under normal d.f. F but which is able to endure mild perturbations from normality. The latter requirement is not satisfied by the classicial least-squares estimator (LSE).

In the location submodel ($p = 1$, $x_{i1} = 1$, $x_{ij} = 0$; $i = 1,\ldots,n$, $j = 2,\ldots,p$), three broad classes of robust estimators, less sensitive to deviations in the distribution shape, namely M-, R- and L-estimators, took the most interest in the statistical practice. From these three classes, M- and R-estimators extend in a straightforward way to the linear model. Their detailed study may be found, e.g. in Huber (1981).

L-estimators have a simple structure in the location model, being just linear combinations of order statistics, yet they have good efficiency and robustness properties. Well-known is the sample median $\tilde{Y}_n$, the α-trimmed mean,

$$L_n(\alpha) = \frac{1}{n-2[n\alpha]} \sum_{i=[n\alpha]+1}^{n-[n\alpha]} Y_{n:i}, \tag{1.2}$$

or the α-Winsorized mean,

$$L_n^*(\alpha) = n^{-1}\{[n\alpha]Y_{n:[n\alpha]} + \sum_{i=[n\alpha]+1}^{n-[n\alpha]} Y_{n:i} + [n\alpha]Y_{n:n-[n\alpha]+1}\}, \quad (1.3)$$

where $0 < \alpha < 1/2$, $[x]$ is the largest integer k satisfying $k \leq x$ and $Y_{n:1} \leq \dots \leq Y_{n:n}$ are the order statistics corresponding to $Y_1,\dots,Y_n$. An extension of L-estimators to the linear model is not as straightforward as in the case of other estimators. One possible extension was suggested by Bickel (1973); but Bickel's estimators are complicated and are not invariant to the reparametrization of design.

Koenker and Bassett (1978) introduced the concept of regression quantile as an extension of the sample quantile to the linear model. This concept seems to provide the right basis for the inference in linear models based on the ordered residuals. While the earlier attempts ordered the residuals from some preliminary estimate, the present procedures order the residuals from the regression quantiles and hence are not effected by the choice of initial estimator.

The present paper provides a review of some estimators and tests based on regression quantiles. The area is under development and thus the review is not exhaustive; but even the partial results indicate that the procedures are not only theoretically interesting but that they are also easily applicable to the practical problems.

2. Regression quantiles and trimmed least-squares estimators

Let $Y_1,\dots,Y_n$ be independent observations, Y_i distributed according to the d.f. $F(y-\sum_{j=1}^p x_{ij}\beta_j)$, $i = 1,\dots,n$, where F is an absolutely continuous d.f. and $\underset{\sim}{X} = (x_{ij})_{i=1,\dots,n}^{j=1,\dots,p}$ is a known design matrix. Fix $\alpha \in (0,1)$ and consider the function

$$\rho_\alpha(x) = \begin{cases} \alpha x & \text{if } x \geq 0 \\ (\alpha-1)x & \text{if } x < 0. \end{cases} \quad (2.1)$$

Denote $\underset{\sim}{\hat\beta}(\alpha) = (\hat\beta_1(\alpha),\dots,\hat\beta_p(\alpha))'$ the solution of the minimization problem

$$\sum_{i=1}^n \rho_\alpha(Y_i - \sum_{j=1}^p x_{ij}t_j) := \min \quad (2.2)$$

with respect to $\underset{\sim}{t} \in R^p$. Notice that, in the location submodel, solution of (2.2) leads to the sample α-quantile. Koenker and Bassett (1978) called $\hat{\underset{\sim}{\beta}}(\alpha)$ the <u>regression α-quantile</u>.

If $\underset{\sim}{X}$ contains an intercept, i.e. $x_{i1} = 1$, $i = 1,\dots,n$, if

$$\frac{1}{n}\underset{\sim}{X}'\underset{\sim}{X} \to \underset{\sim}{Q} \quad \text{(positive definite } p\times p \text{ -matrix)} \tag{2.3}$$

as $n \to \infty$, and if F has a positive and finite derivative in a neighborhood of $F^{-1}(\alpha)$, then (Koenker and Bassett (1978))

$$\mathscr{L}\{n^{1/2}(\hat{\underset{\sim}{\beta}}(\alpha) - \underset{\sim}{\beta}(\alpha))\} \to N_p(\underset{\sim}{O},[\alpha(1-\alpha)/\, f^2(F^{-1}(\alpha))]\underset{\sim}{Q}^{-1}) \tag{2.4}$$

as $n \to \infty$, where

$$\underset{\sim}{\beta}(\alpha) = (\beta_1 + F^{-1}(\alpha),\ \beta_2,\dots,\beta_p)'. \tag{2.5}$$

The simultaneous asymptotic distribution of several regression quantiles is analogous to that of several sample quantiles: if $\hat{\underset{\sim}{\xi}} = (\hat{\underset{\sim}{\beta}}(\alpha_1),\dots,\hat{\underset{\sim}{\beta}}(\alpha_m))'$ and $\underset{\sim}{\xi} = (\underset{\sim}{\beta}(\alpha_1),\dots,\underset{\sim}{\beta}(\alpha_m))'$ then the asymptotic distribution of $n^{1/2}(\hat{\underset{\sim}{\xi}} - \underset{\sim}{\xi})$ is mp-dimensional normal $N_p(\underset{\sim}{O},\underset{\sim}{\Omega} \times \underset{\sim}{Q}^{-1})$, where $\underset{\sim}{\Omega} = (\omega_{ij})_{i,j=1}^m$ with

$$\omega_{ij} = \alpha_i(1-\alpha_j)/f(F^{-1}(\alpha_i))f(F^{-1}(\alpha_j)),\ i,j = 1,\dots,m, \tag{2.6}$$

provided F has positive bounded derivative in a neighborhood of $F^{-1}(\alpha_1),\dots,F^{-1}(\alpha_m)$.

If we put $\alpha = 1/2$, we get an l_1-estimator, which is an extension of the sample median to the regression model.

Koenker and Bassett (1978) also suggested the <u>trimmed least-squares estimator</u> (trimmed LSE) in the following way: fix α_1, α_2, $0 < \alpha_1 < \alpha_2 < 1$, and calculate the regression quantiles $\hat{\underset{\sim}{\beta}}(\alpha_1)$ and $\hat{\underset{\sim}{\beta}}(\alpha_2)$. The solution of (2.2) need not be uniquely determined but we could always determine a rule which selects one of the possible solutions (the choice of a single solution does not effect the asymptotic results). Let $\underset{\sim}{A}$ be the diagonal $n\times n$ matrix with the diagonal

$$a_{ii} = a_i = \begin{cases} 0 & \text{if } Y_i \le \sum_{j=1}^{p} x_{ij}\hat{\beta}_j(\alpha_1) \text{ or } Y_i \ge \sum_{j=1}^{p} x_{ij}\hat{\beta}_j(\alpha_2) \\ 1 & \text{otherwise, } i = 1,\dots,n. \end{cases} \tag{2.7}$$

Trimm-off all Y_i with $a_i = 0$, $i = 1,\dots,n$ and calculate the ordinary LSE from the remaining observations. The resulting estimator $\underset{\sim}{L}_n(\alpha_1,\alpha_2)$ which could be also expressed as

$$\underset{\sim}{L}_n(\alpha_1,\alpha_2) = (\underset{\sim}{X}'\underset{\sim}{A}\,\underset{\sim}{X})^-(\underset{\sim}{X}'\underset{\sim}{A}\,\underset{\sim}{Y}) \tag{2.8}$$

($\underset{\sim}{C}^-$ denotes the generalized inverse of $\underset{\sim}{C}$) is called the trimmed least-squares estimator of $\underset{\sim}{\beta}$.

This estimator was later studied in details by Ruppert and Caroll (1980) who confirmed the asymptotic distribution conjectured by Koenker and Bassett (1978). Let us formulate their result as a theorem.

<u>Theorem 1</u>. (Ruppert and Caroll (1980)). Let $Y_1,\dots,Y_n$ be independent random variables, Y_i distributed according to the d.f. $F(y-\Sigma_{j=1}^{p} x_{ij}\beta_j)$; let F and $\underset{\sim}{X} = (x_{ij})_{i=1,\dots,n}^{j=1,\dots,p}$ satisfy the following conditions:

1° F has a continuous density f such that $f(y) > 0$ on the support of F.

2° $x_{i1} = 0$, $i = 1,\dots,n$ and $\sum_{i=1}^{n} x_{ij} = 0$, $j = 2,\dots,p$.

3° $\lim_{n\to\infty} \max_{1\le i\le n} \{n^{-1/2}|x_{ij}|\} = 0$, $j = 1,\dots,p$.

4° There exists a positive definite $\underset{\sim}{Q}$ such that $\lim_{n\to\infty} n^{-1}\,\underset{\sim}{X}'\underset{\sim}{X} = \underset{\sim}{Q}$.

Then

$$\underset{\sim}{L}_n(\alpha_1,\alpha_2) - \underset{\sim}{\beta} = n^{-1}(\alpha_2-\alpha_1)^{-1}\underset{\sim}{Q}^{-1}\sum_{i=1}^{n}\underset{\sim}{x}_i\psi(E_i) + o_p(n^{-1/2}), \tag{2.9}$$

as $n\to\infty$, where $\underset{\sim}{x}_i = (x_{i1},\dots,x_{ip})'$, and

$$\psi(x) = \begin{cases} F^{-1}(\alpha_1) & \text{if } x < F^{-1}(\alpha_1) \\ x & \text{if } F^{-1}(\alpha_1) \le x \le F^{-1}(\alpha_2) \\ F^{-1}(\alpha_2) & \text{if } F^{-1}(\alpha_2) < x. \end{cases} \tag{2.10}$$

Consequently,

$$\mathscr{L}\{n^{1/2}(\underset{\sim}{L}_n(\alpha_1,\alpha_2)-\underset{\sim}{\beta})\} \to N_p(\underset{\sim}{O},\sigma^2(\alpha_1,\alpha_2,F)\underset{\sim}{Q}^{-1}), \tag{2.11}$$

where

$$\sigma^2(\alpha_1,\alpha_2,F) = (\alpha_2-\alpha_1)^{-2}\{\int_{\alpha_1}^{\alpha_2}(F^{-1}(u)-\delta)^2du+\alpha_1(F^{-1}(\alpha_1)-\delta)^2+ \tag{2.12}$$

$$+(1-\alpha_2)(F^{-1}(\alpha_2)-\delta)^2-[\alpha_1(F^{-1}(\alpha_1)-\delta)+(1-\alpha_2)(F^{-1}(\alpha_2)-\delta)]^2\}$$

and

$$\delta = \delta(\alpha_1,\alpha_2,F) = (\alpha_2-\alpha_1)^{-1}\int_{\alpha_1}^{\alpha_2}F^{-1}(u)du. \tag{2.13}$$

Notice that $\sigma^2(\alpha_1,\alpha_2,F)$ is the asymptotic variance of the trimmed mean in the location case. If, besides the assumptions of Theorem 1, we assume that F is symmetric and put $\alpha_1 = 1-\alpha_2$ we see that $\delta = 0$ and $\sigma^2(\alpha,F)$ takes on a simpler form. Combining the representation (2.9) with Theorem 4.1 of Jurečková (1977), we get

$$n^{1/2}\|\underset{\sim}{L}_n(\alpha) - \underset{\sim}{M}_n\| \overset{p}{\to} 0, \text{ as } n\to\infty, \tag{2.14}$$

where $\underset{\sim}{M}_n$ is the M-estimator defined as a solution of the system of equations

$$\sum_{i=1}^{n} x_{ik}\psi(Y_i-\sum_{j=1}^{p}x_{ij}t_j) = 0, \quad k=1,\ldots,p \tag{2.15}$$

with ψ of (2.10). The M-estimator generated by the function

$$\psi_k(x) = \begin{cases} x & \text{if } |x| \le k \\ k.\text{sign } x & \text{if } |x| > k \end{cases} \tag{2.16}$$

with $k > 0$ is known as Huber's estimator. Hence, we may conclude that the class of trimmed LSE's is asymptotically equivalent to the class of Huber's estimators.

Comparing with $\underset{\sim}{M}_n$, the trimmed LSE is scale-equivariant, i.e. $\underset{\sim}{L}_n(c\,Y_1,\ldots,c\,Y_n) = c\,\underset{\sim}{L}_n(Y_1,\ldots,Y_n)$, $c > 0$. This is not the case of Huber's M-estimator $\underset{\sim}{M}_n$: to make it scale equivariant, we should either supplement $\underset{\sim}{M}_n$ by an estimator of scale or we could alternatively apply the following adaptive procedure proposed by Jurečková and Sen (1984). Fix $\alpha \in (0, 1/2)$, calculate the regression quantiles $\hat{\underset{\sim}{\beta}}(\alpha)$

and $\hat{\underset{\sim}{\beta}}(1-\alpha)$ and put

$$k_n(\alpha) = \frac{1}{2}(\hat{\beta}_1(1-\alpha) - \hat{\beta}_1(\alpha)). \tag{2.17}$$

Let $\underset{\sim}{M}_n^*$ be an estimator defined as a solution of (2.15) with $\psi = \psi_{k_n(\alpha)}$. Then $\underset{\sim}{M}_n^*$ is scale-equivariant and, under the conditions of Theorem 1 and for symmetric F, the asymptotic distribution of $\underset{\sim}{M}_n^*$ concides with that of (2.11)-(2.13) [with $\alpha_1 = \alpha$, $\alpha_2 = 1-\alpha$, $\delta = 0$]. Moreover, if we put

$$\alpha = (1-\varepsilon)(1-\Phi(k)) + \frac{\varepsilon}{2} \tag{2.18}$$

with k satisfying

$$2(\varphi(k)/k) - 2\Phi(-k) = \varepsilon/(1-\varepsilon) \tag{2.19}$$

[Φ is the d.f. of $N(0,1)$ and $\varphi(x) = \frac{d\Phi}{dx}$] then $\underset{\sim}{M}_n^*$ is the minimax estimator minimizing the maximum asymptotic variance over the family of ε-contaminated normal distributions,

$$\mathcal{F} = \{F = (1-\varepsilon)\Phi + \varepsilon H, \quad H \in \mathcal{H}\}, \tag{2.20}$$

where $\mathcal{H}$ is the family of symmetric d.f.'s (cf.Huber (1964, 1981)).

Ruppert and Caroll (1980) also derived the Bahadur representation of regression quantile $\hat{\underset{\sim}{\beta}}(\alpha)$: under the assumptions of Theorem 1,

$$\hat{\underset{\sim}{\beta}}(\alpha) - \underset{\sim}{\beta}(\alpha) = n^{-1}[f(F^{-1}(\alpha))]^{-1}\underset{\sim}{Q}^{-1}\sum_{i=1}^{n} \underset{\sim}{x}_i\varphi_\alpha(E_i - F^{-1}(\alpha)) + o_p(n^{-1/2}), \tag{2.21}$$

where

$$\varphi_\alpha(x) = \alpha - I[x < 0], \qquad x \in R^1. \tag{2.22}$$

Let us now consider a possible definition and properties of the trimmed LSE in a more general situation when some of the conditions of Theorem 1 are violated:

(i) If the design matrix $\underset{\sim}{X}$ does not satisfy 2° and hence there is no intercept, we suggest to extend $\underset{\sim}{X}$ by a column of units to

$$\underset{\sim}{X}^* = (\underset{\sim}{1} : \underset{\sim}{X}), \tag{2.23}$$

where $\underset{\sim}{1} = (1,\ldots,1)'$ is an $n\times 1$ vector; then we shall consider the extended model

$$\underset{\sim}{Y} = \underset{\sim}{X}^* \underset{\sim}{\beta}^* + \underset{\sim}{E} \tag{2.24}$$

with $\underset{\sim}{\beta}^* = (\beta_0, \beta_1,\ldots,\beta_p)'$; $\beta_0 (=0)$ is an artificial intercept. We calculate the trimmed LSE $\underset{\sim}{L}_n^*(\alpha_1,\alpha_2) = (L_0^*, L_1^*,\ldots,L_p^*)'$ for the extended model in the Koenker-Bassett manner and define $\underset{\sim}{L}_n(\alpha_1,\alpha_2) = (L_1^*,\ldots,L_p^*)'$ as the trimmed LSE for the original model. Then

$$\mathscr{L}\{n^{1/2}(\underset{\sim}{L}_n(\alpha_1,\alpha_2)-\beta)\} \to N_p(\underset{\sim}{0},\sigma^2(\alpha_1,\alpha_2,F)\bar{\underset{\sim}{Q}}^{-1}), \tag{2.25}$$

as $n\to\infty$, where $\sigma^2(\alpha_1,\alpha_2,F)$ concides with (2.12) and $\bar{\underset{\sim}{Q}} = (q_{jk})_{j,k=1}^p$ with

$$\bar{q}_{jk} = \lim_{n\to\infty} \frac{1}{n} \sum_{i=1}^n (x_{ij}-\bar{x}_j)(x_{ik}-\bar{x}_k),\ \bar{x}_j = \sum_{i=1}^n x_{ij},\ j = 1,\ldots,p. \tag{2.26}$$

The asymptotic distribution (2.25) and the representation of the type (2.21) in a more general setup and with the remaining term of order $O_p(n^{-3/4})$ are proved in Jurečková (1984).

(ii) As a special case, consider the situation that $\underset{\sim}{X}$ satisfies

$$\sum_{i=1}^n x_{ij} = 0, \quad j = 1,\ldots,p. \tag{2.27}$$

Then, if we defined $\hat{\underset{\sim}{\beta}}(\alpha)$ as a solution of (2.2) we should come to a surprising solution that the regression quantiles corresponding to various α are asymptotically undistinguishable from each other. More precisely, under 1°, 3° and 4° of Theorem 1,

$$\hat{\underset{\sim}{\beta}}(\alpha) - \underset{\sim}{\beta} = n^{-1}(f(0))^{-1}\underset{\sim}{Q}^{-1} \sum_{i=1}^n \underset{\sim}{x}_i\, \varphi_{\alpha_0}(E_i) + o_p(n^{-1/2}) \tag{2.28}$$

holds for $0 < \alpha < 1$, as $n\to\infty$, with $\alpha_0 = F(0)$ and φ_α of (2.22). This proposition is proved in Jurečková (1984), again strengthened to $O_p(n^{-3/4})$. We get as a consequence that the trimmed LSE is not well defined in such situation. In such and similar cases, we recommend to use the approach described in (i).

(iii) Jurečková (1983a) introduced the Winsorized LSE, studied its asymptotic properties and found its M-estimator counterpart. For $\alpha \in (0,1/2)$ the α-Winsorized LSE is defined as

$$\underset{\sim}{L}_n^*(\alpha) = n^{-1}\{[n\alpha]\hat{\underset{\sim}{\beta}}(\alpha) + (n-2[n\alpha]\underset{\sim}{L}_n(\alpha) + [n\alpha]\hat{\underset{\sim}{\beta}}(1-\alpha)\}, \tag{2.29}$$

where $\underset{\sim}{L}_n(\alpha)$ is the α-trimmed LSE. The asymptotic properties of $\underset{\sim}{L}_n^*(\alpha)$ are analogous to those of the α-Winsorized mean in the location case.

An important aspect of every statistical procedure is that of its computation. A possible computation of regression quantiles and of the trimmed LSE was already considered by Koenker and Bassett. It turns out that the regression quantiles could be calculated by a modified linear programming algorithm and hence the calculation of the trimmed LSE (or that of the Winsorized LSE) merely requires the solution of a linear programming problem in addition to the usual least squares computation. Antoch (1984) elaborated a series of computer programs convenient for a direct practical use. An extensive Monte Carlo study by Antoch, Collomb and Hassani (1984) illustrates that the L-estimators behave well for non-normal distributions and reasonably well for normal distribution, comparing with the LSE.

3. Tests of linear and other hypotheses

The trimmed LSE can be used for testing the linear hypothesis in an analogous way as the classical tests use the ordinary LSE. The pertaining critical regions are based on the asymptotic distributions of the test statistics.

Ruppert and Caroll (1980) constructed a consistent estimator of $\sigma^2(\alpha_1,\alpha_2,F)$ of (2.12) of the form

$$S_n^2 = (\alpha_2-\alpha_1)^{-1}\{(n-p)^{-1}Z_n^2+\alpha_1(\hat{\beta}_1(\alpha_1)-L_1)^2 \tag{3.1}$$

$$+ (1-\alpha_2)(\hat{\beta}_1(\alpha_2)-L_1)^2\}-[\alpha_1(\hat{\beta}_1(\alpha_1)-L_1)+(1-\alpha_2)(\hat{\beta}_1(\alpha_2)-L_1)]^2 ,$$

where

$$Z_n^2 = \underset{\sim}{Y}'\underset{\sim}{A}[\underset{\sim}{I}_p - \underset{\sim}{X}(\underset{\sim}{X}'\underset{\sim}{A}\ \underset{\sim}{X})^-\underset{\sim}{X}]\underset{\sim}{A}\ \underset{\sim}{Y}; \tag{3.2}$$

$\hat{\beta}_1(\alpha_i)$ and L_1 is the first component of $\hat{\beta}(\alpha_i)$, $i = 1,2$, and of $L_n(\alpha_1,\alpha_2)$, respectively. The convergence

$$S_n^2 \xrightarrow{p} \sigma^2(\alpha_1,\alpha_2,F), \quad \text{as} \quad n\to\infty, \tag{3.3}$$

was proved by Ruppert and Caroll (1980) under the conditions of Theorem 1 and by Jurečková (1984) under more general conditions including the polynomial regression.

To illustrate some tests based on the trimmed LSE, it is convenient to rewrite the linear model (1.1) in the form

$$Y = \beta_1 \cdot 1 + \bar{X}\bar{\beta} + E, \tag{3.4}$$

where $1 = (1,\ldots,1)'$ is (nx1) vector of units, $\bar{X} = (x_{ij})_{i=1,\ldots,n}^{j=2,\ldots,p}$ and $\bar{\beta} = (\beta_2,\ldots,\beta_p)'$. Consider the hypothesis

$$H_o:\beta_j = 0, \quad j = p-m+1,\ldots,p \quad (1 \leq m \leq p-1).$$

Let $L = L_n(\alpha_1,\alpha_2)$ denote the trimmed LSE of $\beta = (\beta_1,\ldots,\beta_p)'$ and let $\bar{L} = \bar{L}_n(\alpha_1,\alpha_2)$ denote the same calculated under H_o. Then we may test H_o with the aid of statistic

$$T_n = (L - \bar{L})'X'X(L-\bar{L})/S_n^2. \tag{3.5}$$

Under the assumptions 1° and 4° of Theorem 1, provided

$$\max_{1\leq i\leq n} |x_{ij}| = O(n^{1/4}), \quad \text{as} \quad n\to\infty, \text{ for } j = 2,\ldots,p, \tag{3.6}$$

the test with the critical region

$$T_n \geq \chi_m^2(\gamma), \quad 0 < \gamma < 1 \tag{3.7}$$

has an asymptotic size γ, as $n\to\infty$, $\chi_m^2(\gamma)$ is $(1-\gamma)$-quantile of χ^2 distribution with m degrees of freedom; moreover, its Pitman efficiency with respect to the classical F-test coincides with the ARE of the trimmed LSE to the classical LSE. This is proved in Jurečková (1983b) for the case of polynomial regression but the proofs apply to the present setup as well.

Ruppert and Caroll (1980) mention the possible tests of $H_0: \underset{\sim}{K}\underset{\sim}{\bar{\beta}} = \underset{\sim}{c}$ with $\underset{\sim}{K}$ being an $1x(p-1)$ matrix of the rank 1, under the conditions 1°-4°. The latter conditions do not cover the polynomial regression but hopefully the tests can be extended to more general cases with the aid of methodology developed in Jurečková (1983b, 1984).

Parallel tests based on l_1-estimator were developed by Koenker and Bassett (1982a); the tests represent analogues of the classical Wald's likelihood ratio and Lagrange multiplier test, respectively. In another paper, Koenker and Bassett (1982b) propose a test of heteroscedasticity based on regression quantiles, more precisely, a test of hypothesis $H_1: \underset{\sim}{\gamma} = \underset{\sim}{0}$ in the model

$$Y_i = \sum_{j=1}^{p} x_{ij}\beta_j + \sigma_i E_i, \quad i = 1,\dots,n \tag{3.8}$$

with

$$\sigma_i = 1 + h\Big(\sum_{k=1}^{r} z_{ik}\gamma_k\Big), \quad i = 1,\dots,n, \tag{3.9}$$

where x_{ij} and z_{ik} $(i = 1,\dots,n;\ j = 1,\dots,p;\ k = 1,\dots,r)$ are known constants, $\underset{\sim}{\beta} = (\beta_1,\dots,\beta_p)'$ and $\underset{\sim}{\gamma} = (\gamma_1,\dots,\gamma_r)'$ unknown parameters and h is a known, smooth positive function. The power of the test based on regression quantiles is less sensitive to the kurtosis of E_i (and thus to possibly long-tailed distributions) than the tests earlier proposed for normally distributed E_i.

References

[1] Antoch, J.(1984). Collection of programs for robust regression analysis. Charles University, Prague.

[2] Antoch, J., Collomb, G. and Hassani, S.(1984). Robustness in parametric and nonparametric regression: An investigation by computer simulation. COMPSTAT.

[3] Bickel, P.J.(1973). On some analogues to linear combinations of order statistics in the linear model. Ann.Statist. 1, 597-616.

[4] Huber, P.J.(1964). Robust estimation of a location parameter. Ann. Math. Statist. 35, 73-101.

[5] Huber, P.J.(1981). Robust Statistics. J.Wiley, New York.

[6] Jurečková, J.(1977). Asymptotic relations of M-estimates and R-estimates in linear regression model.Ann.Statist. 5, 464-472.

[7] Jurečková, J.(1983a). Winsorized least-squares estimator and its M-estimator counterpart. Contributions to Statistics: Essays in Honour of Norman L.Johnson (ed. P.K.Sen), pp.237-245. North-Holland, Amsterdam.

[8] Jurečková, J.(1983b). Trimmed polynomial regression. Comment.Math. Univ.Carolinae, 24, 597-607.

[9] Jurečková, J.(1983c). Robust estimators of location and regression parameters and their second order asymptotic relations. Trans. 9th Prague Conf.on Inform. Theory, Random Proc. and Statist.Desic. Functions, 19-32. Academia, Prague.

[10] Jurečková, J.(1984). Regression quantiles and trimmed least-squares estimator under a general design. Kybernetika 20, No 5.

[11] Jurečková, J. and Sen, P.K.(1984). On adaptive scale-equivariant M-estimators in linear models. Statistics & Desicions 2.

[12] Koenker, R.(1982). Robust methods in econometrics. Economic Reviews 1, 213-255.

[13] Koenker, R. and Bassett, G., Jr.(1978). Regression quantiles. Econometrica 46, 33-50.

[14] Koenker, R. and Bassett, G., Jr.(1982a). Tests of linear hypotheses and l_1 estimation. Econometrica 50, 1577-1583.

[15] Koenker, R. and Bassett, G., Jr.(1982b). Robust tests for heteroscedasticity based on regression quantiles. Econometrica 50, 43-62.

[16] Ruppert, D. and Carroll, R.J.(1980). Trimmed least squares estimation in the linear model. Journ.Amer.Statist.Assoc., 75, 828-838.

CONNECTED DESIGNS WITH THE MINIMUM NUMBER OF EXPERIMENTAL UNITS

S. Kageyama
Department of Mathematics
Hiroshima University
Japan

Abstract: This paper discusses the minimum number of experimental units for a usual block design to be connected, when the number of treatments and the number of blocks are given. Block designs having such minimum number are characterized and constructed. For some available classes of block designs , the same problem is treated.

1. Introduction

In all connected designs every elementary treatment contrast is estimable. This is a very nice statistical property. Any complete block design is obviously connected, and most of incomplete block designs with a large number of experimental units may usually be connected. On the other hand, incomplete block design with a small number of experimental units have less chance of being connected.

In this paper, the minimum number of experimental units for an incomplete block design to be connected is determined under given v and b, where v is the number of treatments and b is that of blocks. Furthermore, connected block designs having the minimum number of experimental units are characterized and constructed in some ways. Note, however, that in designs which attain the minimum number there is no degree of freedom to estimate the residual variance. So, to estimate residual errors, we consider another situation in which $r_i \geq 2$ and $k_j \geq 2$ for all $i=1,2,\dots,v$, $j=1,2,\dots,b$. Under this restriction we deal with the same problem. Finally, for some available

---

Key words: Connected, experimental unit.
AMS 1980 subject classifications: Primary 62K, Secondary 05B.
Supported in part by Grant 321-6066-58530013 (Japan).

classes of designs (variance-balanced, efficiency-balanced, and so on), the minimum number is determined along with some construction methods.

PART I

2. Minimum number of experimental units

In a block design, let v treatments be applied to plots arranged in b blocks of sizes $k_1, k_2, \ldots, k_b$, and let the ith treatment occur in r_i blocks $(i = 1,2,\ldots,v)$. Originally, a block design is defined to be connected (Bose (1950)) if, given any two treatments i and j, it is possible to construct a chain of treatments $i = i_o, i_1, i_2, \ldots$, $i_n = j$ such that every consecutive two treatments in the chain occur together in a block. This definition is obviously equivalent to the following statistical statement: A block design is said to be connected if all elementary treatment contrasts are estimable. This fact shows the usefulness of connected block designs in statistical analysis.

A portion of a block design may be said to be connected if it is connected in the above sense with regard to the corresponding submatrix of the $v \times b$ incidence matrix of the original design. Any general block design must break up into a number of connected portions such that any treatment belonging to one portion is unconnected with every treatment belonging to any of the other portions. That is, in terms of the incidence matrix, the original design N may be expressed as

$$N = \begin{bmatrix} N_1 & & & O \\ & N_2 & & \\ O & & \ddots & \\ & & & N_g \end{bmatrix},$$

where N_i's are connected portions, for all $i = 1,2,\ldots,g$ (≥ 2). If one wants to make this N connected, then there need to be at least g-1 units such that any two treatments belonging to two different connected portions are connected (in the case of two portions one unit is enough to do so). The problem of determining the minimum of the total number of units for an experiment to be connected can be restricted to the case of a binary block design, i.e. of an incidence matrix with elements 0 and

1 only. Otherwise, the minimum number necessarily increases anyway. Thus, only binary designs are considered in this paper

Proposition 2.1. In an incomplete block design with parameters v, b, r_i, k_j ($v > k_j \geq 2$) for $i=1,2,\ldots,v$ and $j=1,2,\ldots,b$, the minimum of the total number of experimental units for the design to be connected is $v+b-1$, provided v and b are given.

Proof. Under v and b given, from an incidence pattern of the decomposition of the original design N (its $v \times b$ incidence matrix) into connected portions we can arrive at the form

$$N = \begin{bmatrix} 1 & & & \\ \vdots & & 0 & \\ 1 & 1 & & \\ & \vdots & \ddots & 1 \\ & 1 & & \vdots \\ & & 0 & 1 \end{bmatrix},$$

with block sizes k_j, $j=1,\ldots,b$, after some permutation of the rows and columns of the design. Then it holds that $v=k_1+\Sigma_{j=2}^{b}(k_j-\ell_j)$, where ℓ_j is the number of units overlapping between the $(j-1)$th and jth blocks in the above pattern of N, for $j=2,\ldots,b$. Since the total number of experimental units is always $n=\Sigma_{i=1}^{v}r_i=\Sigma_{j=1}^{b}k_j$, it follows that if $\ell_j \geq 1$ for all j, then

$$n = v + \Sigma_{j=2}^{b} \ell_j \geq v + (b-1),$$

and if some $\ell_{j'} = 0$, then the design becomes disconnected. Thus the proof is completed.

We next give another implication of Proposition 2.1. In a practical situation of treatment comparisons, a block of size zero is useless and a block of size one may no contribution to any comparison of treatment effects. In this sense, we are considering here incomplete block designs with blocks such that $k_j \geq 2$ for $j=1,2,\ldots,b$. This setting yields

$$n = \sum_{j=1}^{b} k_j \geq 2b,$$

the equality holding if and only if $k_1=k_2=\ldots=k_b=2$. For a design of this type attaining the bound it is obvious that if $b \leq v-2$, then the design is disconnected. So, we have $b \geq v-1$ for the connectedness, i.e., $2b \geq v+b-1$. Furthermore, it follows that if $b \geq v$, we cannot construct a connected (equiblock-sized) design attaining the minimum number $v+b-1$. Hence, in the case of $k_1 = \ldots = k_s = 2$ the connectedness implies the minimum $b = v - 1$, i.e. $2b = v + b - 1$.

On the other hand, if $v \geq \Sigma_{j=1}^{b} k_j$, the design is necessarily disconnected. So, we have $v < \Sigma_{j=1}^{b} k_j$ for the connectedness, that is, $n \geq v+1$, which implies that $r_i \geq 2$ for some i. Obviously, $n = v+1$ $(=\Sigma_{i=1}^{v} r_i)$ if and only if $r_1 = \ldots = r_{i-1} = r_{i+1} = \ldots = r_v = 1$ and $r_i = 2$ for some i. In the case of $n = v+1$, we cannot construct any connected block design if $b \geq 3$. In fact, when $b = 2$, there exists a connected block design having the minimum number $v+1$ $(=v+b-1$ with $b = 2)$.

3. Characterization of the connected minimum designs

We shall characterize some connected block desings with parameters v, b, r_i, k_j $(i = 1,2,\ldots,v;\ j = 1,2,\ldots,b)$ having the minimum number $v+b-1$ of experimental units.

At first, we can show the following.

Proposition 3.1. If $b \geq v$, then there does not exist a connected block design with parameters v, b, r_i, k_j $(v > k_j \geq 2)$, $i = 1,2,\ldots,v$, $j = 1,2,\ldots,b$, having the minimum number $v+b-1$ of experimental units.

Proof. If $v \leq b$, there exists a non-negative integer ℓ such that $b = v + \ell$. In this case, the minimum number should be $v+b-1 = 2b-\ell-1$, which is a contradiction to the fact that in general $n \geq 2b$ for n being the total number of experimental units. Hence, the proof is completed.

Proposition 3.1 implies that the number of blocks must be less than the number of treatments to be compared.

Next, we consider three cases according to some behaviour of r_i and k_j.

Case I: $k_j=k(\geq 2), r_i=r(\geq 2)$ for all i and j. In this case, we have $vr = v+b-1$ and $bk = v+b-1$, which show that $r = 1 + (b-1)/v$ and $k = 1 + (v-1)/b$. Since r and k are both integers, it holds that $b-1 \geq v$ and $v-1 \geq b$, which yield that $b-1 \geq v \geq b+1$, and this, by itself, is a contradiction. Thus, we have established the following.

Proposition 3.2. There does not exist a connected block design with parameters v,b,r and k, having the minimum number $v+b-1$ of experimental units.

Case II: $k_j = k$ for all j. In this case, $bk = v+b-1$ which yields $k = 1 + (v-1)/b$ or $b = (v-1)/(k-1)$. Hence we have the following.

Proposition 3.3. The following conditions are necessary for the existence of a connected block design with parameters v,b,r_i and $k\geq 2$, having the minimum number $v+b-1$ of experimental units:

(i) $v-1$ is divisible by b (i.e., $k = 1 + (v-1)/b$);
(i)' $v \geq b + 1$;
(ii) $k - 1$ divides $v - 1$.

Case III: $r = r(\geq 2)$ for all i. In this case, $vr = v+b-1$, so $r = 1+(b-1)/v$ or $v = (b-1)/(r-1)$. Hence we can obtain the following.

Proposition 3.4. The following conditions are necessary for the existence of a connected block design with parameters v,b,r and k_j, having the minimum number $v+b-1$ of experimental units:

(i) $b-1$ is divisible by v (i.e., $r = 1 + (b-1)/v$);
(i)' $b \geq v + 1$;
(ii) $r-1$ divides $b-1$.

Propositions 3.1 and 3.4 can yield the following immediately.

Corollary 3.5. There does not exist a connected block design with parameters $v,b,r,k_j\,(v-1 \geq k_j \geq 2,\ j = 1,2,\ldots,b)$, having the minimum number $v+b-1$ of experimental units.

4. Constructions of minimum connected designs

We shall present construction methods of connected block designs with parameters $v,b,r_i,k_j\,(v > k_j \geq 2)$, $i = 1,2,\ldots,v$, $j = 1,2,\ldots,b$, having the minimum number $v+b-1$ of experimental units, under a given v and b.

From Proposition 3.1 it is sufficient to consider a case of $v > b$ for any construction. In this case there exists a positive integer ℓ such that $v - \ell = b$, which always allows to construct a required design, with $b-1$ blocks of size two and the last block of size $\ell+1$, for which $n = v + b - 1$:

$$\begin{array}{rl}
 & \overbrace{\hphantom{1\quad 1\quad \ddots\quad 1\quad 1}}^{b} \\
\begin{array}{r} b\left\{\vphantom{\begin{array}{c}1\\1\\1\\ \ddots\\1\\1\end{array}}\right. \\ \ell\left\{\vphantom{\begin{array}{c}1\\ \vdots\\1\end{array}}\right.\end{array} &
\left[\begin{array}{ccccc} 1 & & & & \\ 1 & 1 & & 0 & \\ & 1 & \ddots & & \\ & & \ddots & & \\ & 0 & & 1 & \\ & & & 1 & 1 \\ \hline & & & & 1 \\ & & 0 & & \vdots \\ & & & & 1 \end{array}\right]
\end{array}.$$

As seen from Proposition 3.2, there does not exist a connected block design which is equireplicated and equiblock-sized. Then we consider here other cases except for the above pattern.

Case A: Equiblock-sized designs of block size k.

(1) For an integer $k\ (\geq 2)$ the incidence pattern

$$\begin{bmatrix} 1 & & & \\ J_{k-1} & 1 & & 0 \\ & J_{k-1} & & \\ & & \ddots & \\ & 0 & & 1 \\ & & & J_{k-1} \end{bmatrix}$$

yields a connected block design with parameters $v,b,r_i = 1$ or 2, $k = 1+(v-1)/b$, where J_s is an s-dimensional column vector with unit elements everywhere. In this case $n = v+b-1$. Alternatively,

$$\begin{bmatrix} 1 & J'_{b-1} & & \\ J_{k-1} & & & O \\ & J_{k-1} & & \\ & & \ddots & \\ O & & & J_{k-1} \end{bmatrix}$$

yields a connected block design with parameters $v,b,r_i = 1$ or b, $k = 1+(v-1)/b$, such that $n = v+b-1$.

(2) When $b = 2r_{i'}$ for some i', the incidence pattern

$$\left[\begin{array}{cccc|cccc} 1 & J'_{r_{i'}-1} & & & & O & & \\ \hline J_{k-1} & & & & 1 & & & \\ & J_{k-1} & & O & & 1 & & O \\ & & \ddots & & & & \ddots & \\ & O & & J_{k-1} & & O & & 1 \\ \hline & & & & J_{k-1} & & & O \\ & O & & & & & \ddots & \\ & & & & & O & & J_{k-1} \end{array}\right]$$

yields a connected block design with parameters $v,b\ (= 2r_{i'})$, $r_i = 1$, 2 or $r_{i'}$, $k = 1+(v-1)/b$, such that $n = v+b-1$.

<u>Case B</u>: Equireplicated designs with replication number r. We are now considering blocks whose sizes satisfy $v-1 \geq k_j \geq 2$ for all $j = 1,2,\ldots,b$. In this case, from Corollary 3.5, there does not exist a connected block design with parameters v,b,r,k_j having the minimum number $v+b-1$ of experimental units. However, if we violate the restriction $k_j \geq 2$ for <u>all</u> j, that is, if we allow some blocks to be of size one, then the transpose of any of the designs in Case A may yield a minimum design. These designs are useful in later discussions (in the sense of using method (8)).

(3) In particular, the incidence pattern

$$\begin{bmatrix} 1 & J'_{r-1} & & & & \\ & & 1 & J'_{r-1} & & O \\ & & & & \ddots & \\ & O & & & & \\ & & & & 1 & J'_{r-1} \end{bmatrix}$$

yields a connected block design with parameters v,b,r = 1+(b-1)/v, k_j = 1 or 2, such that n = v+b-1.

(4) When $v = 2k_{j'}$, for some j',the incidence pattern

$$\left[\begin{array}{c|cccc|cccc} & J'_{r-1} & & & & & & & \\ & & J'_{r-1} & & O & & & O & \\ J_{k_{j'}} & & O & \ddots & & & & & \\ & & & & J'_{r-1} & & & & \\ \hline & 1 & & & & J'_{r-1} & & & \\ & & 1 & & O & & J'_{r-1} & & O \\ & & & \ddots & & & & \ddots & \\ O & O & & & & & O & & \\ & & & & 1 & & & & J'_{r-1} \end{array}\right]$$

yields a connected block design with parameters v,b,r = 1 + (b-1)/v, k_j = 1,2, or $k_{j'}$ such that n = v+b-1.

<u>Case C:</u> Designs with varying r_i and k_j. Approaches similar to Cases A and B lead to the following constructions:

(5) The incidence pattern

$$\begin{bmatrix} 1 & & & \\ J_{k_1-1} & 1 & & \\ & J_{k_2-1} & & O \\ & & \ddots & \\ & O & & 1 \\ & & & J_{k_b-1} \end{bmatrix}$$

yields a connected block design with parameters $v, b, r_i = 1$ or 2, k_j, such that $n = v+b-1$. Alternatively,

$$\begin{bmatrix} 1 & J'_{b-1} & & \\ J_{k_1-1} & & & \\ & J_{k_2-1} & & 0 \\ & & \ddots & \\ & 0 & & J_{k_b-1} \end{bmatrix}$$

yields a connected block design with parameters $v, b, r_i = 1$ or b, k_j, such that $n = v+b-1$.

(6) When $b = 2r_{i'}$ for some i', the incidence pattern

$$\left[\begin{array}{cccc|cccc} 1 & J'_{r_{i'}-1} & & & & & 0 & \\ J_{k_1-1} & & & & 1 & & & \\ & J_{k_2-1} & & 0 & & 1 & & 0 \\ & & \ddots & & & & \ddots & \\ & 0 & & & & 0 & & \\ & & & J_{k_{r_{i'}}-1} & & & & 1 \\ \hline & & & & J_{k_{r_{i'}+1}} & & & 0 \\ & & 0 & & & & \ddots & \\ & & & & & 0 & & J_{k_b} \end{array}\right]$$

yields a connected block design with parameters $v, b, r_i = 1, 2,$ or $r_{i'}$, k_j, such that $n = v+b-1$.

(7) The incidence pattern

$$\begin{bmatrix} J_{b-1} & I_{b-1} \\ 0 & I_{b-1} \end{bmatrix}$$

yields a connected block design with parameters $v = 2(b-1), b, r_i = 1$ or 2, $k_j = 2$ or $b-1$, such that $n = v+b-1$.

(8) Let N_i be connected block designs such that $n_i = v_i+b_i-1$, for $i = 1,2$. Then

$$\begin{bmatrix} N_1 & E \\ O & N_2 \end{bmatrix} \quad \text{or} \quad \begin{bmatrix} N_1 & O \\ E & N_2 \end{bmatrix}$$

is also a connected block design such that $n = v+b-1$, where $n = n_1+n_2+1$, $v = v_1+v_2$, $b = b_1+b_2$, and where E is a matrix of an appropriate order in which only one element is unit and all other are zero.

In method (8), taking designs (1) to (7) as basic designs N_i, other connected block designs having $n=v+b-1$ are obtainable. In this procedure, as basic designs N_i, we may use connected block designs with some blocks of size one. For example, the following designs can be used:

(i) A connected block design with parameters $v, b, r_i, k_j = 1$ or 2, of the pattern

$$\begin{bmatrix} J'_{r_1-1} & 1 & & & \\ & 1 & J'_{r_2-1} & & O \\ & & & \ddots & \\ & O & & 1 & J'_{r_v-1} \end{bmatrix},$$

is such that $n=v+b-1$.

(ii) A connected block design with parameters v $(= 2k_{j'}$ for some $j')$, $b, r_i, k_j = 1, 2,$ or $k_{j'}$, of the pattern

$$\left[\begin{array}{c|cccc|ccc} & J'_{r_1-1} & & & & & & \\ & & J'_{r_2-1} & & O & & O & \\ J_{k_{j'}} & O & & \ddots & & & & \\ & & & & J'_{r_{k_{j'}}-1} & & & \\ \hline & 1 & & & & J'_{r_{k_{j'}+1}} & & O \\ & & 1 & \ddots & O & & \ddots & \\ O & & & & & & & \\ & O & & & 1 & O & & J'_{r_v-1} \end{array}\right],$$

is such that $n=v+b-1$.

Incidentally, any permutation of rows and/or columns in designs described here yields many other connected block designs, as arrangements.

PART II

5. Designs for evaluation of errors

It is known (cf. Ogawa (1974)) that under the usual model the number of degrees of freedom belonging to the best linear estimates of the parameters in a connected block design is exactly equal to $v+b-1$. Hence there are no residual degrees of freedom when the design has the minimum number $v+b-1$ of experimental units. Thus, the conclusion is that a minimum connected design is quite unanalysable. A further point is that the minimum designs presented in Part I appear, almost if not quite without exception, to have at least one value of the replication numbers or block sizes equal to one. Thus, it seems that the minimum designs are not of much practical use, if we are insistently adherent to the analysis of variance.

To overcome all difficulties in the above situation, we consider block designs satisfying $r_i \geq 2$ and $k_j \geq 2$ for $i = 1,2,...,v$; $j = 1,2,...,b$ and investigate a connected block design with the number $v+b+1$ of experimental units to obtain some estimate of the residual variance. For, if we have a block design with the number $v+b+1$ of experimental units, there remain two additional degrees of freedom beyond $v+b-1$, and hence we can estimate the residual variance.

We will present minimum designs with parameters v,b,r_i,k_j $(i=1,2,...,v;\ j = 1,2,...,b)$, in the above sense, similar to those of Section 4.

<u>Case I</u>: $v > b$. Let $v = b + \ell$ for a positive integer ℓ. Then we can get the incidence pattern

$$\begin{array}{c} b\left\{\begin{array}{c} \\ \\ \\ \\ \\ \end{array}\right. \\ \ell\left\{\begin{array}{c} \\ \\ \\ \end{array}\right. \end{array} \left[\begin{array}{cccc} 1 & 1 & & \\ 1 & 1 & & 0 \\ & 1 & \ddots & \\ & & \ddots & \ddots \\ & 0 & 1 & 1 \\ \hline & & 1 & 1 \\ & 0 & \vdots & \vdots \\ & & 1 & 1 \end{array}\right],$$

which yields a connected block design with parameters v, $b = v-\ell$, $r=2$, $k_j = 2,3,\ell+1$, or $\ell+2$. In this case for the estimate of the residual variance there remain $2v-(v+b-1)=\ell+1$ (≥ 2) degrees of freedom, and the number of units attains $v+b+1$ when $\ell = 1$. In general, when $v = \alpha b+\beta$ $(0 \leq \beta < b)$ we can construct a design of similar type. For example, when $v = 2b+2$, by systematical extension we obtain the following incidence pattern:

$$\overbrace{\begin{bmatrix} 1 & 1 & & & \\ 1 & 1 & & 0 & \\ & 1 & \ddots & & \\ & 0 & \ddots & 1 & \\ & & & 1 & 1 \\ \hline & & & 1 & 1 \\ & 0 & & 1 & \\ & 1 & \cdot^{\cdot^{\cdot}} & & \\ 1 & 1 & & 0 & \\ 1 & 1 & & & \\ \hline 1 & 1 & & 0 & \\ & 1 & 1 & & \end{bmatrix}}^{b} .$$

Case II: $b > v$. Let $b = v + \ell$ for a positive integer ℓ. Then we get the pattern

$$v\left\{\begin{bmatrix} 1 & 1 & & & & \\ 1 & 1 & 1 & 0 & & \\ & & 1 & \ddots & & \\ & & & \ddots & 1 & \\ & 0 & & & 1 & \end{bmatrix}\right. \overbrace{\begin{bmatrix} & & \\ & 0 & \\ 1 & \ldots & 1 \\ 1 & \ldots & 1 \end{bmatrix}}^{\ell},$$

which yields a connected block design with parameters $v, b=v+\ell$, $r_i = 2,3,\ell+1$, or $\ell+2$, $k=2$, such that for the estimate of the residual variance there remain $2b-(v+b-1)=\ell+1$ (≥ 2) degrees of freedom, and the number of units attains $v+b+1$ when $\ell = 1$. In general, when $b=\alpha v+\beta$ $(0 \leq \beta < v)$ we can construct a similar design. For example, when $b=3v+2$, we have the following extension:

$$v\left\{\left[\begin{array}{cccc|cccc|cccc|c} 1 & 1 & & & & & 1 & 1 & 1 & & & & \\ 1 & 1 & 1 & 0 & & 0 & \cdot^{\cdot^{\cdot}} 1 & 1 & 1 & 1 & & 0 & 0 \\ & & 1 & \ddots & & \cdot^{\cdot^{\cdot}} & & & & 1 & \ddots & & 1 \\ & & & \ddots 1 & 1 & \cdot^{\cdot^{\cdot}} & & & & & \ddots 1 & 1 & 1 \quad 1 \\ & 0 & & 1 & 1 & & 0 & & 0 & & 1 & 1 & 1 \end{array}\right]\right. .$$

But this case never attains $v+b+1$, as will be seen from Proposition 5.1.

Case III: $v = b$. The incidence pattern

$$v\left\{\begin{bmatrix} 1 & 1 & & & & \\ 1 & 1 & & & 0 & \\ & 1 & 1 & & & \\ & & \ddots & \ddots & & \\ & 0 & & 1 & 1 & \\ & & & & 1 & 1 \end{bmatrix}\right.$$

yields a connected block design with parameters $v = b$, $r_i = 2$ or 3, $k_j = 2$ or 3, in which case for the estimate of the residual variance there remain exactly $2v+1-(v+b-1) = 2$ degrees of freedom.

Remark: It can be shown that there does not exist a connected block design with parameters $v,b,r_i(\geq 2)$, $k_j(\geq 2)$, $i = 1,2,\ldots,v$, $j = 1,2,\ldots,b$, having the minimum number $v+b-1$ of experimental units. In fact, (i) when $v \geq b$, we have $n \geq 2v$, which is greater than $v+b-1$, and hence $n > v+b-1$; (ii) when $v < b$, we have $n \geq 2b$, which is greater than $v+b-1$, and hence $n > v+b-1$.

As in Section 3, we can characterize a connected block design with parameters $v,b,r_i(\geq 2)$ and $k_j(\geq 2)$, $i=1,2,\ldots,v$, $j = 1,2,\ldots b'$, having the number $v+b+1$ of experimental units.

Proposition 5.1. If $b > v+1$, then there does not exist a connected block design with parameters, v,b,r_i,k_j ($v > k_j \geq 2$, $b > r_i \geq 2$; $i = 1,2,\ldots,v$, $j = 1,2,\ldots,b$), having the number $v+b+1$ of experimental units.

The proof follows from the same procedure as that of Proposition 3.1.

Proposition 5.2. There does not exist a connected block design with parameters v,b,r,k ($v > k \geq 2$, $b > r \geq 2$) having the number $v+b+1$ of experimental units.

Proof. It follows that if $n = vr = bk = v+b+1$, then $r = 1+(b+1)/v$ and $k = 1+(v+1)/b$. Since r and k are both integers, it holds that $b+1 \geq v$ and $v+1 \geq b$, which yield $b+1 \geq v \geq b-1$. Hence we get cases $v = b-1$, b, or $b+1$. When $v = b-1$, $r = 2+2/(b-1)$ and hence $b = 3$ with $r = 3$, which is a contradiction to $b > r$. When $v = b$, $r = 1+(b+1)/b$, which is not an integer. When $v = b+1$, $r = 2$ and $k = 2+2/b$. Hence we have $b = 2$, which is a contradiction to $b > r$ (≥ 2). Thus the proof is completed.

From Proposition 5.1 and the proof of Proposition 5.2,

we can obtain the following.

Proposition 5.3. If there exists a connected block design with parameters v,b,r $(b > r \geq 2)$ and k_j $(v > k_j \geq 2,\ j = 1,2,\ldots,b)$, having the number $v+b+1$ of experimental units, then $b = v-1$ and $r = 2$.

For an example of Proposition 5.3, refer to the first design pattern presented in the case I of this section.

Proposition 5.4. The following conditions are necessary for the existence of a connected block design with parameters v,b,r_i $(b > r_i \geq 2,\ i = 1,2,\ldots,v)$ and k $(v > k \geq 2)$, having the number $v+b+1$ of experimental units:

(i) $v+1$ is divisible by b (i.e., $k = 1+(v+1)/b$);
(i)' $v \geq b-1$;
(ii) $k-1$ divides $v+1$.

The proof is obvious from the relation $bk = v+b+1$.

An example satisfying Proposition 5.4 can be seen in the case II of this section.

PART III

6. Practical classes of block designs

There are several classes of optimal block designs. Most of them satisfy the Fisher inequality $b \geq v$. Such are, for example, balanced incomplete block (BIB) designs, variance-balanced block (VBB) designs, efficiency-balanced block (EBB) design and so on. On account of Proposition 3.1 there do not exist such designs having the minimum number $v+b-1$ of experimental units. Though in a partially balanced incomplete block (PBIB) design the Fisher inequality does not hold in general, Proposition 3.2 shows that there does not exist a PBIB design with the minimum number $v+b-1$ of experimental units. Thus, we cannot consider these block designs as minimum designs in the above sense.

In Part II, we consider a block design with parameters v,b,r_i $(b > r_i \geq 2,\ i = 1,2,\ldots,v)$ and k_j $(v > k_j \geq 2,\ j = 1,2,\ldots,b)$, having the number $v+b+1$ of experimental units, to obtain estimates of the residual variance. Under this situation, on account of Proposition 5.1, we have to confine ourselves to designs with $b \leq v+1$. But for designs validating the Fisher inequality there remains the only possibility of $b = v$ and $v+1$. These cases, together with Propositions 5.2, 5.3 and 5.4, imply that the design must be unequally replicated and have $b = v+1$. It seems that it is difficult to produce such a connected minimum block design with $b=v+1$ among VBB designs and EBB designs.

Furthermore, it follows from Proposition 5.2 that even when searching among PBIB designs not validating the Fisher inequality, we will not find a connected design having $v+b+1$ experimental units only.

From now on, we consider VBB designs and EBB designs among block designs which are unequally replicated and/or unequally blocksized, and investigate lower bounds on n along with some constructions of designs attaining the bounds.

6.1. VBB designs. A block design is said to be variance-balanced if every elementary constrast of treatments is estimated with the same variance (cf. Rao (1958)). A variance-balanced block (VBB) design with parameters v, b, r_i, k_j $(i=1,2,\ldots,v;\ j=1,2,\ldots,b)$ is given by an incidence matrix N satisfying

$$R - NK^{-1}N' = \rho\{I_v - (1/v)G_v\},$$

where $R = \text{diag}\{r_1, r_2, \ldots, r_v\}$, $K = \text{diag}\{k_1, k_2, \ldots, k_b\}$, $G_v = J_vJ_v'$ and $\rho = (n-b)/(v-1)$, with $n = \Sigma_{i=1}^{v} r_i = \Sigma_{j=1}^{b} k_j$. Khatri (1982) supposed that from a point of view of overall A-efficiency, among VBB designs one should choose the design which has the minimum number of experimental units. This suggestion enhances the validity of our problem.

From Theorem 1 due to Kageyama and Tsuji (1980), we have

$$\rho \geq \frac{v}{v-1}(\max r_i)(1 - \frac{1}{\min k_j})$$

which, from $\rho = (n-b)/(v-1)$, yields that

$$n \geq b + v(\max r_i)(1 - \frac{1}{\min k_j}).$$

Since $k_j \geq 2$ for all $j = 1,2,\ldots,b$, we have established the following.

Proposition 6.1. For a VBB design, a minimum number of experimental units is given by $b+v(\max r_i)/2$.

Example. VBB designs attaining the bound in Proposition 6.1 are given by plans with Nos. 1, 4, 7, 10 and 13 of Kageyama (1976).

As a general construction method for VBB designs attaining the bound in Proposition 6.1., we have the following incidence patterns:

(6.1): for positive integers v and m such that $m/(v-1) = 1/2$, the pattern

$$\begin{bmatrix} J'_{v-1} & O \\ I_{v-1} & E_{(v-1)\times m} \end{bmatrix},$$

where $E_{a\times b} = J_a J'_b$;

(6.2): for an incidence matrix N of a BIB design with parameters $v,b,r,k = 2\lambda,\lambda$, the pattern

$$\begin{bmatrix} J'_{v} & O \\ I_{v} & N \end{bmatrix};$$

(6.3): for positive integers v and m such that $2m = v-2$,

$$\begin{bmatrix} J'_{v-1} & O & O \\ & J'_{v-2} & O \\ I_{v-1} & I_{v-2} & E_{(v-2)\times m} \end{bmatrix};$$

(6.4): for an incidence matrix N of a BIB design with parameters $v,b,r,k = 2\lambda,\lambda$,

$$\begin{bmatrix} J'_{v+1} & O & O \\ & J'_{v} & O \\ I_{v+1} & & \\ & I_{v} & N \end{bmatrix};$$

(6.5): for positive integers v and m such that $2m = v-3$,

$$\begin{bmatrix} J'_{v-1} & O & O & O \\ & J'_{v-2} & O & O \\ I_{v-1} & & J'_{v-3} & O \\ & I_{v-2} & & \\ & & I_{v-3} & E_{(v-3)\times m} \end{bmatrix}.$$

6.2. EBB designs. A block design is said to be efficiency- balanced if every treatment contrast is estimated with the same efficiency $1-\mu$ (cf. Caliński (1971)). An efficiency-balanced block (EBB) design with parameters v,b,r_i,k_j $(i = 1,2,\ldots,v;\ j = 1,2,\ldots,b)$ is given by an incidence matrix N satisfying

$$R^{-1}NK^{-1}N' - \frac{1}{n}J_v\underline{r}' = \mu(I_v - \frac{1}{n}J_v\underline{r}'),$$

where $\underline{r}' = (r_1,r_2,\ldots,r_v)$ and $n = \Sigma_{i=1}^{v} r_i = \Sigma_{j=1}^{b} k_j$. When the design is equireplicated, the EBB design becomes a VBB design, and when the design is equiblock-sized, the EBB design becomes a BIB design(cf. Kageyama (1980)). Hence we consider here EBB designs which are unequally-replicated and unequally-blocksized. In this case, if $k_j \geq 2$ for all j, we have immediately

$$n \geq 2b + 1,$$

which is trivial. As another bound on n, from Theorem 2.1 due to Kageyama (1980), we can derive the following.

Proposition 6.2. For an EBB design,

$$n \geq (\max r_i)(1-\mu)\left\{\frac{1}{\min k_j} - \mu\right\}^{-1}.$$

Example: The following series of EBB designs with $\mu = 1/4$ attains the bound in Proposition 6.2:

$$\begin{bmatrix} J'_{v-1} & O \\ I_{v-1} & J_{v-1} \end{bmatrix}.$$

In particular, such design with $v = 4$ attains both bounds in Proposition 6.2, and $n \geq 2b+1$.

Since $k_j \geq 2$ for $j = 1,2,\ldots,b$, Proposition 6.2 yields the following.

Corollary 6.3. For an EBB design,

$$n \geq 2(\max r_i) \frac{1-\mu}{1-2\mu}.$$

Finally, we will give an improvement to a bound on μ, i.e to $\mu < 1/(\min k_j)$, for an EBB design, in some sense.

Proposition 6.4. For an EBB design,

$$\mu \leq \frac{v-2}{2(v-1)}.$$

Proof. It holds from Theorem 4 of Kageyama (1980) that $\mu = \{tr(R^{-1}NK^{-1}N')-1\}/(v-1)$. Now

$$tr(R^{-1}NK^{-1}N') = \Sigma_{i=1}^{v}(1/r_i)[\Sigma_{j=1}^{b}n_{ij}/k_j]$$

$$\leq \Sigma_{i=1}^{v}(1/r_i)(1/2)\Sigma_{j=1}^{b}n_{ij} = v/2.$$

Thus we have the required bound.

Remark. Since $1/2 - (v-2)/[2(v-1)] = 1/[2(v-1)] > 0$, the bound in Proposition 6.4 is more stringent than the known bound $\mu < \frac{1}{\min k_j}$ in Corollary 2.1.1 of Kageyama (1980), when $k_j \geq 2$ for some j.

Acknowledgement. The author is thankful to Dr.R.Mukerjee, Indian Statistical Institute, Calcutta, India, for his suggestion of the problem in this paper.

References

[1] Bose, R.C.(1950). Least Square Aspects of Analysis of Variance. Institute of Statistics Mimeo Series 9, University of North Carolina, Chapel Hill.

[2] Caliński, T.(1971), On some desirable patterns in block desings. Biometrics 27, 275-292.

[3] Kageyama, S.(1976). Constructions of balanced block designs. Utilitas Math. 9, 209-229.

[4] Kageyama, S.(1980). On properties of efficiency-balanced designs. Commun.Statist.-Theor. Meth. A9, 597-616.

[5] Kageyama, S. and Tsuji, T.(1980). Some bounds on balanced block designs. J.Statist. Plann. Inf. 4, 155-167.

[6] Khatri, C.G.(1982). A note on variance balanced designs. J.Statist. Plann. Inf. 6, 173-177.

[7] Ogawa, J.(1974). Statistical Theory of the Analysis of Experimental Designs. Marcel Dekker, Inc., New York.

[8] Rao, V.R.(1958). A note on balanced designs. Ann.Math.Statist. 29, 290-294.

SOME REMARKS ON THE SPHERICAL DISTRIBUTIONS AND LINEAR MODELS

C.G.Khatri
Gujart University
Ahmedabad, India

1. Introduction

Let $O(p)$ denote the space of $p\times p$ orthogonal matrices. The random matrix X of order $p\times n$ is said to have a spherical distribution if X and $\Gamma X\Delta$ are identically distributed for every $\Gamma \in O(p)$ and every $\Delta \in O(n)$. By a singular value decomposition theorem, it is obvious that the distribution of X is a symmetric function of the nonzero eigenvalues of XX' $(n \geq p)$ and its density function can be written as

$$\int_{O(p)} f(\Gamma XX'\Gamma')\,[d\Gamma], \tag{1.1}$$

where $f(XX')$ is a function of X such that $\int f(XX')dX = 1$ and $[d\Gamma]$ denotes the unit invariant Haar measure over $O(p)$. X is assumed to have a continuous distribution so that its density function exists.

Let us consider a general linear model (known as the linear growth curve model by Potthoff and Roy (1964)) given by

$$Y = B\xi A + \Theta\varepsilon V_1, \tag{1.2}$$

where Θ and ξ are $p\times p$ and $q\times m$ unknown parameters such that Θ is nonsingular, ε is a $p\times s$ random matrix having a continuous spherical distribution as defined in (1.1), and Y, B, A, V_1 are known matrices of respective orders $p\times n$, $p\times q$, $m\times n$ and $s\times n$ in which rank $V_1 = s$. Potthoff and Roy (1964), Khatri (1966) and Rao (1966) studied this model when $V_1 = I_n$ and the elements of ε are independent $N(0,1)$. We study the model (1.2) when Rank $V_1 = s(<n)$ and Rank $(V_1', A') >$ rank $A + p$. Let us denote $V = V_1'V_1$ and let V^- be any g-inverse of V in the sense $VV^-V = V$. Then

$$V_1 V^- V_1' = I_s, \tag{1.3}$$

and

$$M=(Y-B\xi A)V^-(Y-B\xi A)'=\Theta\varepsilon\varepsilon'\Theta', \quad (Y-B\xi A)(I-V^-V)=O. \tag{1.4}$$

Hence, the density of Y is given by

$$|\Theta\Theta'|^{-s/2}\left(\prod_{i=1}^{s} v_i\right)^{-p/2} \int_{O(p)} f(\Gamma\Theta^{-1}M\Theta'^{-1}\Gamma')[d\Gamma],$$

and (1.5)

$$(Y-B\xi A)(I-V^-V)=O,$$

where $v_1, v_2, \ldots, v_s$ are the nonzero eigenvalues of V. For example, see Khatri (1968) in the case of singular multivariate normal distribution. Let

$$\Theta^{-1} M\Theta'^{-1} = \Gamma_1 D_w \Gamma_1', \tag{1.6}$$

where Γ_1 is a pxp orthogonal matrix, $D_w = \mathrm{diag}\,(w_1, \ldots, w_p)$, the w_i's are the eigenvalues of $\Sigma^{-1}M$ with $\Sigma = \Theta\Theta'$. If Σ is completely unknown, then the elements of D_w and Γ_1 are unknown. Using (1.5), the likelihood function of Y is given by

$$L=|M|^{-s/2}\left(\prod_{i=1}^{s} v_i\right)^{-p/2}|D_w|^{s/2} \int_{O(p)} f(\Gamma D_w \Gamma')\,[d\Gamma]. \tag{1.7}$$

Notice that the likelihood function L is a symmetric function in $w_1, w_2, \ldots w_p$. If f is a decreasing function, then given ξ, the supremum of L over $w_1, \ldots, w_p$ will be obtained if and only if $w_1 = w_2 = \ldots w_p = w$ (say). Hence

$$\sup_{D_w} L = K|M|^{-s/2}, \tag{1.8}$$

where

$$K=\left(\prod_{i=1}^{s} v_i\right)^{-p/2} \sup_{D_w} \int_{O(p)} |D_w|^{s/2} f(\Gamma D_w \Gamma')\,[d\Gamma]. \tag{1.9}$$

The maximum likelihood estimate of ξ will be obtained by minimizing

$$|M| \quad \text{subject to} \quad (Y - B\xi A)(I_n - V^-V) = O. \tag{1.10}$$

This has been done by Srivastava and Khatri [(1979), Theorem 1.10.3 p. 24] when $V = I_n$. We establish a similar results for Rank $V \leq n$, and Rank $(V,A') >$ Rank $A + p$. It has been shown that the results of Khatri (1966) for testing $H_o(\xi = O)$ vs $H(\xi \neq O)$ and the simultaneous confidence bounds for estimable parametric functions $C\xi W$ remain valid without much change. Thus, the likelihood ratio test is robust under the class of spherical distributions. Recently Dawid (1977) and Sinha and Drygas (1982) obtained some results for some particular cases only. Results of robustness due to Sinha and Drygas (1982) are extended to the general situation.

2. Maximum likelihood estimations for ξ

We observe from (1.4) that M is invariant under the choice of V^- and recently Khatri (1982) obtained some particular types of V^- which simplify (1.4) to a great extend. We give them below with a simple proof.

Lemma 1 (Khatri (1982). Let $V_o = V + A'CA$ and let V_o^- be any symmetric g-inverse of V_o, where C is some mxm p.d. matrix. Let

$$A'(AA')^-A + A_1'A_1 = I_n \text{ and } A_1A_1' = I_{n-r},\ A_1A'=O,\ r = \text{Rank } A.$$

Then

$$V_oV_o^-(V,A') = (V,A') \text{ or } V_oV_o^-(V_1',A') = (V_1',A'),$$

$$V_o = A'(AV_o^-A')^-A + VA_1'(A_1VA_1')^-A_1V \tag{2.1}$$

and

$$V_o^-A'F^-AV_o^- + A_1'(A_1VA_1')^-A_1 \text{ is a g-inverse of } V, \tag{2.2}$$

where

$$F = AV_o^-VV_o^-A',$$

$$\text{Rank } V = \text{Rank } F + \text{Rank } (A_1VA_1') \text{ and}$$

$$\text{Rank } (A_1VA_1') = \text{Rank } (V,A') - r. \tag{2.3}$$

Proof. Since $V_o = (V_1',A')\begin{pmatrix} I_s & O \\ O & C\end{pmatrix}(V_1',A')'$ and $\begin{pmatrix} I_s & O \\ O & C\end{pmatrix}$ is p.d.,

$\mathcal{M}(V_o) = \mathcal{M}(V_1', A') = \mathcal{M}(V, A')$ and $\mathcal{M}$ is the column space. This gives

$$V_oV_o^-(V_1', A') = (V_1', A') \quad \text{or} \quad V_oV_o^-(V, A') = (V, A').$$

Further, $AV_o^-A' = AV_o^-V_oV_o^-A'$ and Rank (AV_o^-A') = Rank $(AV_o^-V_o)$ = Rank A, and using Corollary 1.9.2 of Srivastava and Khatri (1979),

$$\text{Rank}(V_o^-A', A_1') = \text{Rank}(AV_o^-A') + \text{Rank}A_1 = \text{Rank}A + \text{Rank}A_1 = n. \tag{2.4}$$

Now, let $V_o = TT'$ where T is an nxt matrix and t = Rank T = Rank V_o. Hence,

$$t = \text{Rank}\{T'(V_o^-A', A_1')\}$$

and this implies

$$I_t = (T'V_o^-A',\ T'A_1')\left\{\begin{pmatrix} AV_o^-T \\ A_1T \end{pmatrix}'(T'V_o^-A', T'A_1')\right\}^-\begin{pmatrix} AV_o^-T \\ A_1T \end{pmatrix}. \tag{2.5}$$

Thus, we get

$$\begin{aligned} V_o &= (A',\ V_oA_1')\begin{pmatrix} AV_o^-A' & 0 \\ 0 & A_1V_oA_1' \end{pmatrix}^-\begin{pmatrix} A \\ A_1V_o \end{pmatrix} \\ &= A'(AV_o^-A')^-A + V_oA_1'(A_1V_oA_1')^-A_1V_o \end{aligned} \tag{2.6}$$

which proves (2.1). Using (2.4), we have

$$\text{Rank}\,[V_1(V_o^-A', A_1')] = s = \text{Rank}V_1,\ A_1VV_o^-A' = A_1V_oV_o^-A' = A_1A' = 0,$$

and

$$I_s = V_1(V_o^-A', A_1')\ \left[\begin{pmatrix} AV_o^- \\ A_1 \end{pmatrix}V(V_o^-A', A_1')\right]^-\begin{pmatrix} AV_o^- \\ A_1 \end{pmatrix}V_1.$$

This implies

$$V = V[V_o^-A'(AV_o^-VV_o^-A')^-AV_o^- + A_1'(AV_oA_1')^-A_1]V \tag{2.7}$$

which proves (2.2). Noting Rank V_o = Rank A + Rank$(A_1V_oA_1')$ and

$$\text{Rank } V = \text{Rank}[V_1(V_o^-A', A_1')] = \text{Rank}\left[\begin{pmatrix} AV_o^- \\ A_1 \end{pmatrix}V(V_o^-A', A_1')\right] = \text{Rank}\begin{pmatrix} AV_o^-VV_o^-A' & 0 \\ 0 & A_1VA_1' \end{pmatrix},$$

we get (2.3)..

Thus, Lemma 1 is established.

Lemma 2. Let $M = (Y - B\xi A)V^-(Y - B\xi A)'$ and $(Y - B\xi A)(I_n - V^-V) = O$, where ξ is an unknown parametric function while Y, B, A, and V are known matrices. Then $|M|$ is minimized under the variation of ξ subject to $(Y - B\xi A)(I_n - V^-V) = O$ at the point $B\xi A = B(B'S^{-1}B)^-B'S^{-1}YV_o^-A'(AV_o^-A')^-A$ where $V_o = V + A'CA$, C is any $m\times m$ p.d., $S = Y[V_o^- - V_o^-A'(AV_o^-A')^-AV_o^-]Y' = YA_1'(A_1VA_1')^-A_1Y'$ and $\text{Rank}\,(A_1VA_1') = \text{Rank}\,(V,A') - \text{Rank}\,A > p$.

Proof. Since M is invariant under any g-inverse V^- of V, hence using (2.2), $V^- = V_o^-A'FAV_o^- + A_1'(A_1VA_1')^-A_1$, $Z = YV_o^-A'(AV_o^-A')^-$ and $F_1 = (AV_o^-A')F^-(AV_o^-A')$ with $F = AV_o^-VV_o^-A'$, we can write

$$M = (Z - B\xi)F_1(Z - B\xi)' + S, \tag{2.8}$$

where $S = Y[V_o^- - V_o^-A'(AV_o^-A')^-AV_o^-]Y' = YA_1'(A_1VA_1')^-A_1Y'$ and on account of YA_1' having a continuous density function and $\text{Rank}(A_1VA_1') = \text{Rank}(V,A') - \text{Rank}\,A > p$, S is p.d. with probability one.

Let $R = I_p - B(B'S^{-1}B)^-B'S^{-1}$ and $Z_1 = (B'S^{-1}B)^-B'S^{-1}Z = (B'S^{-1}B)^-B'S^{-1}YV_o^-A'(AV_o^-A')^-$. Then,

$$(Z - B\xi) = (B, R)\begin{pmatrix} Z_1 - \xi \\ Z \end{pmatrix}$$

and

$$|M| = |S|\,\left|I_p + (B,R)\begin{pmatrix} Z_1 - \xi \\ Z \end{pmatrix}F_1((Z_1 - \xi)', Z')\begin{pmatrix} B' \\ R' \end{pmatrix}S^{-1}\right|$$

$$= |S|\begin{vmatrix} I_q + (Z_1-\xi)F_1(Z_1-\xi)'B'S^{-1}B & (Z_1-\xi)F_1Z'R'S^{-1}R \\ ZF_1(Z_1-\xi)'B'S^{-1}B & I_p + ZF_1Z'R'S^{-1}R \end{vmatrix}$$

$$= |I_p + ZF_1Z'R'S^{-1}R|\;|S + B(Z_1 - \xi)\,G(Z_1 - \xi)'B'|, \tag{2.9}$$

where F_1 is a symmetric matrix,

$$G = F_1 - F_1Z'R'S^{-1}R\,(I_p + ZF_1Z'R'S^{-1}R)^{-1}ZF_1 = F_1G_1 = G_1'F_1$$

with

$$G_1 = I_m - Z'R'S^{-1}R(I_p + ZF_1Z'R'S'^{-1}R)^{-1}\ ZF_1 = (I+Z'R'S^{-1}RZF_1)^{-1}$$

and G is a p.s.d. matrix. From (2.9), it is obvious that M is minimized if and only if

$$B(Z_1 - \xi)G = O \quad \text{or} \quad B(Z_1 - \xi)F_1 = O \tag{2.10}$$

subject to $(Y - B\xi A)\ (I - V^-V) = O$.
Notice that $YV_o^-V_o = Y$ and $A = AV_o^-V_o$. Hence

$$\begin{aligned} O &= (Y - B\xi A)(I - V^-V) = (Y - B\xi A)V_o^-V_o(I - V^-V) \\ &= (Y - B\xi A)V_o^-A'CA\ (I - V^-V) \\ &= (Z - B\xi)\ AV_o^-A'CA(I - V^-V) = (Z - B\xi)\ AV_o^-V_o\ (I - V^-V) \\ &= (Z - B\xi)\ A(I - V^-V). \end{aligned}$$

Now, using $V^- = V_o^-A'F^-AV_o^- + A_1'(A_1V_oA_1')^-A_1$ and (2.10) we get

$$(Z - B\xi)\ A(I - V^-V) = O \Leftrightarrow O = B(Z_1 - \xi)\ A(I - V_o^-A'F^-AV_o^-V)$$

and $B(Z_1 - \xi)F_1 = O \quad \Leftrightarrow \quad B(Z_1 - \xi)\ AV_o^-A'F^-AV_o^-A' \Leftrightarrow O{=}B(Z_1{-}\xi)AV_o^-A'F^-A.$
Hence, the above equations give

$$O = B(Z_1 - \xi)A.$$

Thus $|M|$ is minimized iff $B\xi A = BZ_1A$; and

$$\min_{\xi} |M| = |S+(I-B(B'S^{-1}B)^-B'S^{-1})YV_o^-A'F^-AV_o^-Y'(I-S^{-1}B(B'S^{-1}B)^-B'|.$$

This proves Lemma 2.

Notice that the maximum likelihood estimate of $B\xi A$ is $B(B'S^{-1}B)^-\ \ B'S^{-1}YV_o^-A'(AV_o^-A')^-A$ which is the same as that for $B\xi A$ when errors are normally distributed. Thus, if $C\xi W$ is an estimable parametric function, then the maximum likelihood estimate of $C\xi W$ is

$C(B'S^{-1}B)^{-}B'S^{-1}YV_o^{-}A'(AV_o^{-}A')^{-}W.$

3. The distribution of the maximum likelihood estimate under H_o

Since we are talking of the estimable parametric functions, we shall assume without loss of generality that Rank $B = q$ and Rank $A = m$, for obtaining the distribution of the maximum likelihood estimate of ξ as $\hat{\xi} = (B'S^{-1}B)^{-1}B'S^{-1}YV_o^{-}A'(AV_o^{-}A')^{-1}$ under H_o $(\xi = O)$, or $C\hat{\xi}W$ under $H_o(C\xi W = O)$, where C and W are $c \times q$ and $m \times m_1$ matrices of ranks c and m_1, respectively.

First of all, we shall consider the hypothesis $H_o(\xi = O)$ vs $H(\xi \neq O)$. The likelihood ratio test procedure is to reject H_o if

$$\lambda = \frac{|S+(I-B(B'S^{-1}B)^{-1}B'S^{-1})YV_o^{-}A'F^{-}AV_o^{-}(I-S^{-1}B(B'S^{-1}B)^{-1}B')|}{|S + YV_o^{-}A'F^{-}AV_o^{-}Y'|} \leq d_\alpha, \quad (3.1)$$

where d_α is a constant such that $P(\lambda \leq d_\alpha | H_o) = \alpha$.

Let $Z_2 = YV_o^{-}A'F_2$ with $F_2F_2' = F^{+}$ and $F_3 = AV_o^{-}A'F_2$. Then from (1.5), the joint density of Z_2 and S (see Srivastava and Khatri (1979)) is given by

$$|\Sigma|^{-\frac{1}{2}s}\{\pi^{\frac{1}{2}pf_1}/\Gamma_p(\tfrac{1}{2}f_1)\}|S|^{\frac{1}{2}(f_1-p-1)}\int_{O(p)} f(\Gamma\Theta^{-1}M\Theta'^{-1}\Gamma')\,[d\Gamma], \quad (3.2)$$

where $\Gamma_p(\frac{1}{2}f_1)$ is multivariate gamma function, $\Sigma = \Theta\Theta'$, $f_1 = \text{Rank}(V,A') - m > p$, and

$$M = (Z_2 - B\xi F_3)(Z_2 - B\xi F_3)' + S.$$

Let $B_o = (B,B_1)$ be a nonsingular matrix such that $(B_o^{-1})' = (B(B'B)^{-1},B_1)$, $\Theta_1 = B_o^{-1}\Theta$,

$$\Sigma_1 = B_o^{-1}\Sigma B_o'^{-1} = \Theta_1\Theta_1', \quad S_1 = B_o^{-1}SB_o'^{-1},$$

$$M_1 = B_o^{-1} M B_o'^{-1} = \begin{pmatrix} Z_3 - \xi F_3 \\ Z_4 \end{pmatrix} ((Z_3 - \xi F_3)', Z_4') + \begin{pmatrix} S_{11,1} & S_{12,1} \\ S_{12,1}' & S_{22,1} \end{pmatrix}, \tag{3.3}$$

where $Z_3 = (B'B)^{-1} B'Y V_o^- A'F_2$, $Z_4 = B_1' Y V_o^- A'F_2$, $S_{11,1} = (B'B)^{-1} B'SB(B'B)^{-1}$, $S_{12,1} = (B'B)^{-1} B'SB_1$ and $S_{22,1} = B_1'SB_1$. Then the joint density of S_1, Z_3 and Z_4 is the same as that in (3.2) by replacing Σ by Σ_1, S by S_1, Θ by Θ_1 and M and M_1 defined in (3.3). Use the transformation

$$M_1 = S_1 + \begin{pmatrix} Z_3 - \xi F_3 \\ Z_4 \end{pmatrix} ((Z_3 - \xi F_3)', Z_4') = \begin{pmatrix} T_{11} & T_{12} \\ 0 & T_{22} \end{pmatrix} \begin{pmatrix} T_{11}' & 0 \\ T_{12}' & T_{22}' \end{pmatrix}$$

and
$$\begin{pmatrix} U_1 \\ U_2 \end{pmatrix} = \begin{pmatrix} T_{11}^{-1} & -T_{11}^{-1} T_{12} T_{22}^{-1} \\ 0 & T_{22}^{-1} \end{pmatrix} \begin{pmatrix} Z_3 - \xi F_3 \\ Z_4 \end{pmatrix}. \tag{3.4}$$

The Jacobian of the transformation is

$$J(S_1, Z_3, Z_4 \to M_1, U_1, U_2) = J(S_1 \to M_1 | Z_3, Z_4) J(Z_3, Z_4 \to U_1, U_2 | M_1) = |M_1|^{\frac{1}{2}(s-f_1)}.$$

Hence, using these in (3.2), it is easy to see that (U_1', U_2') and M_1 are independently distributed and their respective density functions are given by

$$\frac{\Gamma_p(\frac{1}{2}, s)}{\pi^{\frac{1}{2}p(s-f_1)} \Gamma_p(\frac{1}{2} f_1)} \left| I - \begin{pmatrix} U_1 \\ U_2 \end{pmatrix} (U_1', U_2') \right|^{\frac{1}{2}(f_1 - p - 1)} \tag{3.5}$$

and

$$|\Sigma_1|^{-s/2} \{\pi^{\frac{1}{2}ps} / \Gamma_p(s/2)\} |M_1|^{\frac{1}{2}(s-p-1)} \int_{O(p)} f(\Gamma \Theta^{-1} M_1 \Theta_1'^{-1} \Gamma') [d\Gamma]. \tag{3.6}$$

Note that the Jacobian of the transformation $U_{(1)} = U_1 (I_{s-f_1} - U_2'U_2)^{-1/2}$ is $T(U_1 \to U_{(1)}) = |I - U_2'U_2|^{\frac{1}{2}q}$ and so from (3.5), it is easy to see that $U_{(1)} = U_1 (I_{s-f_1} - U_2'U_2)^{-\frac{1}{2}}$ and U_2 are independently distributed and their respective density functions are

$$\frac{\Gamma_q(\frac{1}{2}(s-p+q))}{\pi^{\frac{1}{2}q(s-f_1)}\Gamma_q(\frac{1}{2}(f_1-p+q))}|I_q - U_{(1)}U'_{(1)}|^{\frac{1}{2}(f_1-p-1)} \tag{3.7}$$

$$\{\Gamma_{p-q}(\tfrac{1}{2}s)/(\pi^{\frac{1}{2}(p-q)(s-f_1)}\Gamma_{p-q}(\tfrac{1}{2}f_1))\}|I_{p-q}-U_2U_2'|^{\frac{1}{2}(f_1-p+q-1)} . \tag{3.8}$$

Now, in terms of the original variables

$$U_2'U_2 = Z_4'(T_{22}T_{22}')^{-1}Z_4 = Z_4(S_{22\cdot 1} + Z_4Z_4')^{-1}Z_4$$

or

$$\begin{aligned}(I_{s-f_1} - U_2'U_2)^{-1} &= I_{s-f_1} + F_2'AV_o^-Y'B_1(B_1'SB_1)^{-1}B_1'YV_o^-A'F_2 \\ &= I_{s-f_1} + F_2'AV_o^-Y'(S^{-1}-S^{-1}B(B'S^{-1}B)^{-1}B'S^{-1})YV_o^-A'F_2.\end{aligned} \tag{3.9}$$

Further,

$$\begin{aligned}U_1 &= T_{11}^{-1}[(Z_3 - \xi F_3) - T_{12}T_{22}^{-1}Z_4] \\ &= T_{11}^{-1}[(Z_3 - \xi F_3)(I_{s-f_1} - U_2'U_2) - S_{12,1}(S_{22,1} + Z_4Z_4')^{-1}Z_4] \\ &= T_{11}^{-1}(Z_3 - \xi F_3 - S_{12,1}S_{22,1}^{-1}Z_4)(I_{s-f_1} - U_2'U_2).\end{aligned}$$

Hence,

$$U_{(1)} = T_{11}^{-1}(\hat{\xi} - \xi)F_3(I_{s-f_1} - U_2'U_2)^{\frac{1}{2}}. \tag{3.10}$$

From (3.4),

$$T_{11}T_{11}' = (B'S^{-1}B)^{-1} + (\hat{\xi}-\xi)F_3(I-U'U)F'(\hat{\xi}-\xi).$$

Then it is easy to verify that nonzero eigenvalues of $\{(I-U_{(1)}U'_{(1)})^{-1} - I\}$ are the nonzero eigenvalues of $\{(B'S^{-1}B)(\hat{\xi}-\xi)F_3[I_{s-f} + F_3'Z'(S^{-1} - S^{-1}B(B'S^{-1}B)^{-1}B'S^{-1})ZF_3]^{-1}F_3'(\hat{\xi}-\xi)'\}$, where $Z = YV_o^-A'(AV_o^-A')^{-1}$. This shows that the nonzero eigenvalues of $U_{(1)}U'_{(1)}$ are the nonzero eigenvalues of $\{L(I + L)^{-1}\}$ with

$$L = (B'S^{-1}B)(\hat{\xi} - \xi)H(\hat{\xi} - \xi)' \tag{3.11}$$

and

$$H = AV_o^- A' F^+[I + AV_o^- Y' (S^{-1} - S^{-1}B(B'S^{-1}B)^{-1}B'S^{-1})YV_o^- A'F^+]^{-1}AV_o^- A'.$$

Now, we can use Roy's maximum root criterion for obtaining the simultaneous confidence bounds on ξ and we can refer to Khatri (1966) for details. Further, one can obtain the distribution of the eigenvalues of $U_{(1)}U'_{(1)}$ from (3.7).

The simultaneous confidence bounds on $C\xi W$ can be obtained from (3.7). If $V=I$, then we get exactly the same confidence bounds for $C\xi W$ as given by Khatri (1966). For singular V, we shall have to make some changes, but they are not presented here.

Now, we shall consider the distribution of

$$U = T_{11}^{-1}(\hat{\xi} - \xi)F_3, \tag{3.12}$$

from (3.7), (3.8) and (3.10). Since $U = U_{(1)}(I_{s-f_1} - U_2'U_2)^{-\frac{1}{2}}$, the distribution of U given $(I_{s-f_1} - U_2'U_2)$ is

$$\frac{\Gamma_q(\frac{1}{2}(s-p+q))}{\pi^{\frac{1}{2}q(s-f_1)}\Gamma_q(\frac{1}{2}(f_1-p+q))}|I_{p-q}-U_2U_2'|^{\frac{1}{2}q}|I_q-UU'+UU_2'U_2U'|^{\frac{1}{2}(f_1-p-1)}. \tag{3.13}$$

If we integrate over U_2 using (3.8), we get the density of U. For this purpose, we have to find the value of the integral

$$g(UU') = C_1 \int_{U_2} |I_{p-q}-U_2U_2'|^{\frac{1}{2}(f_1+q-p+q-1)} |I_q-UU'+UU_2'U_2U'|^{\frac{1}{2}(f_1-p-1)} dU_2, \tag{3.14}$$

where $C_1 = \Gamma_{p-q}(\frac{1}{2}s)/\pi^{\frac{1}{2}(p-q)(s-f_1)}\Gamma_{p-q}(\frac{1}{2}f_1)$. One can refer to Muirhead (1982) Chapter 7) for

$$|I_q-UU'+UU_2'U_2U'|^{\frac{1}{2}(f_1-p-1)} = |I_q-UU'|^{\frac{1}{2}(f_1-p-1)} |I_{s-f_1}-(-U'(I_q - UU')^{-1}UU_2'U_2)|^{-\frac{1}{2}(p+1-f_1)} =$$

$$= |I_q - UU'|^{\frac{1}{2}(f_1-p-1)} \sum_{k=0}^{\infty} \sum_{K} (-1)^k \frac{(\frac{p+1-f_1}{2})_K}{k!} C_K(U_2U'(I_q-UU')^{-1}UU_2')$$

in terms of zonal polynomials. For zonal polynomials C_K and $(a)_K$ one may refer to Muirhead (1982). Notice that $|I_{p-q}-U_2U_2'|$ is invariant under orthogonal transformation, $U_2 \to \Delta_1 U_2 \Delta_2$ where Δ_1 and Δ_2 are $(p-q)\times(p-q)$ and $(s-f_1)\times(s-f_1)$ orthogonal matrices. Hence (3.14) can be rewritten as

$$g(UU') = |I_q-UU'|^{\frac{1}{2}(f_1-p-1)} \sum_{k=0}^{\infty} \sum_{K} \frac{(-1)^k (\frac{p+1-f}{2})_K}{k!}$$

$$\int_{U_2} \int_{O(s-f_1)} C_K(\Delta_2 U'(I_q-UU')^{-1} U\Delta_2' U_2' U_2)[d\Delta_2] |I-U_2U_2'|^{\frac{1}{2}(f_1+q-p+q-1)} dU_2$$

$$= |I_q-UU'|^{\frac{1}{2}(F_1-p-1)}$$

$$\sum_{k=0}^{\infty} \sum_{K} \frac{(-1)^k (\frac{p+1-f_1}{2})_K}{k!} \frac{C_K(UU'(I_q-UU')^{-1})}{C_K(I_{s-f_1})} g_{1K}$$

[for integration over Δ_2, see James (1964) or Murihead (1982)] , where

$$g_{1K} = \int_{U_2} |I_{p-q}-U_2U_2'|^{\frac{1}{2}(f_1+q-p+q-1)} C_K(U_2'U_2)\,dU_2$$

$$= \frac{\pi^{\frac{1}{2}(p-q)(s-f_1)} C_K(I_{s-f_1}) \Gamma_{s-f_1}(\frac{s-p+2q}{2}) (\frac{p-q}{2})_K}{\Gamma_{s-f_1}(\frac{s+1}{2}) (\frac{s+q}{2})_K} \quad \text{if } p-q \geq s-f_1$$

$$= \frac{\pi^{\frac{1}{2}(p-q)(s-f_1)} C_K(I_{p-q}) \Gamma_{p-q}(\frac{f_1+q}{2}) (\frac{s-f_1}{2})_K}{\Gamma_{p-q}(\frac{s+q}{2})(\frac{s+q}{2})_K} \quad \text{if } s-f_1 \geq p-q$$

(see Khatri and Pillai (1969)). Using these in (3.14) and (3.13) we get the density of U as

$$C|I_q-UU'|^{\frac{1}{2}(f_1-p-1)} \sum_{k=0}^{\infty} \sum_K \frac{(-1)^k (\frac{p+1-f_1}{2})_K (\frac{p-q}{2})_K}{k!(\frac{s+q}{2})_K} C_K(UU'(I_q-UU')^{-1}), \tag{3.15}$$

where

$$C = \frac{\Gamma_q(\frac{s-p+q}{2})\Gamma_{p-q}(\frac{1}{2}s)\Gamma_{p-q}(\frac{f_1+q}{2})}{\pi^{\frac{1}{2}q(s-f_1)}\Gamma_q(\frac{f_1-p+q}{2})\Gamma_{p-q}(\frac{1}{2}f_1)\Gamma_{p-q}(\frac{s+q}{2})}.$$

This distribution can be utilized in obtaining inference on the parameter ξ.

4. UMPI - property of the likelihood ratio test procedure

When $p=q=1$ and $s=n$, the result was established by Sinha and Drygas (1982). We shall take $p=q$, $B=I$ for the case of simplicity and rewrite the density of (3.2) as

$$|\Sigma|^{-\frac{1}{2}s} \int_{O(p)} f[\Gamma\Theta^{-1}((W_1-\eta)(W_1-\eta)'+W_2W_2')\Theta'^{-1}\Gamma]\,[d\Gamma], \tag{4.1}$$

where $\Sigma=\Theta\Theta'$, $\eta=\xi(AV_o^{-1}A')F_2$, $F_2F_2'=(AV_o^-VV_o^-A')^+$, $W_1=YV_o^-A'F_2$ and $W_2=YA_1'F_4$ with $F_4F_4'=(A_1VA_1')^+$. The oreders of the matrices W_1 and W_2 are $p\mathrm{x}(s-f)$ and $p\mathrm{x}f_1$, respectively. The testing of the hypothesis $H_o(\xi=0)$ vs $H(\xi\neq 0)$ will be reduced to $H_o(\eta=0)$ vs $H(\eta\neq 0)$. The testing problem remains invariant under the following group G of transformations

$$W_1 \to P\,W_1\,\Delta_1 \quad \text{and} \quad W_2 \to P\,W_2\,\Delta_2, \tag{4.2}$$

where $P\in O(p)$, $\Delta_1\in O(s-f_1)$ and $\Delta_2\in O(f_1)$. It can be shown that maximal invariants under G are the eigenvalues of $(W_2W_2')^{-1}W_1W_1'$, and maximum invariant parameters $\delta_1\geq\delta_2\geq\dots\geq\delta_p\geq 0$ are the eigenvalues of $\Sigma^{-1}\eta\eta'$. Let $1>l_1>l_2>\dots>l_p>0$ be the eigenvalues of

$(W_1W_1'+W_2W_2')^{-1}W_1W_1'$ if $s-f_1 \geq p$, and these eigenvalues are the sample maximal invariants. Then the null distribution of $U = (W_1W_1'+W_2W_2')^{-\frac{1}{2}}W_1$ is given by

$$\frac{\Gamma_p(\frac{1}{2}s)}{\pi^{\frac{1}{2}p(s-f_1)}\Gamma_p(\frac{1}{2}f_1)} |I_p - UU'|^{\frac{1}{2}(f_1-p-1)} \tag{4.3}$$

and the null density of $l_1, l_2, \ldots, l_p$ is

$$P_o(D\ ) = \frac{\pi^{\frac{1}{2}p^2}}{B_p(\frac{s-f_1}{2}, \frac{f_1}{2})\Gamma_p(\frac{1}{2}p)}$$

$$(\prod_{i=1}^{p} l_i)^{\frac{1}{2}(s-f_1-p-1)} \prod_{i=1}^{p}(1-l_i)^{\frac{1}{2}(f_1-p-1)} \alpha_p(D_l) \tag{4.4}$$

with $\alpha_p(D_l) = \prod_{i=1}^{p-1}\prod_{j=i+1}^{p}(l_i - l_j)$. To obtain the non-null distribution $P(D_l | D_\delta)$ of $l_1, l_2, \ldots, l_p$, we use Wijsaman's result (1967) and write

$$h(D_l | D_\delta) = P(D_l | D_\delta)/P_o(D_l) =$$

$$\int_{O(p)}\int_{O(s-f_1)}\int_P |P|_+^{s-p} f[\Gamma\Theta^{-1}\{P(W_1W_1'+W_2W_2')P' - (PW_1\Delta_1\eta' + \eta\Delta_1'W_1'P')$$

$$+ \eta\eta'\}\ \Theta'^{-1}\Gamma']\ dP[d\Gamma][d\Delta_1],$$

$$\{\int_{O(p)}\int_{O(s-f_1)}\int_P |P|_+^{s-p} f[\Gamma\Theta^{-1}P(W_1W_1'+W_2W_2')P'\Theta'^{-1}\Gamma']dP[d\Gamma][d\Delta]\}^{-1},$$

where $|P|_+$ means the positive value of the determinant. Let $U = (W_1W_1' + W_2W_2')^{-\frac{1}{2}}W_1$. Transform $P \to \Gamma\Theta^{-1}P(W_1W_1'+W_2W_2')^{-\frac{1}{2}}$, $\Gamma \to \Gamma\ \Gamma_1$ and $\Delta_1 \to \Delta_1\ \Delta_3$, where $\Theta^{-1}\eta = \Gamma_1\ D_\delta\Delta_3'$ and $D_\delta = \text{diag}\ (\delta_1, \ldots, \delta_p)$ with $\delta_1 \geq \delta_2 \geq \ldots \geq \delta_p \geq 0$. Then, the above expression can be rewritten as

$$h(D_l | D_\delta) =$$

$$\int_{O(p)} \int_{O(s-f_1)} \int_P |P|_+^{s-p} f[PP' - (PU\Delta_1 D_\delta \Gamma' + \Gamma D_\delta \Delta_1' U'P') + \Gamma D_\delta^2 \Gamma'][d\Gamma][d\Delta_1]dP$$

$$\{\int_P |P|_+^{s-p} f(PP')dP\}^{-1}. \tag{4.5}$$

Let us write (4.5) as a function of $H(U)$. Notice that $H(U) = H(L_1 U L_2')$, where $L_1 \in O(p)$ and $L_2 \in O(s-f_1)$. It is a function of the eigenvalues of UU' (or $U'U$). Further $H(U) = H(-U)$ for any U. Hence by convexity of f, for any α between $[0,1]$,

$$H(-U) = H(U) = \alpha H(-U) + (1-\alpha) H(U) \geq H((1-2\alpha)U)$$

and since $H(U)$ depends only on the eigenvalues of UU' and $(1-2\alpha)^2$, it follows indeed that $H(U)$ is isotonic (or monotonic increasing in the eigenvalues). This shows that the tests based on the eigenvalues of $U'U$ are UMPI test procedures provided f is a convex function.

5. A robustness property of the best linear unbiased estimator

Consider the model

$$X = B\xi A + U_1 \varepsilon U_2, \; E\varepsilon = 0 \; \text{ and } \; V(\text{vec } \varepsilon) = I_{p_1 m_1}, \tag{5.1}$$

where X, B, A, U_1 and U_2 are known matrices of respective orders $p\text{x}n$, $p\text{x}q$, $m\text{x}n$, $p\text{x}p_1$ and $m_1\text{x}n$, and if $\varepsilon = (\underline{\varepsilon}_1, \underline{\varepsilon}_2, \ldots, \underline{\varepsilon}_{m_1})$, then $(\text{vec } \varepsilon)' = (\underline{\varepsilon}_1', \underline{\varepsilon}_2', \ldots, \underline{\varepsilon}_{m_1}')$. Thus the above model (5.1) is

$$(\text{vec } X) = (A' \times B)(\text{vec } \xi) + (U_2' \times U_1)(\text{vec } \varepsilon), \tag{5.2}$$

where $A' \times B$ denotes the Kronecker product between A' and B. Let the BLUE of $C\xi W$ be $G_1 X G_2$ (or the BLUE of $(W' \times C)$ vec ξ is $(G_2' \times G_1)(\text{vec } X))$, and let any linear unbiased estimate of $C\xi W$ be $L_1 X L_2$.

Now, on account of the BLUE property

$$V[(L_2' \times L_1)\,\mathrm{vec}\; X] - V[(G_2' \times G_1)\;\mathrm{vec}\; X] \tag{5.3}$$

is p.s.d. Hence

$$(L_2'U_2'U_2L_2 \times L_1U_1U_1'L_1') - (G_1'U_2'U_2G_2 \times G_1U_1U_1'G_1') \tag{5.4}$$

is p.s.d. and for any vectors $\underline{a}$ and $\underline{b}$,

$$[\underline{a}'(L_2'\Omega_2L_2)\underline{a}][\underline{b}'L_1\Omega_1L_1'\underline{b}] - (\underline{a}'G_2'\Omega_2G_2\underline{a})(\underline{b}'G_1'\Omega_1G_1'\underline{b}) \geq O, \tag{5.5}$$

where $\Omega_2 = U_2'U_2$ and $\Omega_1 = U_1U_1'$. Now, consider A_1 and A_2 be any p.s.d. matrices and consider

$$T_1 = A_1(G_1 \; X \; G_2 - C\xi W)A_2(G_1XG_2 - C\xi W)'$$

and

$$T_2 = A_1(L_1 \; X \; L_2 - C\xi W)A_2(L_1XL_2 - C\xi W). \tag{5.6}$$

Let $l_1^{(i)}, \; l_2^{(i)}, \; \ldots \; , l_c^{(i)}$ be the eigenvalues of T_i for $i = 1,2$, and let $g(l_1^{(i)}, \; l_2^{(i)}, \; \ldots \; , \; l_c^{(i)}) = h_i$, $i = 1,2$, be the functions of the roots such that g is an increasing function in each root separately. Then we shall show that

$$P(h_1 \leq d^2) \geq P(h_2 \leq d^2). \tag{5.7}$$

The above result (5.7) was established by Sinha and Drygas (1982) with $c = 1$. Here c denotes the number of rows of C. The proof of (5.7) is indicated below. Notice that

$$T_1 = A_1G_1U_1 \; \varepsilon \; U_2G_2A_2G_2'U_2'\varepsilon'U_1'G_1'$$

and

$$T_2 = A_2L_1U_1 \; \varepsilon \; U_2L_2A_2L_2'U_2'\varepsilon'U_1'L_1'. \tag{5.8}$$

Then, $l_1^{(1)}, \; l_2^{(1)}, \; \ldots \; , \; l_c^{(1)}$ are the nonzero eigenvalues of $(U_1'G_1'A_1G_1U_1)\;\varepsilon\;(U_2G_2A_2G_2'U_2')\varepsilon'$ and hence $l_1^{(1)}, \ldots , l_c^{(1)}$ are the non-zero eigenvalues of $D_\alpha^{(1)}(\Delta_1\varepsilon\Delta_2)\; D_\beta^{(1)}(\Delta_1\varepsilon\Delta_2)$, where $\Delta_1'D_\alpha^{(1)}\;\Delta_1 = U_1'G_1'A_1G_1U_1$ and $\Delta_2D_\beta^{(1)}\Delta_2' = U_2G_2A_2G_2'U_2'$. $D_\alpha^{(1)}$ and $D_\beta^{(1)}$ are the diagonal matrices and the diagonal elements are the eigenvalues of

$U_1'G_1'A_1G_1U_1$ (or $A_1^{\frac{1}{2}} G_1\Omega_1G_1'A_1^{\frac{1}{2}}$)' and $U_2G_2A_2G_2'U_2'$ (or $A_2^{\frac{1}{2}}G_2'\Omega_2G_2A_2^{\frac{1}{2}}$), respectively. If the distribution of ε is spherical then ε and $\Delta_1 \varepsilon \Delta_2$ are identically distributed. Hence

$$P(h_1 \le d^2) = P(g({}^{(1)}_1, \ldots, {}^{(1)}_c) \le d^2),$$

where ${}^{(1)}_1, \ldots, {}^{(1)}_c$ are the nonzero eigenvalues of $D^{(1)} \varepsilon D^{(1)} \varepsilon'$. Similarly,

$$P(h_2 \le d^2) = P(g({}^{(2)}_1, \ldots, {}^{(2)}_c) \le d^2),$$

where ${}^{(2)}_1, \ldots, {}^{(2)}_c$ are the nonzero eigenvalues of $D_\alpha^{(2)} \varepsilon D_\beta^{(2)} \varepsilon'$. By (5.5), it is easy to see that

$$\alpha_i^{(1)} \le \alpha_i^{(2)}, \ \beta_j^{(1)} \le \beta_j^{(2)} \quad \text{for all } i \text{ and } j,$$

where $\alpha_1^{(t)} \ge \alpha_2^{(t)} \ge \ldots \ge \alpha_c^{(t)} \ge \ldots \ge 0$ and

$$\beta_1^{(t)} \ge \beta_2^{(t)} \ge \ldots \ge \beta_g^{(t)} \ge \ldots \ge 0 \quad \text{for } t = 1,2,.$$

Hence, taking the positive square roots of $\alpha_i^{(t)}$'s we see that

$$(D_\alpha^{(2)})^{\frac{1}{2}} \varepsilon D_\beta^{(2)} \varepsilon' (D_\alpha^{(2)})^{\frac{1}{2}} - (D_\alpha^{(1)})^{\frac{1}{2}} \varepsilon D_\beta^{(1)} \varepsilon' (D_\alpha^{(1)})^{\frac{1}{2}}$$

is p.s.d. Therefore,

$$l_j^{(2)} \ge l_j^{(1)} \quad \text{for } j = 1,2, \ldots, c$$

provided $l_1^{(t)} \ge l_2^{(t)} \ge \ldots \ge l_c^{(t)} \ge 0$ for $t = 1,2$. Since g is an increasing function of l_j's separately, we see that

$$g(l_1^{(2)}, \ldots, l_2^{(2)}) \ge g(l_1^{(1)}, \ldots, l_c^{(1)}).$$

Therefore,

$$g(l_1^{(2)}, \ldots, l_c^{(2)}) \le c^2 \Rightarrow g(l_1^{(1)}, \ldots, l_c^{(1)}) \le c^2$$

and this proves (5.7).

References

David, A.P. (1977). Spherical Matrix Distributions and a Multivariate model. J. Roy. Statist. Soc. Series B. 39, 254-261.

James, A.T. (1964). Distribution of matrix variates and latent roots derived from normal samples. Ann. Math. Statist. 35, 475-501.

Khatri, C.G. (1966). A note on a MANOVA model applied to problems in growth curve. Ann. Inst. Statist. Math., 18, 75-86.

Khatri, C.G. (1968). Some results for the singular normal multivariate regression models. Sankhyā, Series A, 30, 267-280.

Khatri, C.G. (1982). A representation of a matrix and its use in the Gauss-Markoff model. J. Indian Statist. Assoc. 20 (1982), 89-98.

Khatri, C.G. and Pillai, K.C.S. (1969). Distributions of vector corresponding to the largest roots of three Matrices. In Multivariate Analysis, Vol 2 (editor P.R.Krishnaiah). Academic Press, Inc. New York, 219-240.

Muirhead, R.J. (1982). Aspects of multivariate Statistical Theory. John Wiley and sons, New York.

Potthoff, R.F. and Roy, S.N. (1964). A generalized multivariate analysis of variance model useful especially for growth curve problems. Biometrika, 51, 313-326.

Rao, C.R. (1966). Covariance adjustment and related problems in multivariate analysis. In Multivariate Analysis. I (editor P.R. Krishnaiah) Academic Press, New York, pp. 87-103.

Sinha, Bimal K. and Drygas, H. (1982). Robustness properties of the F-test and best linear unbiased estimators in linear models. Centre of multivariate Analysis. University of Pittsburgh, Technical Report No. 82-28.

Srivastava, M.S. and Khatri, C.G. (1979). An introduction to multivariate Statistics. North Holland, New York.

Wijsmann, R.A. (1967). Cross section of orbits and their application to densities of maximal invariants. Fifth Berkeley Symp. Math. Statist. Prob. 1, 389-400.

ON COMPUTATION OF THE LOG-LIKELIHOOD FUNCTIONS UNDER MIXED LINEAR MODELS

J. Kleffe

Institute of Mathematics
Academy of Sciences of the GDR
Berlin, GDR

Summary. We describe a method of computing the log-likelihood and related functions which applies to a large class of mixed linear models. It results in explicite expressions for arbitrary nested classification models and serves to prove boundedness of the log-likelihood function.

1. Introduction

Computation of maximum likelihood estimators for parameters in mixed linear models is a well discussed topic in statistics. Some of the key references to this field are Hartley and Rao (1967), Hemmerle and Hartley (1973), Patterson and Thompson (1975), Harville (1977) and C.R. Rao (1979). These papers consider different methods mostly based on equating derivatives to zero and solving the resulting equations by means of iteration. But very little is known about convergence of the proposed algorithms and whether a maximum of the likelihood function exists in all cases. It is therefore surprising that not much attention has been paid to the simpler problem of just calculating the log-likelihood function. This might be useful to control the progress made in each step of iteration or to employ general computer programs for maximization of complicated functions. Let us consider the general linear model

$$Ey = X\beta, \text{ Cov } y = \sum_{i=1}^{p} U_i \Sigma_i U_i' = V, \tag{1.1}$$

Key words: Likelihood function, mixed linear models, MANOVA.

where y is a normally distributed random vector and $X, U_1,\ldots,U_p$ are given $n\times m_o$, $n\times m_1,\ldots,n\times m_p$ matrices. Here β is a parameter vector of fixed effects and Σ_i, $i=1,\ldots,p$, are variance-covariance matrices of order m_i. Models of this kind appear in multivariate analysis of variance (MANOVA) and in the analysis of growth curves. There is no loss of generality in assuming that U_p is the identity matrix since Σ_p is arbitrary. The variance-covariance matrix V is assumed to be nonsingular except in Section 4.

We are going to discuss in Section 2 computation of the log-likelihood function

$$l(\beta,\Sigma,y) = -\log|V| - (y-X\beta)'V^{-1}(y-X\beta), \tag{1.2}$$

where Σ stands for the set of variance-covariance matrices $\Sigma_1,\ldots,\Sigma_p$. But mostly, interest focusses on the maximum of (1.2) subject to β. It is given by the function

$$l^*(\Sigma,y) = -\log|V| - y'(MVM)^+y, \tag{1.3}$$

where $M = I - XX^+$. Section 3 deals with computation of this function. Section 4 considers the log-likelihood function of the so called error contrasts, which is given by

$$l_m(\Sigma,y) = -\log|B'VB| - y'(MVM)^+y, \tag{1.4}$$

where B is a $(n-r)\times n$-matrix such that

$$B'B = I_{n-r},\ BB' = M,\ r = \text{rank}(X). \tag{1.5}$$

The maximization of the function (1.4) subject to unknown parameters yields restricted or marginal maximum likelihood estimators as proposed by Patterson and Thompson (1975) in order to reduce estimation bias. It is important to know that (1.5) implies

$$(MVM)^+ = B(B'VB)^{-1}B' \tag{1.6}$$

for any n.n.d. matrix V such that $B'VB$ is nonsingular. Therefore, $l_m(\Sigma,y)$ is the log-likelihood function of $B'y$.

The problem with computation of the above functions is the size of the matrices V, MVM and B'VB and the effort required to derive their inverses (pseudoinverses) or determinants. Therefore, several authors remarked that the special structure of V admits stepwise computation of its inverse or determinant by inverting smaller order matrices in each step (see Kleffe (1980)). LaMotte (1972) described an efficient iterative computational procedure for the inverse and determinant of the covariance matrix in a p-way nested classification model. But it still remains to store the results of each step of computation. This fact alone causes heavy restrictions in the size of the model.

We are going to describe a way of computation that avoides such storage problems. The method works most simple if the design matrices U_j admit hierarchical ordering like

$$R(U_1) \subset R(U_2) \subset \ldots \subset R(U_{p-1}) \subset R(\Sigma_p), \tag{1.7}$$

where R(.) denotes column space. In this case, there are matrices Δ_j such that $U_j = U_{j+1}\Delta_j$ for $j = 1,\ldots,p-1$. Another important consequence of (1.7) is $R(V) = R(\Sigma_p)$ and equivalence of nonsingularity of V and Σ_p. Some more simplification is achieved if in addition to (1.7) also

$$R(X) \subset R(U_1), \qquad X = U_1\Delta_o. \tag{1.8}$$

The impact of properties like (1.7) and (1.8) on computational aspects in linear models was first discovered by LaMotte (1972) and fully explored in a paper by Kleffe and Seifert (1983), which discusses computation of MINQUE estimators for variance components. We are going to use many of the results in this paper and therefore introduce to refer to this paper in shorter form as KS(83). This paper also discusses extensions to more general models which apply to our problem in the same way. It is therefore no essential loss in restricting our presentation to the class of most simple cases specified by assumption (1.7). It contains all unbalanced nested classification models which are discussed in Section 5.

2. Computation of $l(\beta,\Sigma,y)$

The special structure of V, as it is given by (1.1), suggests a recursive scheme of computation for its inverse. Let us start with computing

A_p^{-1}, where $A_p = \Sigma_p$, and define the nonsingular matrices

$$A_j = A_{j+1} + U_j \Sigma_j U_j', \quad j = p-1,\dots,1. \tag{2.1}$$

Then, application of formula (3.6) in Harville (1977) provides A_j^{-1} in terms of A_{j+1}^{-1} and inverses of smaller order matrices. Summarizing these single steps of computation yields:

<u>Lemma 2.1</u>. (see KS(83)) Let model (1.1) satisfy (1.7) and define for $j = 1,\dots,p-1$

$$B_j = A_{j+1}^{-1} U_j, \quad F_j = U_j' B_j, \quad E_j = I + \Sigma_j F_j. \tag{2.2}$$

Then, it holds

$$V^{-1} = \Sigma_p^{-1} - \sum_{j=1}^{p-1} B_j E_j^{-1} \Delta_j B_j', \tag{2.3}$$

where the matrices B_j and E_j result from $B_p = F_p = \Sigma_p^{-1}$, $E_p = I$ and

$$B_j = B_{j+1} E_{j+1}^{-1} \Delta_j \tag{2.4}$$

$$F_j = \Delta_j' F_{j+1} E_{j+1}^{-1} \Delta_j, \quad j = p-1,\dots,1. \tag{2.5}$$

Let us now introduce random vectors $z_j = B_j'(y - X\beta)$ for $j = p,\dots,1$. Then, $z_p = \Sigma_p^{-1}(y - X\beta)$ and it follows from (2.4) that

$$z_j = \Delta_j' (E_{j+1}^{-1})' z_{j+1}, \quad j = p-1,\dots,1. \tag{2.6}$$

The matrices E_j required in (2.6) are obtained as outlined in (2.5) and (2.2) so that we have a nice algorithm for computation of z_j. Our main result is now as follows.

<u>Theorem 2.1.</u> Under the assumption (1.7),

$$\begin{aligned} l(\beta,\Sigma,y) &= l(\beta,\Sigma_p,y) - \sum_{j=1}^{p-1} (\log|E_j| - z_j' E_j^{-1} \Sigma_j z_j) \\ &= l(\beta,\Sigma_p,y) - \sum_{j=1}^{p-1} (\log|E_j| + \bar{z}_j' F_j E_j^{-1} \bar{z}_j) \\ &\quad + \sum_{j=1}^{p-1} \bar{z}_j' F_j \bar{z}_j , \end{aligned} \tag{2.7}$$

where $\bar{z}_j = F_j^- z_j$, and

$$l(\beta,\Sigma_p,y) = -\log|\Sigma_p| - (y-X\beta)'\Sigma_p^{-1}(y-X\beta). \tag{2.8}$$

Proof. By definition of z_j, it follows from (2.3) that

$$(y-X\beta)'V^{-1}(y-X\beta) = z_p'\Sigma_p^{-1}z_p - \sum_{j=1}^{p-1} z_j'E_j^{-1}\Sigma_j z_j. \tag{2.9}$$

Thus, it remains to compute $|V|$. But from (2.1) and (2.2) it follows

$$|A_j| = |A_{j+1} + U_j\Sigma_j U_j'| = |A_{j+1}||I + \Sigma_j U_j'A_{j+1}^{-1}U_j| = |A_{j+1}||E_j|. \tag{2.10}$$

Using this result for $j = 1,2,\dots,p-1$ yields

$$|V| = |\Sigma_p||E_{p-1}|\dots|E_1|. \tag{2.11}$$

Expression (2.7) follows immediately from (1.2).

The second expression for $l(\beta,\Sigma,y)$ in Theorem 2.1 is a simple consequence of the identity (2.2) and may have numerical advantages since $\bar{z}_j$ behaves like an average whereas z_j is some form of a total. Computation of $\bar{z}_j$, however, requires additional effort. Another alternative and sometimes more convenient expression for $l(\beta,\Sigma,y)$ is obtained by expressing (2.9) as

$$\sum_{i=1}^{p} (y-X\beta)'V^{-1}U_i\Sigma_i U_i'V^{-1}(y-X\beta) = \sum_{i=1}^{p} s_i'\Sigma_i s_i, \tag{2.12}$$

where the statistics $s_i = U_i'V^{-1}(y-X\beta)$ are obtained by computing

$$\bar{s}_1 = E_1^{-1}\bar{z}_1, \qquad s_1 = F_1\bar{s}_1, \tag{2.13}$$

$$\bar{s}_j = E_j^{-1}(\bar{z}_j - \Delta_{j-1}(\bar{z}_{j-1} - \bar{s}_{j-1})), \qquad s_j = F_j\bar{s}_j. \tag{2.14}$$

A simple proof of this is given in KS (83), and using (2.12) in (1.2) yields

$$l(\beta,\Sigma,y) = -(\ln|\Sigma_p| + s_p'\Sigma_p s_p) - \sum_{j=1}^{p-1} (\ln|E_j| + s_j'\Sigma_j s_j). \tag{2.15}$$

The advantage of this expression over (2.7) is negativity of the second term so that boundedness of the log-likelihood function (2.15) might be verified by just proving boundedness for the first term. It is shown in KS (83) Corollary 2 that

$$s_p'\Sigma_p s_p \geq (y-X\beta)'(\Sigma_p^{-1} - \Sigma_p^{-1}U_{p-1}(U_{p-1}'\Sigma_p^{-1}U_{p-1})^{-}U_{p-1}'\Sigma_p^{-1})(y-X\beta)$$

which provides further help for verification of boundedness.

An important special case is when (1.8) holds in addition to (1.7). Then, it is shown in KS (83) that

$$B_o = V^{-1}X, \qquad F_o = X'V^{-1}X, \tag{2.16}$$

where B_o and F_o result from expanding computation along (2.4) and (2.5) to the case $j = 0$. We may now express the quadratic form (2.9) as

$$\begin{aligned}(y-X\beta)'V^{-1}(y-X\beta) &= y'V^{-1}y - 2\beta'y_o + \beta'F_o\beta \\ &= y'V^{-1}y + y_o'F_o^{-}y_o - (\bar{y}_o - \beta)'F_o(\bar{y}_o - \beta),\end{aligned} \tag{2.17}$$

where $y_o = B_o y$ and $\bar{y}_o = F_o^{-}y_o$ follows by expanding (2.6) to the case $j = 0$ and using $\beta = 0$. Also $y'V^{-1}y$ is obtained from (2.9) or (2.12) by putting $\beta = 0$. The final result is

$$l(\beta,\Sigma,y) = l(0,\Sigma,y) + y_o'F_o^{-}y_o - (\bar{y}_o - \beta)'F_o(\bar{y}_o - \beta). \tag{2.18}$$

3. Computation of $l^*(\Sigma,y)$

The function $l^*(\Sigma,y)$ is obtained if (2.9) is replaced by $y'(MVM)^+y$ in $l(\beta,\Sigma,y)$. It is therefore sufficient to consider computation of $y'(MVM)^+y$ which goes along the lines of Section 2. Let $\tilde{A}_j = MA_jM$, where the matrices A_j are defined by (2.1). Then, by assumption (1.7) it holds $R(MU_j) \subset R(\tilde{A}_{j+1})$ for all j and we have the following lemma, also proved in KS (83).

Lemma 3.1. Let model (1.1) satisfy (1.7) and define

$$\tilde{B}_j = \tilde{A}_{j+1}^{+}U_j, \quad \tilde{F}_j = U_j'\tilde{B}_j, \qquad \tilde{E}_j = I + \Sigma_j\tilde{F}_j \tag{3.1}$$

for $j = p-1, \ldots, 1$. Then, it holds

$$(MVM)^{+} = (M\Sigma_p M)^{+} - \sum_{j=1}^{p-1} \tilde{B}_j \tilde{E}_j^{-1} \Sigma_j \tilde{B}_j', \tag{3.2}$$

where the matrices $\tilde{B}_j$ and $\tilde{E}_j$ are obtained as outlined by (2.4) and (2.5) but starting with $\tilde{B}_p = \tilde{F}_p = (M\Sigma_p M)^{+}$ and $\tilde{E}_p = I$.

Analogously to (2.6) let us now introduce vectors $\tilde{y}_i = B_j'y$, which are obtained by starting with $\tilde{y}_p = (M\Sigma_p M)^{+}y$ and proceeding as

$$\tilde{y}_j = \Delta_j'(\tilde{E}_{j+1}^{-1})'\tilde{y}_{j+1} \quad \text{for } j = p-1, \ldots, 1. \tag{3.3}$$

Then, it follows from (3.2) and (2.11):

<u>Theorem 3.1</u>. Under the assumption (1.7) holds

$$l^*(\Sigma, y) = l^*(\Sigma_p, y) - \sum_{j=1}^{p-1} (\log|E_j| - \tilde{y}_j'\tilde{E}_j^{-1}\Sigma_j\tilde{y}_j) \tag{3.4}$$

$$= l^*(\Sigma_p, y) - \sum_{j=1}^{p-1} (\log|E_j| + \tilde{\bar{y}}_j'\tilde{F}_j\tilde{E}_j^{-1}\tilde{\bar{y}}_j) + \sum_{j=1}^{p-1} \tilde{\bar{y}}_j'F_j\tilde{\bar{y}}_j,$$

where $\tilde{\bar{y}}_j = \tilde{F}_j^{-}\tilde{y}_j$ and

$$l^*(\Sigma_p, y) = -\log|\Sigma_p| - y'(M\Sigma_p M)^{+}y. \tag{3.5}$$

Unfortunately, computation of $\tilde{E}_j$ and $\tilde{y}_j$ is often less convenient than computation of E_j and z_j defined in Section 2. The reason for this is increased complexity of the starting matrix $\tilde{A}_p = M\Sigma_p M$. It is therefore advisable to explore alternative methods of computing $y'(MVM)^{+}y$ based on the formula

$$(MVM)^{+} = V^{-1} - V^{-1}X(X'V^{-1}X)^{-}X'V^{-1}. \tag{3.6}$$

The most simple case is when $R(X) \subset R(U_1)$. Then, (2.16) implies

$$y'(MVM)^{+}y = y'V^{-1}y - y_o'F_o^{-}y_o, \tag{3.7}$$

where y_o is defined in (2.17). We obtain a simplified expression

$$l^*(\Sigma, y) = l(O, \Sigma, y) + y_o'F_o^{-}y_o, \tag{3.8}$$

which alternatively follows by maximizing (2.18) subject to β. There is again a way of expressing (3.7) as

$$\sum_{j=1}^{p} t_j'\Sigma_j t_j, \qquad t_j = U_j'(MVM)^+ y.$$

If (1.8) holds in addition to (1.7), it is shown in KS (83) that we can find t_j by simply using (2.13) and (2.14) if s_j is replaced by t_j, z_j is replaced by $y_j = B_j'y$ and the starting line (2.13) gets changed to

$$\bar{t}_1 = E_1^{-1}(\bar{y}_1 - \Delta_o\bar{y}_o), \quad t_1 = F_1\bar{t}_1. \tag{3.9}$$

The function $l^*(\Sigma,y)$ now takes the form

$$l^*(\Sigma,y) = -(\log|\Sigma_p| + t_p'\Sigma_p t_p) - \sum_{j=1}^{p-1} (\log|E_j| + t_j'\Sigma_j t_j), \tag{3.10}$$

where the second term is again less than zero.

4. Computation of $l_m(\Sigma,y)$

The new problem with computation of $l_m(\Sigma,y)$ is how to find $|B'VB|$, where B satisfies (1.5). There are several ways to solve this problem.

Lemma 4.1. Let B satisfy (1.5) and let V as well as $X'X$ be nonsingular matrices. Then it holds

$$|B'VB| = |V|\,|X'V^{-1}X| \,/\, |X'X|. \tag{4.1}$$

Proof. If $B'VB$ is of order $n-r$, then

$$|B'VB| = (-1)^{n-r}|I - B'(I+V)B| = (-1)^{n-r}|I - (I+V)BB'|, \tag{4.2}$$

where the matrix on the right hand side of (4.2) is of order $n \times n$ and nonsingular. Let us next use $BB' = I - X(X'X)^{-1}X'$ in order to obtain

$$|B'VB| = (-1)^{2n-r}|V - (I+V)X(X'X)^{-1}X'|$$

$$= (-1)^{2n-r}|V|\,|I-V^{-1}(I+V)X(X'X)^{-1}X'| \tag{4.3}$$

$$= (-1)^{2n-r}|V||-X'V^{-1}X(X'X)^{-1}|.$$

The matrix on the right hand side of (4.3) is of order $r \times r$ and (4.1) follows.

Application of Lemma 4.1 is limited to cases in which V and $X'X$ are both nonsingular matrices. This is not necessary for existence of the function $l_m(\Sigma, y)$. Let us therefore describe an alternative way of computing $|B'VB|$ under the assumption that just $B'VB$ is nonsingular.

Lemma 4.2. Under the assumption (1.7) it holds

$$|B'VB| = |B'\Sigma_p B||\tilde{E}_{p-1}| \cdots |\tilde{E}_1|, \tag{4.4}$$

where $\tilde{E}_j$, $j = 1, \ldots, p-1$, are as in Lemma 3.1.

Proof. Let us define $A_j^* = B'A_jB$, where the matrices A_j are those of expression (2.1). Then, it follows from (2.1)

$$|A_j^*| = |A_{j+1}^*||I + \Sigma_j U_j'B(B'A_{j+1}B)^{-1}B'U_j|.$$

Using (1.6) and the definition of A_j in Section 3 we obtain

$$|A_j^*| = |A_{j+1}^*||I + \Sigma_j U_j \tilde{A}_{j+1}^+ U_j| = |A_{j+1}^*||\tilde{E}_j|.$$

The result (4.4) follows by application of this equation for $j = 1, 2, \ldots, p-1$.

A disadvantage of (4.4) is that we need to compute the more complex matrices $\tilde{E}_j$. But the case $R(X) \subset R(U_1)$ admits again important simplifications. If $|X'X| \neq 0$, we can first use $X'V^{-1}X = F_o$ in Lemma 4.1 and second it follows from (3.1), (3.6) and (2.2) that

$$\tilde{E}_j = I + \Sigma_j U_j' \tilde{A}_{j+1}^+ U_j = E_j - \Sigma_j B_j' X(X'A_{j+1}^+ X)^{-1} X'B_j. \tag{4.5}$$

We remark that (3.6) holds for singular V if V^{-1} is replaced by V^+. The same applies to Lemma 2.1, where we may substitute A_{j+1}^{-1} by A_{j+1}^+ in (2.2) in order to obtain V^+ (see KS (83)). By assumption (1.8), there exists $S_j = \Delta_{j-1} \cdots \Delta_o$ such that $X = U_jS_j$ and (4.5) implies

$$\tilde{E}_j = E_j - \Sigma_j F_j S_j (S_j' F_j S_j)^{-1} S_j' F_j$$

and

$$|\tilde{E}_j| = |E_j| |I - (S_j' F_j S_j)^{-1} S_j' F_j E_j^{-1} \Sigma_j F_j S_j| \qquad (4.6)$$
$$= |E_j| |S_j' F_j E_j^{-1} S_j| / |S_j' F_j S_j|.$$

Summarizing the above results and those of Section 3 on computation of $y'(MVM)^+y$ yields the following result.

Theorem 4.1. Let $B'VB$ be p.d. and assume (1.7).

i) The marginal log-likelihood function is given by

$$l_m(\Sigma, y) = l_m(\Sigma_p, y) - \sum_{j=1}^{p-1} (\log|\tilde{E}_j| - \tilde{y}_j' \tilde{E}^{-1} \Sigma_j \tilde{y}_j), \qquad (4.7)$$

where

$$l_m(\Sigma_p, y) = -\log|B'\Sigma_p B| - y'(M\Sigma_p M)^+ y. \qquad (4.8)$$

ii) If $|X'X| \neq 0$ and (1.8) holds in addition, then

$$l_m(\Sigma, y) = -\log|B'\Sigma_p B| - y'\Sigma_p^+ y + y_o' F_o^- y_o$$
$$- \sum_{j=1}^{p-1} (\log|E_j| |S_j' F_j E_j^{-1} S_j| |S_j' F_j S_j|^{-1} - y_j' E_j^{-1} \Sigma_j y_j). \qquad (4.9)$$

iii) For p.d. V it holds

$$l_m(\Sigma, y) = l^*(\Sigma, y) + \log|X'X| |F_o|^{-1}. \qquad (4.10)$$

Proof. Result i) follows immediately from (4.4), (3.2) and (1.5). The result (4.9) is obtained by using (4.4), (4.6) and noting that (2.3) provides V^+ for singular V if Σ_p^{-1} is replaced by Σ_p^+ (see KS (83)). Result (4.10) follows from (3.8) and (4.1).

5. Nested classification models

The general nested classification random model

$$y_{i_1 \ldots i_p} = \beta + \xi^{(1)}_{i_1} + \xi^{(2)}_{i_1 i_2} + \ldots + \xi^{(p)}_{i_1 \ldots i_p} \tag{5.1}$$

is in vector form

$$y = X\beta + U_1\xi_1 + \ldots + U_p\xi_p, \tag{5.2}$$

where $\xi_1,\ldots,\xi_p$ are independently distributed random vectors with zero expectation and variance-covariance matrices $\Sigma_i = \sigma_i I_{m_i}$. The single terms in (5.2) are defined by

$$y = \mathrm{col}_{i_1} \ldots \mathrm{col}_{i_p}(y_{i_1 \ldots i_p}), \ \xi_j = \mathrm{col}_{i_1} \ldots \mathrm{col}_{i_j}(\xi^{(j)}_{i_1 \ldots i_j}),$$

$$X = \mathrm{col}_{i_1} \ldots \mathrm{col}_{i_p}(1), \ U_j = \mathrm{Diag}_{i_1} \ldots \mathrm{Diag}_{i_j} \mathrm{col}_{i_{j+1}} \ldots \mathrm{col}_{i_p}(1). \tag{5.3}$$

Here, the operator col_{i_j} denotes arranging terms in a vector form as subscript i_j runs through its domain. This domain depends on the subscripts $i_1,\ldots,i_{j-1}$ and may have a complicated structure. But without loss of generality the subscript i_p runs from 1 to $n_{i_1 \ldots i_{p-1}}$ and $n = \sum_{i_1 \ldots i_{p-1}} n_{i_1 \ldots i_{p-1}}$ is the total number of observations. The operator Diag_{i_j} stands for arranging terms in block diagonal form. For the sake of simplicity let us also use shorter notations like

$$y = \mathrm{col}_{i_1 \ldots i_p}(y_{i_1 \ldots i_p}) \text{ or } U_j = \mathrm{Diag}_{i_1 \ldots i_j} \mathrm{col}_{i_{j+1} \ldots i_p}(1).$$

We may now routinely follow the instructions of Sections 2 and 3 in order to derive the matrices F_j and vectors y_j, $\bar{y}_j$ and t_j. First it is easy to establish that

$$U_j = U_{j+1}\Delta_j, \ \Delta_j = \mathrm{Diag}_{i_1 \ldots i_j} \mathrm{col}_{i_{j+1}}(1).$$

Then, routine computation based on block matrix algebra yields

$$F_j = \mathrm{Diag}_{i_1 \ldots i_j}(f^{(j)}_{i_1 \ldots i_j}), \tag{5.4}$$

$$E_j = \text{Diag}_{i_1 \ldots i_j}(e^{(j)}_{i_1 \ldots i_j}), \quad e^{(j)}_{i_1 \ldots i_j} = 1 + \sigma_j f^{(j)}_{i_1 \ldots i_j},$$

where $f^{(p)}_{i_1 \ldots i_p} = \sigma_p^{-1}$, $f^{(p-1)}_{i_1 \ldots i_{p-1}} = \sigma_p^{-1} n_{i_1 \ldots i_{p-1}}$ and

$$f^{(j)}_{i_1 \ldots i_j} = \sum_{i_{j+1}} f^{(j+1)}_{i_1 \ldots i_{j+1}} (e^{(j+1)}_{i_1 \ldots i_{j+1}})^{-1}$$

for $j = p-2, \ldots, 0$. Also

$$\bar{y}_j = \text{col}_{i_1 \ldots i_j}(\bar{y}^{(j)}_{i_1 \ldots i_j}), \tag{5.5}$$

where $\bar{y}^{(p)}_{i_1 \ldots i_p} = y_{i_1 \ldots i_p}$,

$$\bar{y}^{(p-1)}_{i_1 \ldots i_{p-1}} = n^{-1}_{i_1 \ldots i_{p-1}} \sum_{i_p} y_{i_1 \ldots i_p} = \bar{y}_{i_1 \ldots i_{p-1}}$$

and

$$\bar{y}^{(j)}_{i_1 \ldots i_j} = f^{(j)^{-1}}_{i_1 \ldots i_j} \sum_{i_{j+1}} f^{(j+1)}_{i_1 \ldots i_{j+1}} (e^{(j+1)}_{i_1 \ldots i_{j+1}})^{-1} \bar{y}^{(j+1)}_{i_1 \ldots i_{j+1}}$$

for $j = p-2, \ldots, 0$.

It is not necessary to calculate the following expressions unless we want make use of result (3.10)

$$t_j = \text{col}_{i_1 \ldots i_j}(t^{(j)}_{i_1 \ldots i_j}), \tag{5.6}$$

where

$$t^{(1)}_{i_1} = f^{(1)}_{i_1}(e^{(1)}_{i_1})^{-1}(\bar{y}^{(1)}_{i_1} - \bar{y}^{(o)})$$

$$t^{(j)}_{i_1 \ldots i_j} = f^{(j)}_{i_1 \ldots i_j}(e^{(i)}_{i_1 \ldots i_j})^{-1}(\bar{y}^{(j)}_{i_1 \ldots i_j} - \bar{y}^{(j-1)}_{i_1 \ldots i_{j-1}} +$$

$$+ (f^{(j-1)}_{i_1 \ldots i_{j-1}})^{-1} t^{(j-1)}_{i_1 \ldots i_{j-1}})$$

for $j = 2, \ldots, p-1$. The last vector in this series is t_p and has components

$$t^{(p)}_{i_1\ldots i_p} = \sigma_p^{-1}(y_{i_1\ldots i_p} - \bar{y}_{i_1\ldots i_{p-1}}) + n^{-1}_{i_1\ldots i_{p-1}} t^{(p-1)}_{i_1\ldots i_{p-1}}. \quad (5.7)$$

Computation of the vectors s_j in (2.15) follows the same way but by replacing $y_{i_1\ldots i_p}$ by $(y_{i_1\ldots i_p} - \beta)$ and using $\bar{y}^{(0)} = 0$.

All these quantities are obtained by recursive computation without inverting matrices and the general results of Sections 2,3 and 4 imply the following result.

<u>Theorem 5.1</u>. For model (5.1) and positive σ_p it holds

i) $l(\beta,\Sigma,y) = l^*(\Sigma,y) - f^{(o)}(\bar{y}^{(o)} - \beta)^2$,

ii) $l^*(\Sigma,y) = l(0,\Sigma,y) + f^{(o)}(\bar{y}^{(o)})^2$,

iii) $l_m(\Sigma,y) = l^*(\Sigma,y) + \log n - \log f^{(o)}$,

where $\bar{y}^{(o)}$ and $f^{(o)}$ are defined by (5.4) and (5.5) and

$$-l(0,\Sigma,y) = n \log \sigma_p + \sigma_p^{-1} \sum_{i_1\ldots i_p} y^2_{i_1\ldots i_p}$$

$$+ \sum_{j=1}^{p-1} \sum_{i_1\ldots i_j} (\log(e^{(j)}_{i_1\ldots i_j}) + f^{(j)}_{i_1\ldots i_j}(e^{(j)}_{i_1\ldots i_j})^{-1}(\bar{y}^{(j)}_{i_1\ldots i_j})^2)$$

$$- \sum_{j=1}^{p-1} \sum_{i_1\ldots i_j} f^{(i)}_{i_1\ldots i_j}(\bar{y}^{(j)}_{i_1\ldots i_j})^2 .$$

Proof is simple and based on (2.18), (3.11) and (4.11).

It has been questioned by many authors whether the log-likelihood function is bounded as σ_p and other variance components approach zero. This question is equivalent to asking whether $l^*(\Sigma,y)$ is bounded. From (3.14) it follows that we need to investigate boundedness of

$$-(\log|\Sigma_p| + t_p'\Sigma_p t_p) = -(n \log \sigma_p + \sigma_p \|t_p\|^2). \quad (5.8)$$

The components of the random vector t_p are given by (5.7) and we find (5.8) to be less than $-(n \log \sigma_p + \frac{1}{\sigma_p} \text{SSR})$, where

$SSR = \sum_{i_1 \ldots i_p} (y_{i_1 \ldots i_p} - \bar{y}_{i_1 \ldots i_{p-1}})^2$. This function is bounded if $SSR > 0$. It even tends to $-\infty$ if $\sigma_p \to 0$ and $\sigma_{p-1}, \ldots, \sigma_1$ are fixed. The log likelihood function approaches finite limits as $\sigma_j \to 0$ for some $j = 1, \ldots, p-1$ and σ_p remains away from zero. Therefore, maximum likelihood (m.l.) estimators for σ_j, $j = 1, \ldots, p-1$, may be located at the boundary but not the m.l. estimator for σ_p. The same holds true for multivariate models of the form (5.1), where observations $y_{i_1 \ldots i_p}$ are vectors and the variance components are replaced by matrices of variance-covariance components. Then, a sufficient condition for boundedness of the log-likelihood function is nonsingularity of

$$SSR = \sum_{i_p, \ldots, i_p} (y_{i_1 \ldots i_p} - \bar{y}_{i_1 \ldots i_{p-1}})(y_{i_1 \ldots i_p} - \bar{y}_{i_1 \ldots i_{p-1}})'.$$

Further investigation of the log likelihood function based on Theorem 5.1 seems attractive and may lead to more effective ways of deriving m.l. estimators. This should first be tried for the one-way random model. In this case it holds

$$1^*(\Sigma, y) = -((n-m)\log \sigma_2 + \sigma_2^{-1} SSR) + \sum_{i=1}^{m} (\log(\sigma_2 + \sigma_1 n_i) + \frac{n_i}{\sigma_2 + \sigma_1 n_i} (\bar{y}_{i.} - \bar{y}_o)^2)), \tag{5.9}$$

where $SSR = \sum_{ij} (y_{ij} - \bar{y}_{i.})^2$, $\bar{y}_o = y_o / f_o$

and $f_o = \sum_{i=1}^{m} n_i (\sigma_2 + \sigma_1 n_i)^{-1}$, $y_o = \sum_{i=1}^{m} n_i (\sigma_2 + \sigma_1 n_i)^{-1} \bar{y}_{i.}$.

m is the number of cells and n_i observations are made at the i-th level of the random factor. For the balanced model holds $n_i = n$, $i = 1, \ldots, m$, and $\bar{y}_o$ becomes independent of σ_1 and σ_2. This makes it considerably easier to maximize (5.9) leading to the well-known result mentioned by Klotz, Milton and Zacks (1969).

References

[1] Hartley. H.O. and Rao, J.N.K. (1967). Maximum likelihood estimation for the mixed analysis of variance model, Biometrika 54, 93-108.

[2] Harville, D.A. (1977). Maximum likelihood approaches to variance component estimation and to related problems. Journ. Amer. Statist. Assoc. 72, 320-340.

[3] Hemmerle, W.I. and Hartley, H.O. (1973). Computing maximum likelihood estimates for the mixed AOV model using the W-transformation. Technometrics 15, 819-831.

[4] Kleffe, J.(1980). On recent progress of MINQUE theory - nonnegative estimation, consistency, asymptotic normality and explicite formulae. Math. Operationsforsch. Statist., Ser. Statistics 11, 4,563-588.

[5] Kleffe, J. and Seifert, B. (1983). Computation of variance components by the MINQUE method, Journ. Mult. Analysis (in print).

[6] Klotz, J.H., Milton, R.C. and Zacks, S. (1969). Mean square efficiency of estimators of variance components. Journ. Amer. Statist. Assoc. 64, 1383-1402.

[7] LaMotte, L.R. (1972). Notes on the covariance matrix of a random nested ANOVA model, Ann.Math. Statist. 43, 2,659-662.

[8] Patterson, H.D. and Thompson, R. (1975). Maximum likelihood estimation of components of variance. Proc. of 8th International Biometric Conference, pp. 197-207.

[9] Rao, C.R. (1970). Estimation of variance components - MINQE theory and its relation to ML and MML estimation of variance components, Sankhya Ser.B 41, 3-4 , 138-153.

SOME REMARKS ON IMPROVING UNBIASED ESTIMATORS BY MULTIPLICATION WITH A CONSTANT

J.Kleffe
Institute of Mathematics
Academy of Sciences of the GDR
Berlin, GDR

Summary. Using a general result due to Perlman (1972), we show that no invariant unbiased estimator for variance components (not necessarily being quadratic) qualifies to admissibility under the mean squared error risk function. Uniformly better estimates are easily given by multiplication of an unbiased estimator with some constant less than but close to one. The optimal choice of such a multiplier is investigated and more specific results are derived for quadratic estimation.

1. Introduction

Perlman (1972) published a general result on improving unbiased estimators under the mean squared error risk function by just multiplying it with a suitable constant. A simple generalization of his result to vector valued estimating functions with a possibly singular dispersion matrix is as follows.

Let $\hat{\gamma}$ be a vector of estimating functions with expectation $\gamma = \gamma(\theta)$ and variance-covariance matrix W_θ, where $\theta \in \mathcal{T}$ is the unknown parameter.

The mean squared error risk function for $\hat{\gamma}$ as an estimator of γ is

$$MSE_\theta(\hat{\gamma}) = E_\theta(\hat{\gamma} - \gamma)(\hat{\gamma} - \gamma)' \tag{1.1}$$

and it holds.

Lemma 1. Assume $\gamma \in R(W_\theta)$ for all $\theta \in \mathcal{T}$, where $R(\cdot)$ denotes column space.

(i) There exists a multiplier $\alpha \neq 1$ such that

$$MSE_\theta(\alpha\hat{\gamma}) \leq W_\theta \quad \text{for all} \quad \theta \in \mathcal{T} \tag{1.2}$$

iff there is a constant c such that

$$\gamma\gamma' \leq cW_\theta \quad \text{for all} \quad \theta \in \mathcal{T}. \tag{1.3}$$

(ii) The inequality (1.2) holds for all α with

$$\max(0, \frac{c-1}{c+1}) \leq \alpha \leq 1 . \tag{1.4}$$

Proof is simple along lines of Perlman. The minimal choice of c is

$$c = \sup_{\theta \in \mathcal{T}} \gamma' W_\theta^- \gamma, \tag{1.5}$$

where W_θ^- is any g-inverse of W_θ.

Perlman also investigated the optimal choice of the constant α, when it is possible, by proving the following result.

Lemma 2. Let γ be an s-vector with $\gamma \in R(W_\theta)$.

(i) For $s = 1$, there exists α_* with

$$MSE_\theta(\alpha_*\hat{\gamma}) \leq MSE_\theta(\alpha\hat{\gamma}) \tag{1.6}$$

for all α and $\theta \in \mathcal{T}$ iff $\gamma^2 W_\theta^{-1}$ is independent of $\theta \in \mathcal{T}$ and α_* is given by

$$\alpha_* = 1/(1 + W_\theta/\gamma(\theta)^2). \tag{1.7}$$

(ii) There exists no α_* with the property (1.6) if $s > 1$ and rank $(W_\theta) > 1$ for at least one $\theta \in \mathcal{T}$.

The quantity (1.5) is not only bounded but constant in $\theta \in \mathcal{T}$ if there exists a nonsingular linear transformation T_θ such that the distribution of $T_\theta\hat{\gamma}$ is independent of $\theta \in \mathcal{T}$. This follows from $\gamma = W_\theta T_\theta' \delta$ for some δ and

$$c = (ET_\theta\hat{\gamma})'(\text{Cov } T_\theta\hat{\gamma})^-(ET_\theta\hat{\gamma}) = \gamma' T_\theta'(T_\theta W_\theta T_\theta')^- T_\theta \gamma = \tag{1.8}$$

$$\delta' T_\theta W_\theta T_\theta' (T_\theta W_\theta T_\theta')^- T_\theta W_\theta T_\theta' \delta = \delta' T_\theta W_\theta T_\theta' \delta = \gamma' W_\theta^- \gamma .$$

An interesting example to which this fact applies is estimation of the covariance matrix in the multivariate linear regression model

$$Y = XB + \varepsilon ,$$

where the rows of the $N \times p$ random matrix ε are i.i.d. normally distributed with expectation zero and nonsingular variance-covariance matrix Σ. The usual estimator $\hat{\Sigma} = k^{-1} Y'MY$ for Σ has the Wishart distribution $W_p(k,\Sigma)$ with $k = \mathrm{tr} M$ degrees of freedom, where $M = I - XX^+$. However,

$$\Sigma^{-\frac{1}{2}} \hat{\Sigma} \Sigma^{-\frac{1}{2}} \sim W_p(k, I)$$

and has a distribution which is independent of the unknown parameter Σ. Routine computation implies that $c = kp/2$ and that $\alpha\hat{\Sigma}$ has uniformly smaller mean squared error risk function if

$$\frac{kp-2}{kp+2} \leq \alpha \leq 1 .$$

If $p = 1$, then $\alpha_* = k/(k+2)$ is the best choice of α by Lemma 2. Note that the set of p.d. matrices is dense in the class of all n.n.d. matrices so that the improvement extends to singular covariance matrices too. Other improvements of $\hat{\Sigma}$ are given by Olkin and Selliah (1975).

Unfortunately, since some properties of the Wishart distribution play a fundamental role in obtaining the above results, it is not easy to construct such multipiers α for estimation of variance components in the general mixed linear model. But under fairly general assumptions we are able to describe a simple upper bound for c.

Lemma 3. Let $\hat{\gamma}$ be an unbiased estimator for $\gamma = F\theta$ with covariance matrix W_θ which satisfies

$$R(W_\theta) = R(FJ_\theta^{-1}F') \quad \text{and} \quad W_\theta \geq FJ_\theta^{-1}F', \tag{1.9}$$

where J_θ is Fisher's information matrix, assumed to be nonsingular (Cramèr-Rao inequality).

(i) For every $\theta \in \mathcal{T}$ holds

$$\gamma' W_\theta^- \gamma \leq \theta'(J_\theta - J_\theta F^*(F^{*\prime} J_\theta F^*)^- F^{*\prime} J_\theta)\theta, \tag{1.10}$$

where F^* is any matrix with $R(F^*) = N(F')$.

(ii) An upper bound for c which does not depend on F is given by

$$c = \sup_{\theta \in \mathcal{T}} \theta' J_\theta \theta .$$

<u>Proof</u>. From (1.8) it follows $W_\theta^+ \leq (FJ_\theta^{-1}F')^+$ and

$$\sup_{\theta \in \mathcal{T}} \gamma' W_\theta^+ \gamma \leq \sup_{\theta \in \mathcal{T}} \theta' F'(FJ_\theta^{-1}F')^+ F\theta .$$

But a simple extension of result 33 at page 77 in Rao (1973) yields the identy

$$F'(FJ_\theta^{-1}F')^+ F = J_\theta - J_\theta F^*(F^{*\prime} J_\theta F^*)^- F^{*\prime} J_\theta \tag{1.11}$$

and proves Lemma 3.

The most important conclusion we can draw from Lemma 3 is that boundedness of $\theta' J_\theta \theta$ on $\theta \in \mathcal{T}$ implies improvability of each unbiased estimator of a linear parametric function. Further, the bounds for c given in parts (i) and (ii) of Lemma 3 coincide for each nonsingular F since $F^* = O$ in this case.

2. <u>Estimation of variance components</u>

The general mixed linear model can be presented as

$$y = X\beta + \varepsilon, \quad E\varepsilon = O, \quad E\varepsilon\varepsilon' = \sum_{i=1}^{p} \theta_i V_i = V_\theta, \tag{2.1}$$

where $X, V_1, \ldots, V_p$ are given matrices, while β and $\theta = (\theta_1, \ldots, \theta_p)'$ are the unknown parameters. The matrices V_i are n.n.d. and symmetric. It is a common practice to base estimation of linear functions of θ on the statistic $t = My$, $M = I - XX^+$ which follows the linear model

$$Et = O, \quad Ett' = \sum_{i=1}^{p} \theta_i MV_i M . \tag{2.2}$$

If y is normally distributed, then it is known from Patterson and

Thompson (1975) that Fisher's information matrix for model (2.2) is the $p \times p$ symmetric matrix

$$J_\theta = \frac{1}{2}(\mathrm{tr}\; V_i(MV_\theta M)^+ V_j(MV_\theta)^+). \tag{2.3}$$

It is nonsingular iff the matrices $MV_1M,\ldots,MV_pM$ are linearly independent and if $\theta \in \mathcal{T}$, where

$$\mathcal{T} = \{\theta \in R^p : R(MV_\theta M) \supset R(MV_iM),\quad i = 1,\ldots,p\}. \tag{2.4}$$

Using arguments as given in Lemma 2 by LaMotte (1977) the set $\mathcal{T}$ is easily seen to be open and dense in the set of all parameter vectors θ for which $MV_\theta M$ is nonnegative definite (n.n.d.) and therefore, uniform improvement of an estimator on $\theta \in \mathcal{T}$ is equivalent to improvement on the entire parameter space.

<u>Lemma 4</u>. Let y be a normally distributed random vector and $\hat{\gamma} \in R(F)$ be an unbiased estimator for $\gamma = F\theta$ based on the statistic $t = My$.*) Then $\alpha\hat{\gamma}$ satisfies (1.2) for all α such that

$$\frac{\mathrm{rank}\ (M) - 2}{\mathrm{rank}\ (M) + 2} \leq \alpha < 1. \tag{2.5}$$

<u>Proof</u>. For $\theta \in \mathcal{T}$ and by the Cramer-Rao inequality if follows $W_\theta \geq FJ_\theta^{-1}F'$ and $R(W_\theta) \supset R(FJ_\theta^{-1}F')$. The relation $\hat{\gamma} \in R(F)$ implies $R(W_\theta) = R(FJ_\theta^{-1}F')$ and therefore $\gamma \in R(W_\theta)$ so that the assumptions of Lemmae 1 and 3 are satisfied. Now

$$\theta' J_\theta \theta = \frac{1}{2}\mathrm{tr}\,(MV_\theta M)^+ V_\theta = \mathrm{tr}\ M/2$$

proves the desired result.

A simple and direct proof of (2.5) for $\hat{\gamma} = ((t'A_it))$, $i = 1,\ldots,s$, being a vector of quadratic forms is only based on Cauchy-Schwarz inequality and was given by LaMotte (1979) for $s = 1$.

For any vector x put and $A_x = \sum_{i=1}^{s} x_iA_i$. Then,

$$x'(E_\theta\hat{\gamma})(E_\theta\hat{\gamma})'x = (\mathrm{tr}\ A_xMV_\theta M)^2 = (\mathrm{tr}\,(MV_\theta M)^{\frac{1}{2}}A_x(MV_\theta M)^{\frac{1}{2}})^2 \tag{2.6}$$

*) Estimates for $\gamma = F\theta$ with $\hat{\gamma} \notin R(F)$ or $\Pr_\theta(\hat{\gamma} \in R(F)) < 1$ for some $\theta \in \mathcal{T}$ are inadmissible.

$$\leq (\mathrm{tr}\ M)\ \mathrm{tr}\ A_x MV_\theta M A_x MV_\theta M = (\mathrm{tr}\ M) x' W_\theta x/2$$

implies

$$\gamma\gamma' \leq (\mathrm{tr}\ M) W_\theta/2 \quad \text{and} \quad \gamma' W_\theta^- \gamma \leq \mathrm{tr}\ M/2 .$$

The result of Lemma 4 cannot be improved by means of Lemma 3 if F is a nonsingular matrix. But there might be some improvement of (2.5) if rank $F < p$.

Under the assumptions of Lemma 4 holds

$$\sup_{\theta \in \mathcal{T}} \gamma' W_\theta^- \gamma \leq \frac{\mathrm{rank}\ (M) - 2}{\mathrm{rank}\ (M) + 2} - \inf_{\theta \in \mathcal{T}} \theta' J_\theta F^* (F^{*\prime} J_\theta F^*)^- F^{*\prime} J_\theta \theta . \tag{2.7}$$

The infimum on the right hand side need not be zero but often is. To see this fact let us note that the bound (2.7) does not depend on the choice of a particular estimator for $\gamma = F\theta$, so that if

$$\sup_{\theta \in \mathcal{T}} \gamma' W_\theta^- \gamma = \frac{\mathrm{rank}\ (M) - 2}{\mathrm{rank}\ (M) + 2} \tag{2.8}$$

for some choice of an estimate, then necessarily

$$\inf_{\theta \in \mathcal{T}} \theta' J_\theta F^* (F^{*\prime} J_\theta F^*)^- F^{*\prime} J_\theta \theta = 0. \tag{2.9}$$

This argument applies to the example discussed in Section 1. More generally, if equality holds in (2.6) for some $\theta \in \mathcal{T}$, then (2.8) follows and $(MV_\theta M)^{\frac{1}{2}} A_x (MV_\theta M)^{\frac{1}{2}} = \delta_x M$ for all x and some constant δ_x.

By computation of the expectation of $\hat{\gamma}$ follows that

$$\Gamma = \{\gamma = F\theta : F = (\delta_j\ \mathrm{tr}(MV_\theta M)^+ V_i),\ \theta \in \mathcal{T},\ \delta_j \text{ arbitrary}\}$$

is a set of linear parametric functions to which (2.9) applies. It is easily seen to be the class of n.n.d. estimable linear function of θ which was characterized by Pukelsheim (1981).

On the other hand, a constant α smaller than the lower bound in (2.5) if possible, does not necessarily provide a better choice for improving the unbiased estimator. From Lemma 2 it follows that the best choice of α, if it exists, is not the smallest one.

Let us therefore prove some general results on the existence of a best choice of α.

Lemma 5. Let $\hat{\gamma} = t'At$ be a quadratic unbiased estimator for $\gamma \neq 0$ such that

$$\hat{\gamma} \sim \frac{\gamma}{k}\chi_k^2 \quad \text{for all} \quad \theta \in \mathcal{T} . \tag{2.10}$$

Then $\frac{k}{k+2}$ satisfies condition (1.5) so that $k\hat{\gamma}/(k+2)$ is the best improvement of $\hat{\gamma}$ in the sense of Lemma 2. It is known as Hodges-Lehmann modification in special cases.

Proof. The χ^2 distribution (2.10) implies

$$\frac{k}{\gamma} A(MV_\theta M)A = A \quad \text{for all} \quad \theta \in \mathcal{T}$$

and hence

$$\text{var } \hat{\gamma} = 2 \text{ tr}(MV_\theta M)A(MV_\theta M)A = \frac{2\gamma}{k} \text{ tr}(MV_\theta M)A = 2\,\frac{\gamma^2}{k}.$$

Therefore $\gamma^2(\text{var } \hat{\gamma})^{-1} = k/2$ is constant and by Lemma 2 it follows that $\alpha_* = k/(k+2)$ is the best choice of α.

Estimating functions with property (2.10) exist in a number of balanced ANOVA models for which the matrices $MV_1M,\ldots,MV_pM$ generate a p-dimensional commutative quadratic subspace.
Following Seely (1972) such a space has a basis $A_1,\ldots,A_p$ of idempotent and orthogonal ($A_iA_j = 0$ for $i \neq j$) matrices and it is easy to prove that

$$t'A_it \sim \frac{\text{tr } A_iV_\theta}{\text{tr } A_i}\,\chi^2_{\text{tr } A_i} \quad \text{for all} \quad \theta \in \mathcal{T} \quad .$$

These matrices are known for a number of balanced classification models and given in Pukelsheim (1978), Kleffe (1980) and Rao, Kleffe (1983).
We are able to prove necessity of the χ^2 distribution if the matrices $MV_1M,\ldots,MV_pM$ generate a p-dimensional quadratic subspace $\mathcal{H}_M$.

Lemma 6. Let $\mathcal{H}_M$ be a quadratic space and let $\hat{\gamma} = t'At$, $A \in \mathcal{H}_M$, be an unbiased estimator for γ, where $t = My$. There exists a number α_* with (1.6) iff

$$\hat{\gamma} \sim \frac{\gamma}{r(A)} \chi^2_{r(A)} \quad \text{for all} \quad \theta \in \mathscr{T} . \tag{2.11}$$

Proof. Following Lemma 2 we have to investigate the existence of a constant $c \geq 0$ such that

$$\mathrm{var}_\theta \hat{\gamma} = c\gamma^2 \quad \text{for all} \quad \theta \in \mathscr{T} .$$

Since $\mathscr{T}$ is an open set, this condition is equivalent to

$$\mathrm{tr}\, A(MV_iM)A(MV_jM) = c\ \mathrm{tr}\, A(MV_iM) * \mathrm{tr}\, A(MV_jM), \quad i,j = 1,\dots,p, \tag{2.12}$$

or

$$\mathrm{tr}(MV_jM)[A(MV_iM)A - c(\mathrm{tr}\, A(MV_iM))A] = 0, \quad i,j = 1,\dots,p. \tag{2.13}$$

But A_M and $\mathscr{H}_M$ being a quadratic subspace implies

$$A(MV_iM)A = c(\mathrm{tr}\, A(MV_iM))A, \quad i = 1,\dots,p, \tag{2.14}$$

and

$$A(MV_\theta M)A = c\gamma A \quad \text{for all} \quad \theta \in \mathscr{T} .$$

This proves the χ^2 distribution for $\hat{\gamma}/2\gamma$, since

$$(MV_\theta M)\left(\frac{1}{\gamma c} A\right)(MV_\theta M)\left(\frac{1}{\gamma c} A\right)(MV_\theta M) = (MV_\theta M)\left(\frac{1}{c\gamma} A\right)(MV_\theta M)$$

which by Ogasawara and Takahashi (1951) is a necessary and sufficient condition.

We did not require the quadratic subspace to be commutative in order to prove Lemma 6 and it would be interesting to search for corresponding examples. There are plenty of examples for commutative quadratic subspaces of which we discuss two.

a) Let the model be

$$y = XB + \varepsilon, \ E\varepsilon\varepsilon' = \theta_1 I + \theta_2 vv',$$

where v is given vector. Then, $\mathscr{H}_M$ is a commutative quadratic subspace and it is spanned by the following idempotent and orthogonal matrices

$$\frac{1}{v'Mv} Mvv'M, \qquad M - \frac{1}{v'Mv} Mvv'M.$$

The two χ^2-variables are

$$\frac{1}{r(M)-1}\, y'(M-\frac{1}{v'Mv}\, Mvv'M)y \sim \frac{\theta_1}{r(M)-1}\,\chi^2_{r(M)-1} \tag{2.15}$$

and

$$\frac{1}{v'Mv}\,(y'Mv)^2 \sim (\theta_1 + v'Mv\,\theta_2)\chi^2_1 \tag{2.16}$$

and their improvements according to Lemma 5 are

$$\frac{1}{r(M)+1}\, y'(M-\frac{1}{v'Mv}\, Mvv'M)y \tag{2.17}$$

as an estimator for θ_1 and

$$\frac{1}{v'Mv+2}\,(y'Mv)^2 \tag{2.18}$$

as an estimator for $\theta_1 + v'Mv\theta_2$.

b) The balanced nested classification model is given by matrices

$$X = (1_{n_1} \otimes \ldots \otimes 1_{n_p}),$$

$$V_i = (I_{n_1} \otimes \ldots \otimes I_{n_i} \otimes J_{n_{i+1}} \otimes \ldots \otimes J_{n_p}), \quad i = 1,\ldots,p-1,$$

$$V_p = (I_{n_1} \otimes \ldots \otimes I_{n_p}),$$

where J_{n_i} is the $n_i \times n_i$ matrix with all entries equal to 1. The symbol $\otimes$ denotes Kronecker product of matrices. It is shown in Kleffe, Rao (1983) that the matrices

$$H_i = (\prod_{j=i+1}^{p} n_j)^{-1}(V_i - n_i^{-1}V_{i-1}), \quad i = 1,\ldots,p-1,$$

and

$$H_p = V_p - n_p^{-1}V_{p-1},$$

where $V_o = XX'$, form an orthogonal basis for the commutative quadratic subspace $\mathscr{H}_M$. Hence, we have the χ^2 variables

$$\frac{y_i H_1 y}{n_1 - 1} \sim \frac{\gamma_1}{n_1-1}\,\chi^2_{n_1-1} \tag{2.19}$$

and

$$\frac{y'H_i y}{n_1 \ldots n_{i-1}(n_i-1)} \sim \frac{\gamma_i}{n_1 \ldots n_{i-1}(n_i-1)}\,\chi^2_{n_1 \ldots n_{i-1}(n_i-1)}, \tag{2.20}$$

where $i = 2,\ldots,p$, for estimation of

$$\gamma_i = n_{i+1} \cdots n_p\theta_1 + \ldots + n_p\theta_{p-1} + \theta_p, \quad i = 1,\ldots,p-1,$$

and

$$\gamma_p = \theta_p.$$

According to Lemma 5 their improvements are

$$\frac{y'H_1y}{n_1+1}, \quad \frac{y'H_iy}{n_1\cdots n_{i-1}(n_i-1)+2}, \quad i = 2,\ldots,p. \tag{2.21}$$

Both examples a) and b) suggest to consider linear combinations of the improved estimators (2.17) and (2.18) or (2.21) as new estimators for the corresponding linear combinations of their expectations. Do these linear combinations provide improvements over the unbiased estimator ?

<u>Lemma 7</u>. Let $\hat{\gamma} = (\hat{\gamma}_1,\ldots,\hat{\gamma}_p)'$ be a vector of independent χ^2 variables such that

$$\hat{\gamma}_i \sim \frac{\gamma_i}{k_i}\chi^2_{k_i}, \quad i = 1,\ldots,p,$$

and let $\hat{\hat{\gamma}} = (\hat{\hat{\gamma}}_1,\ldots,\hat{\hat{\gamma}}_p)'$ be the vector of their improvements $\hat{\hat{\gamma}}_i = k_i\hat{\gamma}_i/(k_i+2)$ as estimator for γ_i. Then it holds

$$\mathrm{MSE}_\theta(\hat{\hat{\gamma}}) \leq \mathrm{MSE}_\theta(\hat{\gamma}) \quad \text{for all} \quad \theta \in \mathscr{T} \tag{2.22}$$

iff

$$\sum_{i=1}^{p} \frac{k_i}{k_i+1} \leq 2. \tag{2.23}$$

<u>Proof</u>. Using the second moments of the χ^2 distribution implies

$$\mathrm{MSE}_\theta(\hat{\gamma}) - \mathrm{MSE}_\theta(\hat{\hat{\gamma}}) = 4\ \mathrm{Diag}\,(\gamma_i)K\ \mathrm{Diag}\,(\gamma_i),$$

where

$$K = 2\ \mathrm{Diag}\left(\frac{1+k_i}{k_i(k_i+2)^2}\right) - \mathrm{col}\left(\frac{1}{k_i+2}\right)\mathrm{col}'\left(\frac{1}{k_i+2}\right).$$

Here, Diag (•) denotes the diagonal matrix formed by the terms in parentheses and col (•) denotes the vector. The condition (2.22) is now equivalent to $K \geq 0$. It is a routine exercise to find

$$|K| = 2^{p-2} \prod_{i=1}^{p} \frac{1+k_i}{k_i(k_i+2)^2}\left(2 - \sum_{i=1}^{p} \frac{k_i}{k_i+1}\right),$$

and $|K| \geq 0$ holds iff (2.23). The special structure of K also implies nonnegativity of the determinants of all main sections of K and therefore $K \geq 0$ iff $|K| \geq 0$ (see Rao (1973), p.36 (iv)).

The condition (2.23) holds for $p = 2$ and arbitrary $k_i \geq 1$, so that in fact linear combinations of two improved χ^2 variables provide an improvement over the unbiased estimator. Condition (2.23) is never satisfied for $p \geq 5$. There are some values for k_i which satisfy (2.23) in case $p = 3$ or $p = 4$.

Acknowledgement. The author thanks Professor L.R. LaMotte and a referee for several valuable suggestions regarding presentation of this paper.

References

[1] Kleffe, J.(1980). On recent progress of MINQUE theory-non-negative estimation, consistency, asymptotic normality and explicite formulae. Math.Operationsforsch. Statist. Ser.Statistics, 11, 563-588.

[2] Kleffe, J. and Rao, J.N.K. (1983). On the existence of asymptotically unbiased nonnegative quadratic estimates of variance components in ANOVA models, Technical Report, Carleton University, Ottawa Dept. of Mathematics and Statistics.

[3] LaMotte, L.R.(1977). A canonical form for the general linear model. Ann. Statist. 5, 787-789.

[4] LaMotte, L.R. (1979). On admissible invariant quadratic estimators of variance components, in: Variance Components and Animal Breeding. Proc. of a Conference in Honor of C.R. Henderson, L.Dale Vleck and S.R.Searle ed.

[5] Ogasawara, T. and Takahashi, M.(1951). Independence of quadratic forms in normal systems. J.Sci.Hiroshima University, 15, 1-9.

[6] Olkin, I. and Selliah, J.B. (1975). Estimating covariances in a multivariate normal distribution, Technical Report No. 100, Dept. of Statistics, Stanford University.

[7] Patterson, H.D. and Thompson, R.(1975). Maximum likelihood estimation of components of variance. In: Proc. of 8th International Biometric Conference, 197-207.

[8] Perlman, M.D. (1972). Reduced mean squared error estimation for several parameters. Sankhya Ser. B, 89-92.

[9] Pukelsheim, F.(1981). On the existence of unbiased nonnegative estimates of variance components. Ann. Statist. 9, 293-299.

[10] Pukelsheim, F. (1978). Examples for unbiased nonnegative estimation in variance component models. Technical Report No.113, Dept. of Statistics, Stanford University.

[11] Rao, C.R.(1973). Linear Statistical Inference and its Applications. Second Edition. John Wiley, New York.

[12] Seely, J.(1971). Quadratic subspaces and completeness. Ann.Math. Statist. 42, 710-721.

ON IMPROVING ESTIMATION IN A RESTRICTED GAUSS-MARKOV MODEL

K. Kłaczyński and P. Pordzik
Department of Mathematical and
Statistical Methods
Academy of Agriculture
Poznań, Poland

Abstract. In a Gauss-Markov model with linear restrictions, two estimators of the vector parameters are compared with respect to the matrix risk function and with respect to a weighted quadratic risk function. The first from the competitory estimators is based on the sample and linear restrictions contained in the model, while the second estimator additionally utilizes some linear restrictions which are not necessary fulfilled by the vector parameters. The criteria of the superiority of one of the estimators over the other are expressed by the conditions on parameters of a well defined probability density function.

1. Introduction

Consider a restricted Gauss-Markov model

$$\mathcal{M}_1 = \{\underline{y},\ \underline{X\beta} \mid \underline{R}_1'\underline{\beta} = \underline{r}_1,\ \ \sigma^2\underline{V}\}, \tag{1}$$

in which $\underline{y}$ is an $n \times 1$ observable random vector with expectation $E(\underline{y}) = \underline{X\beta}$ and with dispersion matrix $D(\underline{y}) = \sigma^2\underline{V}$, where $\underline{X}$ is an $n \times p$ known matrix of rank $r(\underline{X}) = p$, $\underline{V}$ is a known positive definite matrix of order n, σ^2 is an unknown positive scalar, and $\underline{\beta}$ is a $p \times 1$ vector of unknown parameters. These parameters satisfy linear restrictions

$$\underline{R}_1'\underline{\beta} = \underline{r}_1, \tag{2}$$

Key words: Gauss-Markov model, linear restrictions, comparisons of estimators, matrix risk function, weighted quadratic risk function.

expressed by known $p \times m_1$ matrix $\underline{R}_1$ and $m_1 \times 1$ vector $\underline{r}_1$ such that the equations (2) are consistent, i.e. $\underline{r}_1 \in \mathscr{C}(\underline{R}_1')$, where $\mathscr{C}(\cdot)$ denotes the column space of a matrix argument. It is well known that the minimum dispersion linear unbiased estimator (MDLUE) of $\underline{\beta}$ in model (1) is

$$\underline{b}_1 = \underline{b}_o - \underline{S}^{-1}\underline{R}_1(\underline{R}_1'\underline{S}^{-1}\underline{R}_1)^-(\underline{R}_1'\underline{b}_o - \underline{r}_1), \tag{3}$$

where $\underline{S} = \underline{X}'\underline{V}^{-1}\underline{X}$, $\underline{b}_o = \underline{S}^{-1}\underline{X}'\underline{V}^{-1}\underline{y}$ is the MDLUE of $\underline{\beta}$ in the corresponding unrestricted Gauss-Markov model

$$\mathscr{M}_o = \{\underline{y}, \underline{X\beta}, \sigma^2\underline{V}\}, \tag{4}$$

and the symbol $\underline{A}^-$ denotes a generalized inverse of $\underline{A}$, that is, any matrix satisfying the equation $\underline{AA}^-\underline{A} = \underline{A}$. Moreover, the dispersion matrix of $\underline{b}_1$ is

$$D(\underline{b}_1) = \sigma^2\underline{U} \tag{5}$$

where

$$\underline{U} = \underline{S}^{-1} - \underline{S}^{-1}\underline{R}_1(\underline{R}_1'\underline{S}^{-1}\underline{R}_1)^-\underline{R}_1'\underline{S}^{-1}. \tag{6}$$

The purpose of this note is to develop criteria for comparing, under model (1), the estimator $\underline{b}_1$ as defined in (3) with the estimator $\underline{b}_{12}$ which utilizes some additional linear restrictions

$$\underline{R}_2'\underline{\beta} = \underline{r}_2 , \tag{7}$$

expressed by known $p \times m_2$ matrix $\underline{R}_2$ and $m_2 \times 1$ vector $\underline{r}_2$ such that $\underline{r}_2 \in \mathscr{C}(\underline{R}_2')$ and

$$\mathscr{C}(\underline{R}_1) \cap \mathscr{C}(\underline{R}_2) = \{\underline{O}\}. \tag{8}$$

It should be emphasized that contrary to equations (2), $\underline{\beta}$ need not fulfill equations (7) and therefore $\underline{b}_{12}$ does not in general dominate $\underline{b}_1$ uniformly over the parameter space. The estimators $\underline{b}_1$ and $\underline{b}_{12}$ will be compared with respect to the matrix risk function and with respect to a weighted quadratic risk function, which, for an estimator $\widehat{\underline{b}}$ of $\underline{\beta}$, are defined as

$$M(\widehat{\underline{b}}\ ;\ \underline{\beta}) = E[(\widehat{\underline{b}} - \underline{\beta})(\widehat{\underline{b}} - \underline{\beta})']$$

and

$$\rho_W(\tilde{\underline{b}}\,;\,\underline{\beta}) = E[(\tilde{\underline{b}}-\underline{\beta})'\underline{W}(\tilde{\underline{b}}-\underline{\beta})],$$

respectively, $\underline{W}$ being a known positive definite matrix of order p. The criteria established in this paper generalize rezults of Toro -Vizcarrondo and Wallace (1968), and Yancey, Judge and Bock (1973), concerning the matrix risk and a quadratic risk comparisons between estimators $\underline{b}_o$ and $\underline{b}_1$ under model (4), and also supplement and extend results of Guilkey and Price (1981), concerning the matrix risk comparison between $\underline{b}_1$ and $\underline{b}_{12}$ under model (1).

2. Results

The equations (2) and (7) can jointly be written in the form

$$\underline{R}_{12}'\underline{\beta} = \underline{r}_{12}, \tag{9}$$

where $\underline{R}_{12} = (\underline{R}_1 : \underline{R}_2)$ and $\underline{r}_{12}' = (\underline{r}_1' : \underline{r}_2')$. Suppose, for a while, that $\underline{\beta}$ satisfies equations (9). Then, on account of (3) and (5), the MDLUE of $\underline{\beta}$ in the model $\mathscr{M}_{12} = \{\underline{y},\ \underline{X\beta} \mid \underline{R}_{12}'\underline{\beta} = \underline{r}_{12},\ \sigma^2\underline{V}\}$ and its dispersion matrix are of the forms

$$\underline{b}_{12} = \underline{b}_o - \underline{S}^{-1}\underline{R}_{12}(\underline{R}_{12}'\underline{S}^{-1}\underline{R}_{12})^-(\underline{R}_{12}'\underline{b}_o - \underline{r}_{12}), \tag{10}$$

and

$$D(\underline{b}_{12}) = \sigma^2[\underline{S}^{-1} - \underline{S}^{-1}\underline{R}_{12}(\underline{R}_{12}'\underline{S}^{-1}\underline{R}_{12})^-\underline{R}_{12}'\underline{S}^{-1}] \tag{11}$$

Alternative expressions for $\underline{b}_{12}$ and $D(\underline{b}_{12})$ are given in the following.

<u>Lemma</u> 1. The MDLUE of $\underline{\beta}$ in the model $\mathscr{M}_{12}$ and its dispersion matrix are expressible in the forms

$$\underline{b}_{12} = \underline{b}_1 - \underline{U}\underline{R}_2(\underline{R}_2'\underline{U}\underline{R}_2)^-(\underline{R}_2'\underline{b}_1 - \underline{r}_2) \tag{12}$$

and

$$D(\underline{b}_{12}) = \sigma^2[\underline{U} - \underline{U}\underline{R}_2(\underline{R}_2'\underline{U}\underline{R}_2)^-\underline{R}_2'\underline{U}], \tag{13}$$

where $\underline{b}_1$ and $\underline{U}$ are as defined in (3) and (6), respectively.

<u>Proof</u>. It is known (cf. Searle (1971), p.27) that one of the generalized inverse of a nonnegative definite (n.n.d.) matrix

$$\underline{N} = \begin{pmatrix} \underline{A} & \underline{B} \\ \underline{B}' & \underline{D} \end{pmatrix}$$

is

$$\underline{N}^- = \begin{pmatrix} \underline{A}^- + \underline{A}^-\underline{B}(\underline{D} - \underline{B}'\underline{A}^-\underline{B})^-\underline{B}'\underline{A}^- & -\underline{A}^-\underline{B}(\underline{D} - \underline{B}'\underline{A}^-\underline{B})^- \\ -(\underline{D} - \underline{B}'\underline{A}^-\underline{B})^-\underline{B}'\underline{A}^- & (\underline{D}-\underline{B}'\underline{A}^-\underline{B})^- \end{pmatrix}. \tag{14}$$

Expressing $(\underline{R}_{12}'\underline{S}^{-1}\underline{R}_{12})^-$ as in (14), with $\underline{A} = \underline{R}_1'\underline{S}^{-1}\underline{R}_1$, $\underline{B} = \underline{R}_1'\underline{S}^{-1}\underline{R}_2$ and $\underline{D} = \underline{R}_2'\underline{S}^{-1}\underline{R}_2$, formulae (12) and (13) follow from (10) and (11) in a straightforward manner.

In the case where restrictions (7) are not fulfilled, the estimator $\underline{b}_{12}$ is no longer the MDLUE of $\underline{\beta}$, but it can be noticed that, because of the nonnegative definiteness of the matrix $D(\underline{b}_1) - D(\underline{b}_{12})$, it is still possible that $\underline{b}_{12}$ dominates $\underline{b}_1$ with respect to the matrix risk function.

Theorem 1. Under model (1), the estimator $\underline{b}_{12}$ given in (12) dominates the estimator $\underline{b}_1$ given in (3) with respect to the matrix risk function if and only if $\lambda \le \frac{1}{2}$, and $\underline{b}_1$ dominates $\underline{b}_{12}$ if and only if simultaneously $r(\underline{R}_2) = 1$ and $\lambda \ge \frac{1}{2}$, where

$$\lambda = (\underline{R}_2'\underline{\beta} - \underline{r}_2)'(\underline{R}_2'\underline{U}\underline{R}_2)^-(\underline{R}_2'\underline{\beta} - \underline{r}_2)/(2\sigma^2). \tag{15}$$

Proof. Denoting $\underline{R}_2'\underline{\beta} - \underline{r}_2 = \underline{\delta}$, we have

$$M(\underline{b}_1 ; \underline{\beta}) - M(\underline{b}_{12} ; \underline{\beta}) = \underline{U}\underline{R}_2(\underline{R}_2'\underline{U}\underline{R}_2)^-[\sigma^2\underline{R}_2'\underline{U}\underline{R}_2 - \underline{\delta}\underline{\delta}'](\underline{R}_2'\underline{U}\underline{R}_2)^-\underline{R}_2'\underline{U}.$$

Now, the relations $\underline{\delta} \in (\underline{R}_2')$ and $r(\underline{R}_2'\underline{U}\underline{R}_2) = r(\underline{R}_2)$*) imply that

$$\underline{\delta} \in \mathscr{C}(\underline{R}_2'\underline{U}\underline{R}_2) \tag{16}$$

or, equivalently, $(\underline{R}_2'\underline{U}\underline{R}_2)(\underline{R}_2'\underline{U}\underline{R}_2)^-\underline{\delta} = \underline{\delta}$ (cf. Rao and Mitra (1971), p.22). Hence it follows that $\underline{b}_{12}$ dominates $\underline{b}_1$ if and only if $\underline{T}=\sigma^2\underline{R}_2'\underline{U}\underline{R}_2-\underline{\delta}\underline{\delta}'$

---

*) This equality can be seen as follows. Notice that $\mathscr{C}(\underline{R}_1) \in \mathscr{C}^{\perp}(\underline{U})$ and $r(\underline{U}) = p - r(\underline{R}_1)$. Hence $\dim \mathscr{C}(\underline{R}_1) = \dim \mathscr{C}^{\perp}(\underline{U})$ and, using the assumption that $\mathscr{C}(\underline{R}_1) \cap \mathscr{C}(\underline{R}_2) = \{\underline{O}\}$, we obtain $r(\underline{R}_2') \ge r(\underline{R}_2'\underline{U}) = r(\underline{R}_2') - \dim\{\mathscr{C}(\underline{R}_2) \cap \mathscr{C}^{\perp}(\underline{U})\} = r(\underline{R}_2') - \dim\{\mathscr{C}(\underline{R}_2) \cap \mathscr{C}(\underline{R}_1)\} = r(\underline{R}_2')$. Consequently, due to the nonnegative definiteness of $\underline{U}$, it follows that $r(\underline{R}_2') = r(\underline{R}_2'\underline{U}) = r(\underline{R}_2'\underline{U}\underline{R}_2)$.

is n.n.d. The first part of the theorem is now obvious when applying a result of Baksalary and Kala (1980) stating that, given a n.n.d. matrix $\underline{A}$, a vector $\underline{b}$ such that $\underline{b} \in \mathscr{C}(\underline{A})$ and a positive scalar d, the matrix $d\underline{A} - \underline{bb}'$ is n.n.d. if and only if $\underline{b}'\underline{A}^{-}\underline{b} \leq d$. On the other hand, $\underline{b}_1$ dominates $\underline{b}_{12}$ if and only if $-\underline{T} = \underline{\delta\delta}' - \sigma^2 \underline{R}_2'\underline{UR}_2$ is n.n.d. On account of a well known property of n.n.d. matrices $\underline{A}$ and $\underline{B}$ that a necessary condition for $\underline{A}-\underline{B}$ to be n.n.d. is that $r(\underline{A}) \geq r(\underline{B})$ it follows that $-\underline{T}$ cannot be n.n.d. whenever $r(\underline{R}_2) \geq 2$. Now, it can easily be established that, given a n.n.d. matrix $\underline{A}$ of rank $r(\underline{A}) = 1$, a vector $\underline{b}$ such that $\underline{b} \in \mathscr{C}(\underline{A})$ and a positive scalar d, the matrix $\underline{bb}' - d\underline{A}$ is n.n.d. if and only if $\underline{b}'\underline{A}^{-}\underline{b} \geq d$. An application of this proposition completes the proof of the second part of the theorem.

Let us now consider the problem of comparing the estimators $\underline{b}_1$ and $\underline{b}_{12}$ with respect to a weighted quadratic risk function. Corresponding conditions, weaker than those in Theorem 1 as being only sufficient , are stated in the following.

Theorem 2. Under model (1), the estimator $\underline{b}_{12}$ given in (12) dominates the estimator $\underline{b}_1$ given in (3) with respect to a weighted quadratic risk function with $\underline{W}$ being the matrix of weights if $\lambda \leq (2\xi_L)^{-1}\mathrm{tr}(\underline{K})$, whereas $\underline{b}_1$ dominates $\underline{b}_{12}$ if $\lambda \geq (2\xi_S)^{-1}\mathrm{tr}(\underline{K})$, where λ is as defined in (15), and ξ_L, ξ_S are the largest and the smallest nonzero eigenvalues of $\underline{K} = (\underline{R}_2'\underline{UR}_2)^{-}\underline{R}_2'\underline{UWUR}_2$, respectively.

Proof. Since $\rho_W(\underline{\tilde{b}}\,;\,\underline{\beta}) = \mathrm{tr}[M(\underline{\tilde{b}}\,;\,\underline{\beta})\underline{W}]$ it follows from the proof of Theorem 1 that $\underline{b}_{12}$ dominates $\underline{b}_1$ in the sense that $\rho_W(\underline{b}_1\,;\,\underline{\beta}) \geq \rho_W(\underline{b}_{12}\,;\,\underline{\beta})$ if and only if $\gamma \leq \mathrm{tr}(\underline{K})/2$, where

$$\gamma = (2\sigma^2)^{-1}\underline{\delta}'(\underline{R}_2'\underline{UR}_2)^{-}\underline{R}_2'\underline{UWUR}_2(\underline{R}_2'\underline{UR}_2)^{-}\underline{\delta}.$$

Similarly, it follows that $\underline{b}_1$ dominates $\underline{b}_{12}$ if and only if $\gamma \geq \mathrm{tr}(\underline{K})/2$. Consequently, the proof reduces to showing that $\lambda\xi_S \leq \gamma \leq \lambda\xi_L$.

Using the notation $\underline{\eta} = \underline{U}^{\frac{1}{2}}\underline{R}_2(\underline{R}_2'\underline{UR}_2)^{-}\underline{\delta}$, and the relation (16), it can be seen that

$$\gamma/\lambda = \underline{\eta}'\underline{U}^{\frac{1}{2}}\underline{R}_2(\underline{R}_2'\underline{UR}_2)^{-}\underline{R}_2'\underline{UWUR}_2(\underline{R}_2'\underline{UR}_2)^{-}\underline{R}_2'\underline{U}^{\frac{1}{2}}\underline{\eta}/(\underline{\eta}'\underline{\eta}).$$

Now observe, basing on the spectral decomposition (cf. Rao (1973), p.39)

$$\underline{A} = (\underline{Q}_1 \quad \underline{Q}_2)\begin{pmatrix} \underline{\Delta} & \underline{O} \\ \underline{O} & \underline{O} \end{pmatrix}\begin{pmatrix} \underline{Q}_1' \\ \underline{Q}_2' \end{pmatrix}$$

of a n.n.d. matrix $\underline{A}$, that

$$\sup_{\underline{O} \neq \underline{x} \in \mathscr{C}(\underline{A})} \frac{\underline{x}'\underline{A}\underline{x}}{\underline{x}'\underline{x}} = \sup_{\underline{1} \neq \underline{O}} \frac{(\underline{Q}_1 \underline{1})'\underline{A}(\underline{Q}_1 \underline{1})}{(\underline{Q}_1 \underline{1})'(\underline{Q}_1 \underline{1})} = \sup_{\underline{1} \neq \underline{O}} \frac{\underline{1}'\underline{\Delta}\underline{1}}{\underline{1}'\underline{1}} = \lambda_L$$

and similarly, that

$$\inf_{\underline{O} \neq \underline{x} \in \mathscr{C}(\underline{A})} \frac{\underline{x}'\underline{A}\underline{x}}{\underline{x}'\underline{x}} = \lambda_S$$

where λ_L and λ_S are the largest and the smallest nonzero eigenvalues of $\underline{A}$, respectively.

This result implies the inequalities $\xi_S \leq \gamma/\lambda \leq \xi_L$, where the eigenvalues ξ_S and ξ_L are related here to the matrix $\underline{A} = \underline{U}^{\frac{1}{2}}\underline{R}_2(\underline{R}_2'\underline{U}\underline{R}_2)^- \underline{R}_2'\underline{U}\underline{W}\underline{U}\underline{R}_2(\underline{R}_2'\underline{U}\underline{R}_2)^-\underline{R}_2'\underline{U}^{\frac{1}{2}}$, and thus are the same as those of $(\underline{R}_2'\underline{U}\underline{R}_2)^-\underline{R}_2'\underline{U}\underline{W}\underline{U}\underline{R}_2$.

3. Remarks

Under additional assumption that $\underline{y}$ is normally distributed, all criteria specified in Theorems 1 and 2 are operational, i.e., are stated as conditions on parameters of a well defined probability density function. Applying some known results from distribution theory of quadratic forms (cf. Rao and Mitra (1971), p. 171-176), it can be proved that, under model (1) the statistic

$$F = \frac{(\underline{R}_2'\underline{b}_1 - \underline{r}_2)'(\underline{R}_2'\underline{U}\underline{R}_2)^-(\underline{R}_2'\underline{b}_1-\underline{r}_2)}{(\underline{y}-\underline{X}\underline{b}_1)'\underline{V}^{-1}(\underline{y}-\underline{X}\underline{b}_1)} \quad \frac{n+r(\underline{R}_1)-p}{r(\underline{R}_2)} \tag{17}$$

follows the $F_{n_1,n_2}(\lambda)$ distribution with $n_1 = r(\underline{R}_2)$ and $n_2 = n+r(\underline{R}_1)-p$ degrees of freedom and with the noncentrality parameter λ as specified in (15) and thus yields the common test procedure for $H_o : \lambda = 0$. But following the results of Toro-Vizcarrondo and Wallace (1968), it can be shown that basing on statistic (17) the UMP test procedure for testing $H_o : \lambda \leq \lambda_o$ against $H_1 : \lambda > \lambda_o$ can also be constructed. Roughly, the hypothesis H_o is rejected if $F \geq C_o$, where C_o is determined for a given significance level α by the equality $\int_{C_o}^{\infty} dF_{n_1,n_2}(\lambda_o) = \alpha$. Some tables for making use of such test are given in Wallace and Toro-Vizcarrondo (1969), Goodnight and Wallace (1972).

The criteria obtained in this note are extensions of some results given in the literature. More precisely substituting $\underline{R}_1 = \underline{0}$, and $\underline{R}_2 = \underline{R}$ in Theorems 1 and 2 leads to results established by Toro-Vizcarrondo and Wallace (1968) and Yancey, Judge and Bock (1973), respectively. Also, similar approach can be found in the paper of Guilkey and Price (1981), where some general results demonstrate limitation on comparisons between the estimators utilizing linear restrictions. However, the results obtained by them are not operational in the sense mentioned in the present note.

Acknowledgements

The authors are grateful to Dr.J.K.Baksalary for his constant guidance and for his help in reviewing the manuscript for publication.

References

[1] J.K.Baksalary, R.Kala, 1980, Two properties of a nonnegative definite matrix. Bull.Acad.Polon.Sci., Ser.Sci.Math. 28, 233-235.

[2] J.Goodnight, T.D.Wallace, 1972, Operational techniques and tables for making weak tests for restrictions in regressions. Econometrica 40, 699-709.

[3] D.K.Guilkey, J.M.Price, 1981, On comparing restricted least squares estimators. Journal of Econometrics 15, 397-404.

[4] C.R.Rao, 1973, Linear Statistical Inference and its Applications. John Wiley and Sons, Inc., New York.

[5] C.R.Rao, S.K.Mitra 1971, Generalized Inverse of Matrices and its Applications. John Wiley and Sons, Inc., New York.

[6] S.R.Searle, 1971, Linear Models, John Wiley and Sons, Inc., New York.

[7] C.Toro-Vizcarrondo, T.D.Wallace, 1968, A test of the mean square error criterion for restrictions in linear regressions. J.Amer. Statist.Assoc.63, 558-572.

[8] T.D.Wallace, C.Toro-Vizcarrondo, 1969, Tables for the mean square error test for exact linear restrictions in regression. J.Amer. Statist.Assoc. 64, 1649-1663.

[9] T.A.Yancey, G.G.Judge, M.F.Bock, 1973, Wallace's mean squared error criterion for testing linear restrictions in regression: a tighter bound. Econometrica 41, 1203-1209.

DISTRIBUTION OF THE DISCRIMINANT FUNCTION

M. Krzyśko
Institute of Mathematics
Mickiewicz University
Poznań, Poland

1. Introduction

The classical problem of discrimination is associated with the identification of objects and phenomena described by a set of features. In many practical situations, however, problems arise in which it is necessary to identify objects and phenomena characterized by functions of time, that is, continuous or discrete processes.

This paper is concerned with discrimination among multivariate autoregressive processes with known parameters by the Bayes method. The assumption that the parameters are known is not fully realistic from the practical point of view. However, investigation of the theoretical structure of the discriminant function is fully justified. It is specially usefull when we want to compare an estimator of the discriminant function with the theoretical function.

To calculate the value of the Bayesian risk we need to know the distribution of the discriminant function. Hence, the exact and asymptotic distributions of the discriminant function are investigated.

2. Bayes method of classification

Let π_i, $i = 1,2,\dots,k$, denote a class of p-dimensional stochastic processes $\{\underline{X}_i(t),\ t = 0,\pm 1,\pm 2,\dots\}$ with expected value $E\{\underline{X}_i(t)\} = \underline{\mu}_i$ and covariance matrix

$$E\{(\underline{X}_i(m) - \underline{\mu}_i)(\underline{X}_i(n) - \underline{\mu}_i)'\} = \underline{\Gamma}_i(n-m) = \underline{\Gamma}_i'(m-n)$$

generated by an autoregressive equation of order q_i

$$\sum_{u=0}^{q_i} \underline{B}_i(u)(\underline{X}_i(t-u) - \underline{\mu}_i) = e_i(t). \tag{1}$$

Here $\underline{B}_i(0) = \underline{I}_p$ and

(i) $\underline{B}_i(u)$, $u = 1,2,\ldots,q_i$, $i = 1,2,\ldots,k$, are $p \times p$ matrices of parameters,

(ii) $\{\underline{e}_i(t),\ t = 0,\pm 1,\ldots\}$ is a sequence of independently normally distributed random vectors with zero mean and p.d. covariance matrix $\underline{V}$, $i = 1,2,\ldots,k$.

An important class of vector time series is that generated by a linear mechanism of form (1). Such a mechanism should often be a reasonable approximation to reality, since the idea that $\underline{X}(t)$ is determined by immediate past values of itself, together with past disturbances, is an appealing one. Experience suggests that a model of form (1) will fit a wide range of data (see e.g. Hannan (1970)).

Assume that

$\Theta_i = (\underline{B}_i(1),\ldots,\underline{B}_i(q_i),\ \underline{\mu}_i,\underline{V})$, $i = 1,2,\ldots,k$, are known which also implies that the covariance matrices $\underline{\Gamma}_1(t)$, $\underline{\Gamma}_2(t),\ldots,\underline{\Gamma}_k(t)$ are known.

Let us assume that we have a realization $\underline{X} = (\underline{X}(1),\ \underline{X}(2),\ldots,\underline{X}(T))$ of a process belonging to one of the classes $\pi_1,\ \pi_2,\ldots,\pi_k$. We consider k alternative hypotheses $H_1,H_2,\ldots,H_k$ forming a complete system of disjoint events, i.e. if P_i ($P_i > 0$), $i = 1,2,\ldots,k$, is the a priori probability of the acceptability of hypothesis H_i, then $P_1 + P_2 + \ldots + P_k = 1$. The hypothesis H_i states that $\underline{X}$ is a realization of a process belonging to class π_i, $i = 1,2,\ldots,k$. Our task is to verify the acceptability of one of these k alternative hypotheses. In the statistical literature (see, for example, Anderson (1958)) this problem is known as classification problem. We will make use of the Bayesian method of classification dividing $\mathcal{R}^N$ ($N = pT$) into non-intersecting classification regions $\mathcal{R}_1,\ \mathcal{R}_2,\ldots,\ \mathcal{R}_k$ defined so as to minimize the Bayesian risk

$$r = 1 - \sum_{i=1}^{k} P_i \int_{\mathcal{R}_i} f(\underline{x}|\Theta_i)\,d\underline{x}, \tag{2}$$

where $f(\underline{x}|\underline{\theta}_i) = f(\underline{x}(1),\underline{x}(2),\ldots,\underline{x}(T)|\underline{\theta}_i)$ is the density function of the joint distribution of $\underline{X}$ in class π_i, $i = 1,2,\ldots,k$. By (1) and condition (ii)

$$f(\underline{x}|\underline{\theta}_i) = (2\pi)^{-\frac{N}{2}}|\underline{G}_i|^{-\frac{1}{2}} \exp\{-\tfrac{1}{2}\text{vec}'(\underline{x}-\underline{m}_i)\underline{G}_i^{-1}\text{vec}(\underline{x}-\underline{m}_i)\}, \tag{3}$$

where $m_i = \underline{\mu}_i 1'$, $1' = (1,1,\ldots,1)$, $i = 1,2,\ldots,k$, while vec $\underline{A}$ is a vector formed by stacking the columns of $\underline{A}$, one on top of the other, in order from left to right.

The matrix $\underline{G}_i$ is a block matrix of T^2 blocks, where block (u,v) contains the matrix $\underline{\Gamma}_i(v-u)$, $i = 1,2,\ldots,k$.

The region $\mathscr{R}_i$ minimizing the Bayesian risk (2) is

$$\mathscr{R}_i = \{\underline{x}\colon v_{ij}(x) \geq \ln(P_j/P_i),\ j = 1,2,\ldots,k,\ \ j \neq i\}, \tag{4}$$

where v_{ij} is a discriminant function of the following form

$$v_{ij}(\underline{x}) = \ln(f(\underline{x}|\underline{\theta}_i)/f(\underline{x}|\underline{\theta}_j)) = -\tfrac{1}{2}[\text{vec}'(\underline{x}-\underline{m}_i)\underline{G}_i^{-1}\text{vec}(\underline{x}-\underline{m}_i) - \tag{5}$$

$$-\text{vec}'(\underline{x}-\underline{m}_j)\underline{G}_j^{-1}\text{vec}(\underline{x}-\underline{m}_j) - \ln(|\underline{G}_j|/|\underline{G}_i|)],\ i,j=1,2,\ldots,k,\ j\neq i.$$

According to the Bayesian method, H_i is accepted if $\underline{x} \in \mathscr{R}_i$, $i = 1,2,\ldots,k$.

When the hypothesis H_i is true the probability α_i of rejecting it is equal to

$$\alpha_i = 1 - P(v_{ij} \geq \ln(P_j/P_i),\ j = 1,2,\ldots,k,\ k \neq i|\underline{\theta} = \underline{\theta}_i) =$$

$$= 1 - \int_{\mathscr{R}_i} f(\underline{x}|\underline{\theta}_i)dx,\quad i = 1,2,\ldots,k,$$

and the minimum value of the Bayesian risk is

$$r = \sum_{i=1}^{k} P_i\alpha_i = 1 - \sum_{i=1}^{k} P_i P(v_{ij} \geq \ln(P_j/P_i),\ j = 1,2,\ldots,k,\ \ j \neq i|\underline{\theta} = \underline{\theta}_i).$$

The probability α_i is very difficult to evaluate in this case, because of the very complicated shapes of the classification regions. An alternative way of obtaining a lower bound for α_i is by using the Bonferroni inequality which yields

$$\alpha_i \leq \sum_{\substack{j=1\\ j\neq 1}}^{k} P(v_{ij} < \ln(P_j/P_i)), \quad i = 1,2,\ldots,k, \tag{6}$$

and

$$r \leq \sum_{i=1}^{k} \sum_{\substack{j=1\\ j\neq 1}}^{k} P_i P(v_{ij} < \ln(P_j/P_i)). \tag{7}$$

If $k = 2$, then equality holds in the above formulas. To calculate the minimum value of the Bayesian risk we need to know the distribution of $v_{ij}(\underline{X})$, $i,j = 1,2,\ldots,k$, $j \neq i$.

3. Distribution of the discriminant function

Since $\underline{G}_i^{-1}$ is a symmetric p.d. matrix, there exists a nonsingular matrix $\underline{A}_i$ such that $\underline{G}_i^{-1} = \underline{A}_i'\underline{A}_i$, $i = 1,2,\ldots,k$. For $n = 1,\ldots,N$ let $\lambda_{ij,n}$ be the eigenvalue and $\underline{P}_{ij,n}$ be the corresponding eigenvector of $\underline{R}_{ij} = (A_i^{-1})'\underline{G}_j^{-1}\underline{A}_i^{-1}$, and let $\underline{P}_{ij} = (\underline{P}_{ij,1},\ldots,\underline{P}_{ij,N})$ and $\underline{\lambda}_{ij} = (\lambda_{ij,1},\ldots,\lambda_{ij,N})'$, where $i,j = 1,2,\ldots,k$, $j \neq i$.
Observe that the eigenvalues of $\underline{R}_{ij}$ are the roots of $|\underline{G}_i - \lambda G_j| = 0$.

<u>Theorem 1</u>. The distribution function of the discriminant function $v_{ij}(\underline{X})$ is

$$P\{v_{ij}(\underline{X}) < y \,|\, \underline{\theta} = \underline{\theta}_i\} = \frac{1}{2} - \frac{1}{\pi}\int_0^{\infty} \frac{\sin[s_{ij}(u,y)]}{u t_{ij}(u)}\, du, \tag{8}$$

where

$$s_{ij}(u,y) = \frac{1}{2}\sum_{n=1}^{N}[\tan^{-1}(b_{ij,n}u) + \delta_{ij,n}^2 b_{ij,n}u(1+b_{ij,n}^2u^2)^{-1}] - \frac{1}{2}(y-a_{ij})u,$$

$$t_{ij}(u) = \prod_{n=1}^{N}(1+b_{ij,n}^2u^2)^{\frac{1}{4}}\exp\{\frac{1}{2}\sum_{n=1}^{N}(\delta_{ij,n}b_{ij,n}u)^2/(1+b_{ij,n}^2u^2)\}.$$

The remaining quantities $b_{ij,n}$, a_{ij} and $\delta_{ij,n}^2$ are given by (13) and (14), respectively.

<u>Proof</u>. Writing

$$\text{vec}(\underline{X} - \underline{m}_j) = \text{vec}(\underline{X}-\underline{m}_i) + \text{vec}(\underline{m}_i-\underline{m}_j)$$

we have

$$v_{ij}(\underline{X}) = -\frac{1}{2}[(\underline{A}_i\mathrm{vec}(\underline{X}-\underline{m}_i))'(\underline{I}-\underline{R}_{ij})(\underline{A}_i\mathrm{vec}(\underline{X}-\underline{m}_i))-\underline{\rho}'_{ij}(\underline{A}_i\mathrm{vec}(\underline{X}-\underline{m}_i))]+k_{ij},$$

where

$$\underline{\rho}_{ij} = 2(\underline{A}'_i)^{-1}\underline{G}_j^{-1}\mathrm{vec}(\underline{m}_i-\underline{m}_j), \tag{9}$$

while

$$k_{ij} = \frac{1}{2}[\mathrm{vec}'(\underline{m}_i-\underline{m}_j)\underline{G}_j^{-1}\mathrm{vec}(\underline{m}_i-\underline{m}_j) + \ln(|\underline{G}_j|/|\underline{G}_i|)], \tag{10}$$

$$i,j = 1,2,\dots,k, j \neq i.$$

Because

$$\underline{R}_{ij} = \underline{P}_{ij}\,\underline{\Lambda}_{ij}\underline{P}'_{ij},$$

where $\underline{\Lambda}_{ij} = \mathrm{diag}(\lambda_{ij,1},\dots,\lambda_{ij,N})$, we obtain

$$v_{ij}(\underline{X}) = -\frac{1}{2}[(\underline{P}'_{ij}\underline{A}_i\mathrm{vec}(\underline{X}-\underline{m}_i))'(\underline{I}-\underline{\Lambda}_{ij})(\underline{P}'_{ij}\underline{A}_i\mathrm{vec}(\underline{X}-\underline{m}_i))-$$

$$-\underline{\gamma}'_{ij}(\underline{P}'_{ij}\underline{A}_i\mathrm{vec}(\underline{X}-\underline{m}_i))]+k_{ij},$$

where

$$\underline{\gamma}_{ij} = (\gamma_{ij,1},\dots,\gamma_{ij,N})' = \underline{P}_{ij}^{-1}\,\underline{\rho}_{ij}, \tag{11}$$

while $\underline{\rho}_{ij}$ is given by (9), $i,j = 1,2,\dots,k,\ j \neq i$.

The distribution of $v_{ij}(\underline{X})$ is the same as the distribution of

$$v_{ij}(\underline{Z}_{ij}) = -\frac{1}{2}[\underline{Z}'_{ij}(\underline{I}-\underline{\Lambda}_{ij})\underline{Z}_{ij} - \underline{\gamma}'_{ij}\underline{Z}_{ij}] + k_{ij} = \tag{12}$$

$$= \frac{1}{2}[\sum_{n=1}^{N}(\lambda_{ij,n}-1)Z_{ij,n}^2 + \sum_{n=1}^{N}\gamma_{ij,n}Z_{ij,n}] + k_{ij},$$

where

$$\underline{Z}_{ij} = (Z_{ij,1},\dots,Z_{ij,N})' = \underline{P}'_{ij}\underline{A}_i\mathrm{vec}(\underline{X}-\underline{m}_i).$$

Now, under condition $\underline{\Theta} = \underline{\Theta}_i$, $\mathrm{vec}(\underline{X}-\underline{m}_i)$ has a normal distribution with zero mean and covariance matrix $\underline{G}_i$ and $\underline{Z}_{ij}$ has a normal distribution with zero mean and the identity covariance matrix so that

$$v_{ij}(\underline{Z}_{ij}) = \sum_{n=1}^{N} b_{ij,n}(Z_{ij,n} + \delta_{ij,n})^2 + a_{ij},$$

where

$$b_{ij,n} = \frac{1}{2}(\lambda_{ij,n}-1), \quad a_{ij} = k_{ij} - \sum_{n=1}^{N} \gamma_{ij,n}^2/4b_{ij,n}, \tag{13}$$

while k_{ij} and $\gamma_{ij,n}$ are given by (10) and (11), respectively; $(Z_{ij,n} + \delta_{ij,n})^2$ has the noncentral chi-square distribution with one degree of freedom and noncentrality parameter

$$\delta_{ij,n}^2 = \gamma_{ij,n}^2/16\, b_{ij,n}^2, \tag{14}$$

$$i,j = 1,2,\ldots,k, \quad j \neq i, \quad n = 1,2,\ldots,N.$$

Applying formula (3.2) given by Imhof (1961) we get (8).

Remark 1. Since the distribution function of $v_{ij}(\underline{X})$ is very complicated we will find its asymptotic distribution.

4. Asymptotic distribution of the discriminant function

Expressing $v_{ij}(\underline{X})$ as

$$2v_{ij}(\underline{X}) = \sum_{t=q_j+1}^{T} y_j^2(t) - \sum_{t=q_i+1}^{T} y_i^2(t) + w_{q_j}^2 - w_{q_i}^2 + 2A_{ij},$$

where

$$y_i^2(t) = \Big(\sum_{u=0}^{q_i} \underline{B}_i(u)(\underline{x}(t-u) - \underline{\mu}_i)\Big)'\underline{V}^{-1}\Big(\sum_{u=0}^{q_i} \underline{B}_i(u)(\underline{x}(t-u)-\underline{\mu}_i)\Big),$$

$$w_{q_i}^2 = \operatorname{vec}(\underline{x}_{q_i}-\underline{m}_i)'\underline{G}_{q_i}^{-1}\operatorname{vec}(\underline{x}_{q_i}-\underline{m}_i), \quad \underline{x}_{q_i} = (\underline{x}(1),\ldots,\underline{x}(q_i)),$$

$$A_{ij} = \frac{1}{2}\ln(|\underline{G}_{q_j}|/|\underline{G}_{q_i}|) + \frac{1}{2}(q_i-q_j)\ln|\underline{V}|.$$

Here $\underline{G}_{q_i}$ is a block matrix of q_i^2 blocks, where the block (u,v) contains the matrix $\underline{\Gamma}_i(v-u)$, $i,j = 1,2,\ldots,k$, $j \neq i$.

We will need the following notation

$$Z_{ij}(t) = y_j^2(t) - y_i^2(t), \quad n = T-q, \quad q = \max_{1\le i\le k} q_i,$$

$$S_{ij}(n) = \sum_{t=1}^{n} Z_{ij}(t), \quad m_{ij} = E\{Z_{ij}(t) \mid \underline{\Theta} = \underline{\Theta}_i\},$$

$$\sigma_{ij}^2(n) = E\{(S_{ij}(n) - nm_{ij})^2 \mid \underline{\Theta} = \underline{\Theta}_i\}.$$

Using a central limit theorem for dependent random variables due to Ibragimov (1975) one may show the following result (see Krzyśko (1983)).

<u>Theorem 2</u>. For all pairs (i,j), $i \neq j$, as $T \to \infty$,

$$\left(v_{ij}(\underline{X}) - A_{ij} - \frac{T-q}{2} m_{ij}\right) / \tfrac{1}{2}\sigma_{ij}(T-q)$$

is asymptotically, normally distributed with zero mean and unit variance.

5. <u>Deviation of the distribution of the discriminant function $v_{ij}(X)$ from the asymptotic normal distribution</u>

Theorem 2 shows that the limit distribution of $v_{ij}(\underline{X})$ is normal. Now, we will estimate the precision of this approximation according to the lenght T of the time series. Notice that $v_{ij}(\underline{X})$ and $v_{ij}(\underline{Z}_{ij})$ have the same distribution. Thus the expected value and the variance of $v_{ij}(\underline{Z}_{ij})$, $i,j = 1,2,\ldots,k$, $j \neq i$, are

$$m_{ij}(N) = E\{v_{ij}(\underline{Z}_{ij}) \mid \underline{\Theta} = \underline{\Theta}_i\} = k_{ij} + \sum_{n=1}^{N} b_{ij,n},$$

$$\sigma_{ij}^2(N) = \mathrm{Var}\{v_{ij}(\underline{Z}_{ij}) \mid \underline{\Theta} = \underline{\Theta}_i\} = 2\sum_{n=1}^{N} b_{ij,n}^2 + \frac{1}{4}\sum_{n=1}^{N} \gamma_{ij,n}^2,$$

respectively.

<u>Theorem 3</u>. The following inequality holds

$$\sup_y \left| P\{v_{ij}(\underline{X}) < y \mid \underline{\Theta} = \underline{\Theta}_i\} - \Phi\left(\frac{y - m_{ij}(N)}{\sigma_{ij}(N)}\right)\right| \le c(\underline{\Theta}_i, \underline{\Theta}_j, T),$$

where

$$c(\underline{\Theta}_i, \underline{\Theta}_j, T) = \frac{8.1 \max_{i\le n\le N} |\lambda_{ij,n} - 1|}{\sqrt{2\pi}\,\sigma_{ij}^3(N)} \left\{1.8\sigma_{ij}^2(N) - \frac{1}{6}\sum_{n=1}^{N} (\lambda_{ij,n} - 1)^2\right\} \tag{15}$$

$i, j = 1, 2, \ldots, k,\ j \neq i.$

Remark 2. Theorem 3 with $p = 1$ (i.e. N=T) reduces to a result due to Misiukas (1978). One should observe that the two constants which appear in the expression $c(\underline{\theta}_i, \underline{\theta}_j, T)$ of the Misiukas theorem are miscalculated. Instead of 8.455 and 1.68 there should be 8.1 and 1.8, respectively.

To prove this theorem, we shall require a result of Zolotarev (1965).

Let $L(x)$ be a distribution function and let $H(x)$ be a function with bounded variation such that

$$q = \sup|H'(x)| < \infty \ , \text{ and } \ H(-\infty) = 1 - H(\infty) = 0.$$

Let $l(t)$ and $h(t)$ denote the Fourier-Stieltjes transformations of $L(x)$ and $H(x)$, respectively. Next, let $p(x)$ be a density function of a symmetric distribution and let $\omega(t)$ be its characteristic function.
Let

$$\Delta = \sup|L(x) - H(x)|, \qquad \delta = l-h,$$

$$S(x) = x\int_0^x p(u)\,du, \qquad Q(T) = \frac{T}{2\pi q}\int_0^\infty |\omega(t)\,\delta(tT)|\frac{dt}{t},$$

and let A denote the positive solution of the equation $4S(x) = x$.

Lemma 1 (Zolotarev (1965)). For every $T > 0$ and $x > A$ holds

$$\Delta \leq 2q\,\frac{x[S(x)+Q(T)]}{T[4S(x)-x]}\ .$$

Remark 3. If $H(x)$ is a distribution function, then the assertion holds also when $|l(t)-h(t)|$ is replaced by $|l^*(t) - h^*(t)|$, where $l^*(t)$ and $h^*(t)$ are the characteristic functions of L and H, respectively.

To prove Theorem 3 we shall also require some properties of the (semiinvariants) cumulants.

Let X be a real random variables and $\varphi_X(t)$ its characteristic function.
The cumulant of order k of X is

$$\Gamma_k(X) = \frac{1}{i^k}\left[\frac{d^k}{dt^k} \operatorname{Ln} \varphi_X(t)\right]_{t=0}, \quad k = 1,2,\ldots .$$

Here $\operatorname{Ln} z$ states for the principal branch of the function $\ln z$ so that $\operatorname{Ln} \varphi_X(0) = 0$.

Proof of Theorem 3. To estimate the expression

$$\sup_x |P\{v_{ij}(Z_{ij}) < x | \underline{\Theta} = \underline{\Theta}_i\} - \Phi(x)|$$

we use the above mentioned Zolotarev's lemma.
Put

$$\bar{v}_{ij}(Z_{ij}) = \frac{v_{ij}(Z_{ij}) - m_{ij}(N)}{\sigma_{ij}(N)}, \quad i,j = 1,2,\ldots,k,\ j \neq i,$$

and define

$$L(x) = P\{\bar{v}_{ij}(Z_{ij}) < x | \underline{\Theta} = \underline{\Theta}_i\},\ H(x) = \Phi(x),$$

$$h^*(t) = \exp\{-t^2/2\}, \quad q = 1/\sqrt{2\pi},$$

$$l^*(t) = E\{\exp[it\ \bar{v}_{ij}(Z_{ij})]\},$$

and

$$p(x) = \frac{1-\cos x}{\pi x^2}, \quad -\infty < x < \infty.$$

The characteristic function of $p(x)$ is

$$\omega(t) = \begin{cases} 1-|t| & \text{for } |t| \leq 1, \\ 0 & \text{for } |t| > 1. \end{cases}$$

Hence

$$S(x) = x\int_0^x \frac{1-\cos u}{\pi u^2} du = \frac{1}{\pi}[\cos x + x\ \operatorname{Si}(x) - 1],$$

where $\operatorname{Si}(x)$ denotes the sine integral.

To estimate the expression $|l^*(t) - \exp(-t^2/2)|$ we apply the cumulants of $\bar{v}_{ij}(\underline{Z}_{ij})$. We have

$$\Gamma_k(\bar{v}_{ij}(\underline{Z}_{ij})) = \frac{(k-1)!}{2\sigma_{ij}^k(N)} \sum_{n=1}^N (\lambda_{ij,n} - 1)^k \left(1 + k\frac{\gamma_{ij,n}^2}{4(\lambda_{ij,n}-1)^2}\right),$$

$$\Gamma_1(\bar{v}_{ij}(\underline{Z}_{ij})) = 0, \quad \Gamma_2(\bar{v}_{ij}(\underline{Z}_{ij})) = 1,$$

$$\operatorname{Ln} 1^*(t) = \sum_{k=1}^{\infty} \frac{\Gamma_k(\bar{v}_{ij}(\underline{Z}_{ij}))}{k!}(it)^k$$

$$= -t^2/2 + \sum_{k=3}^{\infty} \frac{\Gamma_k(\bar{v}_{ij}(\underline{Z}_{ij}))}{k!}(it)^k$$

$$= -t^2/2 + \sum_{n=1}^{N} \frac{1}{8(\lambda_{ij,n}-1)^2} \sum_{k=3}^{\infty} a_{nk}\left(\frac{i}{b_n}t\right)^k,$$

where

$$a_{nk} = \gamma_{ij,n}^2 + \frac{4(\lambda_{ij,n}-1)^2}{k}, \qquad b_n = \frac{\sigma_{ij}(N)}{\lambda_{ij,n}-1}.$$

Also

$$\sum_{k=3}^{\infty} a_{nk}\left(\frac{i}{b_n}t\right)^k = a_{n3}\sum_{k=3}^{\infty}\frac{a_{nk}}{a_{n3}}\left(\frac{i}{b_n}t\right)^k = a_{n3}\sum_{k=3}^{\infty} z_k\left|\frac{t}{b_n}\right|^k$$

$$= a_{n3} z \sum_{k=3}^{\infty}\left|\frac{t}{b_n}\right|^k,$$

where Θ and z are complex numbers and

$$i\frac{t}{b_n} = \Theta\left|\frac{t}{b_n}\right|, \qquad z_k = \frac{a_{nk}}{a_{n3}}\Theta^k, \qquad |z| \le 1;$$

here i is the imaginary unit.

Hence for $|t| < \Delta_N$

$$\operatorname{Ln} 1^*(t) = -t^2/2 + \frac{1}{8}\sum_{n=1}^{N}(\lambda_{ij,n}-1)^{-2} a_{n3}\, z \sum_{k=3}^{\infty}\left|\frac{t}{b_n}\right|^k$$

$$= -t^2/2 + \frac{z}{8}\sum_{n=1}^{N}(\lambda_{ij,n}-1)^{-2} a_{n3}\frac{|t|^3}{|b_n|^3\left(1-\left|\frac{t}{b_n}\right|\right)}$$

$$= -t^2/2 + \frac{z|t|^3}{24\sigma_{ij}^2(N)}\sum_{n=1}^{N}\frac{3\gamma_{ij,n}^2 + 4(\lambda_{ij,n}-1)^2}{|b_n|-|t|} = -t^2/2 + \frac{w|t|^3 c_N}{24\sigma_{ij}^2(N)(\Delta_N-|t|)},$$

where

$$c_N = \sum_{n=1}^{N}[3\gamma_{ij,n}^2 + 4(\lambda_{ij,n}-1)^2],$$

$$\Delta_N = \frac{\sigma_{ij}(N)}{\max_{1\le n\le N}|\lambda_{ij,n}-1|},$$

while w is a complex number with $|w| \leq 1$.
We see that $c_N \geq 0$, but the interesting case is that of $c_N > 0$. If $c_N = 0$, then $v_{ij}(\underline{Z}_{ij}) = k_{ij}$, $i,j = 1,2,\ldots,k$, $j \neq i$, and we have the trivial case of the discriminant function.
Therefore, for all $|t| < \Delta_N$ we have

$$1^*(t) = \exp\{-t^2/2 + \frac{w|t|^3 c_N}{24\sigma_{ij}^2(N)(\Delta_N - |t|)}\}.$$

Now let $T_\varepsilon = \dfrac{24\sigma_{ij}^2(N)\Delta_N}{c_N + 24\sigma_{ij}^2(N)\varepsilon}$, where $\varepsilon > 0$.

Then, for all $\varepsilon > 0$, the condition $|t| < \Delta_N$ may be replaced by $|t| \leq \varepsilon T_\varepsilon$. For each $\varepsilon > 0$ we have $0 < \varepsilon T_\varepsilon < \Delta_N$. Since $|e^z - 1| \leq e^{|z|} - 1$ for any complex number z therefore, for $c_N > 0$ and $|t| \leq \varepsilon T_\varepsilon$, where $\varepsilon > 0$, we have

$$|1^*(t) - \exp(t^2/2)| \leq \exp(-t^2/2)\,[\exp(\frac{|t|^3 c_N}{24\sigma_{ij}^2(N)(\Delta_N - |t|)}) - 1)]$$

$$\leq \exp(-t^2/2)[\exp(\frac{|t|^3 c_N}{24\sigma_{ij}^2(N)(\Delta_N - \varepsilon T_\varepsilon)}) - 1]$$

$$= \exp(-t^2/2)[\exp(\frac{|t|^3}{T_\varepsilon}) - 1] = \exp(-t^2/2)\sum_{k=1}^{\infty}\frac{|t|^{3k}}{T_\varepsilon^k k!}$$

$$= \exp(-t^2/2)\frac{|t|}{T_\varepsilon}\sum_{k=1}^{\infty}\frac{|t|^{k-1}(t^2)^k}{T_\varepsilon^{k-1}k!} \leq \exp(-t^2/2)\frac{|t|}{T_\varepsilon}\sum_{k=1}^{\infty}\frac{(\varepsilon T_\varepsilon)^{k-1}(t^2)^k}{T_\varepsilon^{k-1}k!}$$

$$= \exp(-t^2/2)\frac{|t|}{\varepsilon T_\varepsilon}[\exp(\varepsilon t^2) - 1].$$

Hence for $c_N > 0$ and $|t| \leq \varepsilon T_\varepsilon$, where $\varepsilon > 0$,

$$|1^*(t) - \exp(-t^2/2)| \leq \frac{|t|}{\varepsilon T_\varepsilon}\{\exp[-(\tfrac{1}{2} - \varepsilon)t^2] - \exp(-t^2/2)\}.$$

Now put $T = \varepsilon T_\varepsilon$, where $0 < \varepsilon < \frac{1}{2}$.
Then

$$Q(\varepsilon T_\varepsilon) \leq \frac{\varepsilon T_\varepsilon}{\sqrt{2\pi}}\int_0^{\varepsilon T_\varepsilon}|1^*(t) - \exp(-t^2/2)|\frac{dt}{t}$$

$$< \frac{1}{\sqrt{2\pi}}\left(\int_0^\infty \exp[-(\tfrac{1}{2}-\varepsilon)t^2]dt - \int_0^\infty \exp(-t^2/2)dt\right) = \frac{1}{2}\left(\frac{1}{\sqrt{1-2\varepsilon}} - 1\right).$$

By the least inequality and Lemma 1 we have that for all $0 < \varepsilon < \frac{1}{2}$ and for all $\tau > 1.7$, where 1.7 is the positive solution of $4S(x)=x$,

$$\sup_x |\Pr\{\bar{v}_{ij} < x | \underline{\theta} = \underline{\theta}_i\} - \Phi(x)| \leq \frac{c_N + 24\varepsilon\sigma_{ij}^2(N)}{12\sqrt{2\pi}\Delta_N\sigma_{ij}^2(N)} K_1(\varepsilon,\tau),$$

where

$$K_1(\varepsilon,\tau) = \frac{\tau[S(\tau) + \frac{1}{2}((1-2\varepsilon)^{-\frac{1}{2}}-1)]}{\varepsilon[4S(\tau)-\tau]}.$$

It is easy to verify that

$$\frac{c_N + 24\varepsilon\sigma_{ij}^2(N)}{12\sqrt{2\pi}\Delta_N\sigma_{ij}^2(N)} = \frac{\max_{1\leq n\leq N}|\lambda_{ij,n}-1|}{\sqrt{2\pi}\,\sigma_{ij}^3(N)}\left[\sigma_{ij}^2(N)(1+2\varepsilon) - \frac{1}{6}\sum_{n=1}^{N}(\lambda_{ij,n}-1)^2\right].$$

Hence

$$\sup_x |\Pr\{\bar{v}_{ij} < x | \underline{\theta} = \underline{\theta}_i\} - \Phi(x)| \leq \frac{\max_{1\leq n\leq N}|\lambda_{ij,n}-1|}{\sqrt{2\pi}\,\sigma_{ij}^3(N)} K_2(\underline{\theta}_i,\underline{\theta}_j,T),$$

where

$$K_2(\underline{\theta}_i,\underline{\theta}_j,T) = \inf_{\substack{0<\varepsilon<\frac{1}{2}\\ 1.7<\tau}} \left\{\left[\sigma_{ij}^2(N)(1+2\varepsilon) - \frac{1}{6}\sum_{n=1}^{N}(\lambda_{ij,n}-1)^2\right]K_1(\varepsilon,\tau)\right\}. \quad (16)$$

The expression $K_1(\varepsilon,\tau)$ reaches its minimal value approximatly for $(\varepsilon = 0.4,\ \tau = 3.6)$ and it is not greater than 8.1. Hence we have

$$K_2(\underline{\theta}_i,\underline{\theta}_j,T) \leq 8.1\left[1.8\sigma_{ij}^2(N) - \frac{1}{6}\sum_{n=1}^{N}(\lambda_{ij,n}-1)^2\right].$$

After the transformation $y = \sigma_{ij}(N)x + m_{ij}(N)$ we get the assertion of Theorem 3.

<u>Remark 4</u>. The constant c given by (15) may be replaced by $K_2(\underline{\theta}_i,\underline{\theta}_j,T)$ given by (16).

<u>Corollary 1</u>. From Theorem 3 we conclude that

$$P(v_{ij}(\underline{X}) < \ln(P_j/P_i)) = \Phi\left(\frac{\ln(P_j/P_i) - m_{ij}(N)}{\sigma_{ij}(N)}\right) + r_{ij},$$

where

$$|r_{ij}| \leq c(\underline{\theta}_i, \underline{\theta}_j, T), \quad i,j = 1,2,\ldots,k, \quad j \neq i.$$

6. Numerical example

Let us consider the three classes of the two-dimensional second-order autoregressive series with the parameters

$$\underline{\theta}_1 = \left(\begin{pmatrix} -1.0 & -0,3 \\ 3.3 & 1.0 \end{pmatrix}, \begin{pmatrix} -0.02 & 0.0 \\ 0.0 & -0.02 \end{pmatrix}, \begin{pmatrix} 0.0 \\ 0.0 \end{pmatrix}, \begin{pmatrix} 1.0 & 0.5 \\ 0.5 & 1.25 \end{pmatrix}\right),$$

$$\underline{\theta}_2 = \left(\begin{pmatrix} -1.0 & -1.0 \\ 0.24 & 0.0 \end{pmatrix}, \begin{pmatrix} 0.0 & -0.05 \\ 0.048 & 0.11 \end{pmatrix}, \begin{pmatrix} 20.0 \\ 30.0 \end{pmatrix}, \begin{pmatrix} 1.0 & 0.5 \\ 0.5 & 1.25 \end{pmatrix}\right)$$

and

$$\underline{\theta}_3 = \left(\begin{pmatrix} -1.0 & -2.0 \\ 0.3 & 0.6 \end{pmatrix}, \begin{pmatrix} -0.04 & 0.0 \\ 0.035 & 0.03 \end{pmatrix}, \begin{pmatrix} -20.0 \\ 60.0 \end{pmatrix}, \begin{pmatrix} 1.0 & 0.5 \\ 0.5 & 1.25 \end{pmatrix}\right).$$

The values of $c(\underline{\theta}_i, \underline{\theta}_j, T)$ given by (15) according to the length T of the time series are given in Table 1.

Table 1

The values of $c(\underline{\theta}_i, \underline{\theta}_j, T)$

T	$c(\underline{\theta}_1, \underline{\theta}_2, T)$	$c(\underline{\theta}_1, \underline{\theta}_3, T)$	$c(\underline{\theta}_2, \underline{\theta}_3, T)$
10	0.184	0.107	0.073
20	0.129	0.075	0.056
30	0.105	0.061	0.047
40	0.090	0.052	0.041
50	0.081	0,047	0.037

References

[1] Anderson, T.W. (1958), An Introduction to Multivariate Statistical Analysis, Wiley, New York.

[2] Hannan, E.J. (1970), Multiple Time Series, Wiley, New York.

[3] Ibragimov, I.A. (1975), A note on the central limit theorem for dependent random variables , Teor. Verojatnost. i Primenen. 20, 134-139 (in Russian).

[4] Imhof, J.P. (1961), Computing the distribution of quadratic forms in normal variables, Biometrika 48, 419-426.

[5] Krzysko, M.(1983), Asymptotic distribution of the discriminant function, Statistics and Probability Letters 1, 243-250.

[6] Misiukas, R.(1978), Estimate of the probability of misclassification of autoregressive sequences, in: Statistical Problems of Control 30 (Institute of Mathematics and Cybernetics of the Academy of Sciences of the Lithuanian SSR, Vilnius), 29-52. (in Russian).

[7] Zolotarev, V.M.(1965), On the closeness of the distributions of the two sums of independent random variables, Teor. Verojatnost. i Primenen. 10, 519-526 (in Russian).

ADMISSIBILITY, UNBIASEDNESS, AND NONNEGATIVITY IN THE BALANCED, RANDOM, ONE-WAY ANOVA MODEL

L.R. LaMotte
University of Houston
U.S.A

1. Introduction

Problems encountered in estimating variance components in mixed-effects analysis of variance (ANOVA) models have stimulated research in linear estimation. Pukelsheim (1976) noted that quadratics in the observations which are invariant to fixed effects follow a linear model in the variance components, and that results in linear estimation theory can then be applied directly to this derived linear model. However, this model has characteristics which do not permit application of earlier results on admissibility of linear estimators by Cohen (1966) and Rao (1976). Olsen, Seely and Birkes (1976) provided results on admissibility and completeness among unbiased linear estimators in a very general setting which encompasses the peculiar structure of linear models for invariant quadratics. In models with two variance components, they described the essentially complete class of admissible unbiased invariant quadratic estimators of linear combinations of the variance components. These results when combined with existing results on unbiased linear estimation, fairly completed the story on unbiased invariant quadratic estimation of variance components.

Several authors have noted the existence of biased estimators of variance components which are better than the best unbiased invariant quadratic estimator (Hodges and Lehmann 1951; Klotz, Milton and Zacks 1969; LaMotte 1976). Assuming a normally-distributed observation vector, LaMotte (1979, 1980) showed that the derived linear model in invariant quadratics satisfies conditions established by Perlman (1972) and hence that no invariant quadratic estimator (except 0) is admissible among invariant quadratic estimators of its expected value. That is, no (nontrivial) unbiased invariant quadratic estimator is admissible among the class of invariant quadratic estimators (IQEs). Given an unbiased IQE, there exist positive scalar multiples which have uniformly less risk. Identifying a better estimator is easy, but identifying one which also is admissible is not easy.

Extending the work of Olsen, Seely and Birkes (1976) to include biased linear estimators, LaMotte (1982) established a characterization of admissible linear estimators in a general linear model. This result enables one to describe the class of admissible linear estimators in any linear model, and hence the admissible IQEs in a variance-component model. However, the resulting description often does not give a clear picture of the geometrical and topological structure of the admissible estimators. For example, the simple structure of the admissible linear estimators in the standard regression model, though deriveable from LaMotte's (1982) characterization, would be difficult to see, had Cohen (1966) not already described it.

Aside from their inadmissibility, unbiased IQEs of variance components may take negative values. LaMotte (1973) showed that nonnegative unbiased IQEs exist only for a restricted class of positive linear combinations of variance components. Pukelsheim (1981) characterized the linear combinations of variance components for which nonnegative IQEs exist. Hartung (1981) derived and examined estimators constructed by first restricting the class to nonnegative IQEs, then finding in that class one with minimum bias. Whether a nonnegative minimum bias estimator is better (in terms of risk) than an unbiased estimator is a question not considered in Hartung (1981).

This short survey of work on variance component estimation provides the background for the present paper and raises the following questions.

1. Which invariant quadratic estimators are admissible among IQEs ?
2. Which IQEs are better than a given unbiased IQE ?
3. For estimating a positive linear combination of variance components, which IQEs (if any) are nonnegative, admissible among IQEs, and better than a given unbiased IQE ?

Relations to unbiased estimators are important because there is a strong preference for unbiased estimators in applications. As a particular point in the third question, we shall see whether minimum bias estimators provide any improvement over unbiased estimators.

We shall investigate these questions in the balanced, one-way, random ANOVA model. This model is important in its own right in applications. However, it is particularly appealing here because it is the simplest and most tractable of variance component models. In particular, the IQEs can be reduced to an essentially complete class in two dimensions so that graphs can be drawn.

2. Preliminaries

In the balanced, random, one-way ANOVA model, it may be shown that the within-groups mean-square Q_1 and the between groups mean-square Q_2 have a completeness property: for any invariant quadratic estimator, there is a linear combination of Q_1 and Q_2 which is as good. For our purposes, therefore, it is sufficient to begin with the derived model for Q_1 and Q_2. Denoting the error variance component by σ_e^2, and the treatment effect variance component by σ_b^2, we have

$$E(Q_1) = \sigma_e^2, \quad E(Q_2) = \sigma_e^2 + n\sigma_b^2, \tag{2.1}$$

$$\mathrm{Var}(Q_1) = \frac{2(\sigma_e^2)^2}{m(n-1)}, \quad \mathrm{Var}(Q_2) = \frac{2(\sigma_e^2+n\sigma_b^2)^2}{m-1},$$

$$\mathrm{Cov}\ (Q_1,Q_2) = 0,$$

assuming that n observations of the response variable are taken under each of m treatment levels. Let $\gamma_1 = \sigma_e^2$, $\gamma_2 = \sigma_e^2+n\sigma_b^2$, $c_1=2/(m(n-1))$, $c_2 = 2/(m-1)$, so that, with $Q=(Q_1,Q_2)'$,

$$E(Q) = \gamma, \quad \mathrm{Var}(Q) = \mathrm{diag}(c_i\gamma_i^2,\ i = 1,2). \tag{2.2}$$

For purposes of comparison we shall consider two parameter sets:

$$\Omega = \{\gamma_1,\gamma_2: \gamma_1 \geq 0,\ \gamma_2 \geq 0\}, \tag{2.3}$$

$$\Omega_+ = \{\gamma_1,\gamma_2: \gamma_1 \geq 0,\ \gamma_2 \geq \gamma_1\}.$$

The set Ω_+ is the parameter set corresponding to the balanced one-way model; Ω is the parameter set which results by dropping the restriction $\gamma_2 \geq \gamma_1$.

We shall consider estimators $\ell_1 Q_1 + \ell_2 Q_2$ of $p_1\gamma_1 + p_2\gamma_2$. We shall use mean squared error (MSE) to compare estimators:

$$\begin{aligned} MSE_\ell(\gamma) &= E[(\ell'Q - p'\gamma)^2] \\ &= \ell'Var(Q)\ell + [(\ell-p)'E(Q)]^2 . \end{aligned} \tag{2.4}$$

We shall use standard definitions of the relations "as good as" and "better than" between estimators. All discussion here is restricted to linear estimators $\ell'Q$. Thus to say that $\ell'Q$ is admissible shall be taken to mean that there exists no $\ell_* \in R^2$ such that $\ell_*'Q$ is better than $\ell'Q$ in estimating $p'\gamma$.

3. The admissible estimators

Given $p = (p_1, p_2)'$, we shall use LaMotte's (1982) characterization to identify all the admissible estimators of $p'\gamma$ in $\mathscr{L} = R^2$. We shall begin with Ω as the parameter set.

Let

$$\begin{aligned} \mathscr{T} &= \{(Var(Q),\ E(Q)E(Q)') : \gamma\varepsilon\Omega\} \\ &= \{(diag(c_i\gamma_i^2),\ \gamma\gamma') : \gamma\varepsilon\Omega\}. \end{aligned} \tag{3.1}$$

Denote by $[\mathscr{T}]$ the minimal closed convex cone containing $\mathscr{T}$. Then

$$\begin{aligned} [\mathscr{T}] = \{(diag(c_iM_{ii}), M) : (M)_{ij} = M_{ij}, \\ M_{11} \geq 0,\ M_{12} \geq 0,\ M_{22} \geq 0,\ M_{12}^2 \leq M_{11}M_{22}\}. \end{aligned} \tag{3.2}$$

In the linear subspace $\mathscr{W}$ of 2x2 symmetric matrices, the set

$$\mathscr{M} = \{M: M_{11} \geq 0,\ M_{12} \geq 0,\ M_{22} \geq 0,\ M_{12}^2 \leq M_{11}M_{22}\}$$

is generated as the minimal closed convex cone containing the set $\mathscr{C}$ shown in Figure 1. $\mathscr{C}$ is a half-disk in the affine set generated by $\begin{pmatrix}1&0\\0&0\end{pmatrix}$, $\begin{pmatrix}0&0\\0&1\end{pmatrix}$ and $1/2\begin{pmatrix}1&1\\1&1\end{pmatrix}$ (the plane of Figure 1), with the line segment joining $\begin{pmatrix}1&0\\0&0\end{pmatrix}$ and $\begin{pmatrix}0&0\\0&1\end{pmatrix}$ as its base. Ω is generated by the

Figure 1

2×2 Symmetric matrices M in the flat $\mathrm{tr}(M)=1$

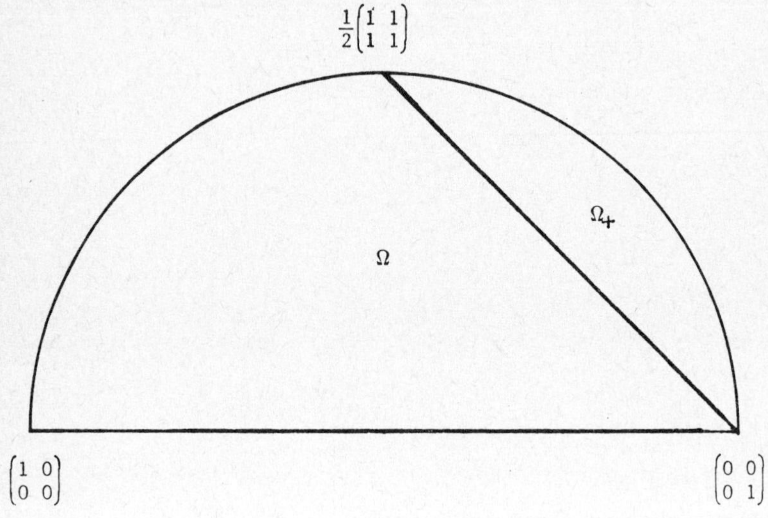

half-disk, while Ω_+ is generated by the part of the disk on and above the line segment joining $\begin{pmatrix}0 & 0\\ 0 & 1\end{pmatrix}$ and $1/2\begin{pmatrix}1 & 1\\ 1 & 1\end{pmatrix}$.

In the developments to follow it is convenient to use $\mathscr{M}$ in place of $[\mathscr{T}]$. There is no loss of generality in doing so, due to Theorem 9.1 in Rockafellar (1970). Extended to $\mathscr{W}$,

$$\mathrm{MSE}_\ell(M) = \ell'[\mathrm{diag}(c_i M_{ii})]\ell + (\ell-p)'M(\ell-p).$$

If $\ell = (\ell_1, \ell_2)'$ is an admissible estimator of $p'\gamma$, it must be best at some point M in $\mathscr{C}$. That is, ℓ must satisfy

$$(1+c_1)M_{11}\ell_1 + M_{12}\ell_2 = p_1 M_{11} + p_2 M_{12} \tag{3.3}$$

$$M_{12}\ell_1 + (1+c_2)M_{22}\ell_2 = p_1 M_{12} + p_2 M_{22}$$

for some point in $\mathscr{C}$. If $M_{11} > 0$ and $M_{22} > 0$ then (3.3) has a unique solution and that solution is an admissible estimator of $p'\gamma$. If $M_{11} = 0$ or $M_{22} = 0$ then (3.3) has many solutions: some are admissible and some are not. We shall deal with the two singular points first, then come back to describe the set of unique solutions to (3.3).

Consider those estimators which are best at $M_{11}=0$ in $\mathscr{C}$. With $M_{11}=0$, (3.3) becomes

$$(1+c_2)\ell_2=p_2, \quad \text{or} \quad \ell_2=\frac{p_2}{1+c_2}\,. \tag{3.4}$$

In order that such an estimator be admissible it must be admissible among the set of all such estimators, namely $\mathscr{L}_1=\{\ell\in R^2 : \ell_2=p_2/(1+c_2)\}$. It may be seen that, at any point in $\mathscr{W}$ with $M_{11}=0$ and $p_2M_{12}=0$ every estimator in $\mathscr{L}_1$ is best among $\mathscr{L}_1$. Such points are called trivial points for $\mathscr{L}_1$. Denote by $\mathscr{S}_1$ the set of all such trivial points in $\mathscr{W}$. If $p_2\neq 0$, $\mathscr{S}_1$ is the linear subspace generated by $\binom{0\ 0}{0\ 1}$. If $p_2=0$, $\mathscr{S}_1$ is the linear subspace generated by $\binom{0\ 0}{0\ 1}$ and $\binom{0\ 1}{1\ 0}$. In order that $\ell\in\mathscr{L}_1$ be admissible among $\mathscr{L}_1$, ℓ must be best among $\mathscr{L}_1$ at a nontrivial point in $[\mathscr{M}+\mathscr{S}_1]$. If $p_2\neq 0$, $[\mathscr{M}+\mathscr{S}_1]$ intersects the plane of Figure 1 in the set of points above and to the left of $\binom{0\ 0}{0\ 1}$. If $p_2=0$, $[\mathscr{M}+\mathscr{S}_1]$ intersects the plane of Figure 1 in the set of points to the left of $\binom{0\ 0}{0\ 1}$. To be best among $\mathscr{L}_1$ at a point $M\in[\mathscr{M}+\mathscr{S}_1]$, ℓ_1 must satisfy

$$(1\ 0)[\mathrm{diag}(c_iM_{ii})+M][\ell_1,p_2/(1+c_2)]'=(1\ 0)Mp \tag{3.5}$$

or

$$(1+c_1)M_{11}\ell_1=p_1M_{11}+c_2p_2M_{12}/(1+c_2)\,. \tag{3.6}$$

The nontrivial points in $[\mathscr{M}+\mathscr{S}_1]$ which permit best estimators (points with $M_{11}=0$ and $M_{12}\neq 0$ do not unless $p_2=0$, in which case such points are trivial points for $\mathscr{L}_1$) all have $M_{11}>0$. Thus

$$\ell_1=\frac{p_1}{1+c_1}+\frac{M_{12}}{M_{11}}\,\frac{c_2p_2}{(1+c_1)(1+c_2)}\,. \tag{3.7}$$

If $p_2\neq 0$, $M_{12}/M_{11}\geq 0$ in $[\mathscr{M}+\mathscr{S}_1]$; if $p_2=0$, the second term in (3.7) is zero. Thus we may conclude that the admissible estimators which are in $\mathscr{L}_1$ consist of all ℓ such that

$$\ell_2=\frac{p_2}{1+c_2},\quad \ell_1=\frac{p_1}{1+c_1}+\alpha p_2,\quad \alpha\geq 0. \tag{3.8}$$

All these estimators are admissible among $\mathscr{L}$ on Ω and these are the only estimators in $\mathscr{L}_1$ which are admissible among $\mathscr{L}$ on Ω. By noting that Ω_+ is characterized by $M_{12}/M_{11}\geq 1$, the estimators in $\mathscr{L}_1$ which are admissible among $\mathscr{L}$ on Ω_+ are all ℓ such that

$$\ell_2 = \frac{p_2}{1+c_2}, \quad \ell_1 = \frac{p_1}{1+c_1} + \alpha\frac{c_2p_2}{(1+c_1)(1+c_2)}, \quad \alpha \geq 1. \tag{3.9}$$

A symmetrical analysis yields the set of all estimators which are best at $\binom{1\ 0}{0\ 0}$ and admissible among $\mathscr{L}$ on Ω:

$$\ell_1 = \frac{p_1}{1+c_1}, \quad \ell_2 = \frac{p_2}{1+c_1} + \alpha p_1, \ \alpha \geq 0. \tag{3.10}$$

That was the hard part. All that remains is to describe the set of unique solutions of (3.3) for $M_{11} > 0$ and $M_{22} > 0$. Let $k_1 = M_{12}/M_{11}$ and $k_2 = M_{12}/M_{22}$. It is helpful to visualize k_1 and k_2 in terms of Figure 1. Draw straight lines from $\binom{1\ 0}{0\ 0}$ and $\binom{0\ 0}{0\ 1}$ intersecting at a point in $\mathscr{C}$: k_1 is the tangent of the angle the line out of $\binom{0\ 0}{0\ 1}$ makes with the base line, and k_2 is the tangent of the angle the other line makes with the base line. We have $k_1k_2 \leq 1$. The only estimator best in the relative interior of the base line is $\ell_1 = p_1/(1+c_1)$, $\ell_2 = p_2/(1+c_2)$. With $k_1 > 0$ and $k_2 > 0$, (3.3) becomes

$$(1+c_1)\ell_1 + k_1\ell_2 = p_1+p_2k_1 \tag{3.11}$$

$$k_2\ell_1 + (1+c_2)\ell_2 = p_1k_2 + p_2.$$

(It is worth noting that (3.11) has a unique solution for any nonnegative values of k_1 and k_2. With $k_1 = 0$, (3.11) generates the admissible estimators which are best at $\binom{0\ 0}{0\ 1}$; with $k_2 = 0$, it generates the admissible estimators among those which are best at $\binom{1\ 0}{0\ 0}$; and, with $k_1 = k_2 = 0$, it generates the one estimator which is best at points in the relative interior of the base line.)

With k_1 fixed, it may be shown that the solutions to (3.11), with k_2 running from 0 to $1/k_1$, form the line segment joining the solution at $k_2 = 0$ and the solution at $k_2 = 1/k_1$. The endpoints of this line segment are $\ell_0(k_1)$ and $\ell_*(k_1)$ with

$$\ell_{10} = \frac{p_1}{1+c_1} + \frac{k_1c_2p_2}{(1+c_1)(1+c_2)}, \quad \ell_{20} = \frac{p_2}{1+c_2}; \tag{3.12}$$

$$\ell_{1*} = \frac{c_2(p_1+p_2k_1)}{(1+c_1)(1+c_2)-1}, \quad \ell_{2*} = \frac{c_1(p_1+p_2k_1)}{k_1[(1+c_1)(1+c_2)-1]} .$$

In order to plot the admissible estimators, we have only to plot these line segments. It is particularly worth noting that all these line segments, if extended, would pass through the point $(p_1/(1+c_2), p_2)$. Several plots are shown in Section 5.

4. The estimators better than a linear unbiased estimator

The proof of Proposition 3.6 in Olsen, Seely and Birkes (1976) indicates how to identify all the estimators as good as a given estimator. Let $h \in R^2$ and $\gamma \geq 0$. Given ℓ_*, $\ell_* - \gamma h$ is as good as ℓ_* iff

$$0 \geq MSE_{\ell}(M) - MSE_{\ell*}(M) \tag{4.1}$$

$$= \gamma^2 h'S_{11}h - 2\gamma h'F_*(M) \quad \text{for all} \quad M\varepsilon \mathscr{M},$$

where $S_{11} = \operatorname{diag}(c_iM_{ii}) + M$ and $F_*(M) = S_{11}\ell_* - Mp$. If $\gamma > 0$ then in order that (4.1) be satisfied it is necessary that $h'F_*(M) \geq 0$ for all $M\varepsilon \mathscr{M}$. (Note that no such $h \neq 0$ exists if $F_*(M) = 0$ for some M in the relative interior of $\mathscr{M}$.) Given such an h, then $\ell_* - \gamma h$ is as good as ℓ_* for $0 \leq \gamma \leq m_h$, where

$$m_h = \bigwedge_{M\varepsilon \mathscr{M}} \{2h'F_*(M)/h'S_{11}h\}. \tag{4.2}$$

With a little more argument we may conclude that ℓ is as good as ℓ_* iff $\ell = \ell_* - \gamma h$, $h'F_*(M) \geq 0$ for all $M \in \mathscr{M}$, and $0 \leq \gamma \leq m_h$.

In the present model, for unbiased estimators $\ell_* = p$ and the ratio in (4.2) is

$$g(M) = h'F_*(M)/h'S_{11}h \tag{4.3}$$

$$= \frac{c_1p_1h_1M_{11}+c_2p_2h_2M_{22}}{(1+c_1)h_1^2M_{11}+(1+c_2)h_2^2M_{22}+2h_1h_2M_{12}} .$$

To find m_h we may restrict M to $\mathscr{C}$ (Figure 1). With the numerator nonnegative, $g(M)$ is monotone in M_{12}. Thus if $h_1h_2 > 0$, g takes its minimum on the upper boundary of $\mathscr{C}$, namely, where $M_{12}^2 = M_{11}M_{22}$. If $h_1h_2 \le 0$, g takes its minimum on the lower boundary. Corresponding to Ω the lower boundary is the base line, while in Ω_+ it is the line segment joining $1/2\begin{pmatrix}1 & 1\\ 1 & 1\end{pmatrix}$ and $\begin{pmatrix}0 & 0\\ 0 & 1\end{pmatrix}$.

Following these facts, the following descriptions may be found with respect to the unrestricted parameter set Ω. If $p_1p_2 \le 0$, the estimators as good as $p'Q$ fill a rectangle with corners at

$$((1-c_1)p_1/(1+c_1), p_2), \qquad (p_1, p_2),$$

$$((1-c_1)p_1/(1+c_1),\ (1-c_2)p_2/(1+c_2)),\ (p_1, (1-c_2)p_2/(1+c_2)).$$

If $p_1 > 0$ and $p_2 > 0$, the estimators as good as $p'Q$ form a set with $\ell_1 \le p_1$, $\ell_2 \le p_2$ and bounded on the lower left by the curve $\{p - m_h h\colon h_1 \ge 0,\ h_2 \ge 0\}$ with

$$m_h = 4\{(1+c_1)h_1/c_1p_1 + (1+c_2)h_2/c_2p_2 \tag{4.4}$$

$$+ \sqrt{[(1+c_1)h_1/c_1p_1 - (1+c_2)h_2/c_2p_2]^2 + 4h_1h_2/(c_1p_1c_2p_2)}\}^{-1}.$$

With respect to Ω_+ we may obtain the following description. The "better" estimators are bounded above and below by $\ell_2 = p_2$ and $\ell_2 = (1-c_2)p_2/(1+c_2)$. If $p_1p_2 \le 0$, the rest of the boundary is formed by the ellipse $\{\ell\colon MSE_\ell(M) = MSE_p(M)$ with $M = \begin{pmatrix}1 & 1\\ 1 & 1\end{pmatrix}\}$. This part of the boundary may be expressed as $\{p - m_h h\colon p_2h_2 \ge 0,\ c_1p_1h_1 + c_2p_2h_2 \ge 0\}$ with

$$m_h = \frac{2(c_1p_1h_1 + c_2p_2h_2)}{(1+c_1)h_1^2 + (1+c_2)h_2^2 + 2h_1h_2}. \tag{4.5}$$

If $p_1 > 0$ and $p_2 > 0$, the rest of the boundary may be expressed in the same way, but with m_h as in (4.4) if

$$h_2[c_2p_2 - c_1(1+c_2)p_1] < h_1[c_1p_1 - (1+c_1)c_2p_2]. \tag{4.6}$$

5. Graphs

This section presents graphs of the admissible IQEs and of the IQEs which are better than the best unbiased IQE. All the graphs here are for the case $m=5$, $n=3$, corresponding to $c_1=.2$, $c_2=.5$. However, it may be shown that the basic characteristics of the graphs are essentially unchanged for other values of c_1 and c_2.

Figure 2

Admissible IQEs of $p_1\gamma_1+p_2\gamma_2$ and IQEs as good as (p_1,p_2)

$p_1=1$, $p_2=0$

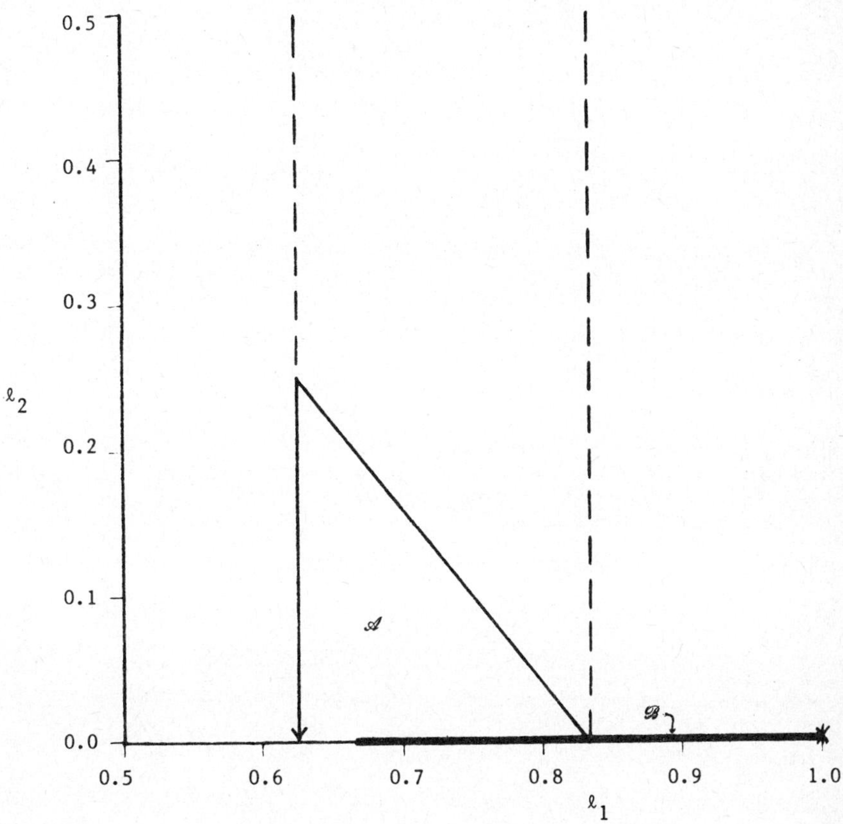

Legend. Solid lines refer to Ω_+; dashed lines show extension to Ω.

* - (p_1,p_2); $\mathscr{A}$-admissible; $\mathscr{B}$ - as good as (p_1,p_2).

Figure 2 shows relations among IQEs of σ_e^2, i.e., $p_1 = 1$, $p_2 = 0$. The best unbiased IQEs has $\ell_1 = 1$, $\ell_2 = 0$. In the unrestricted parameter set Ω, the admissible IQEs are

$$\{(\ell_1,\ell_2): c_2/(c_1+c_2+c_1c_2) \le \ell_1 \le (1+c_1)^{-1},\ \ell_2 > 0\} \cup \{((1+c_1)^{-1},\ 0)\}.$$

Figure 3

Admissible IQEs of $p_1\gamma_1 + p_2\gamma_2$ and IQEs as good as (p_1,p_2)

$p_1 = 0$, $p_2 = 1$

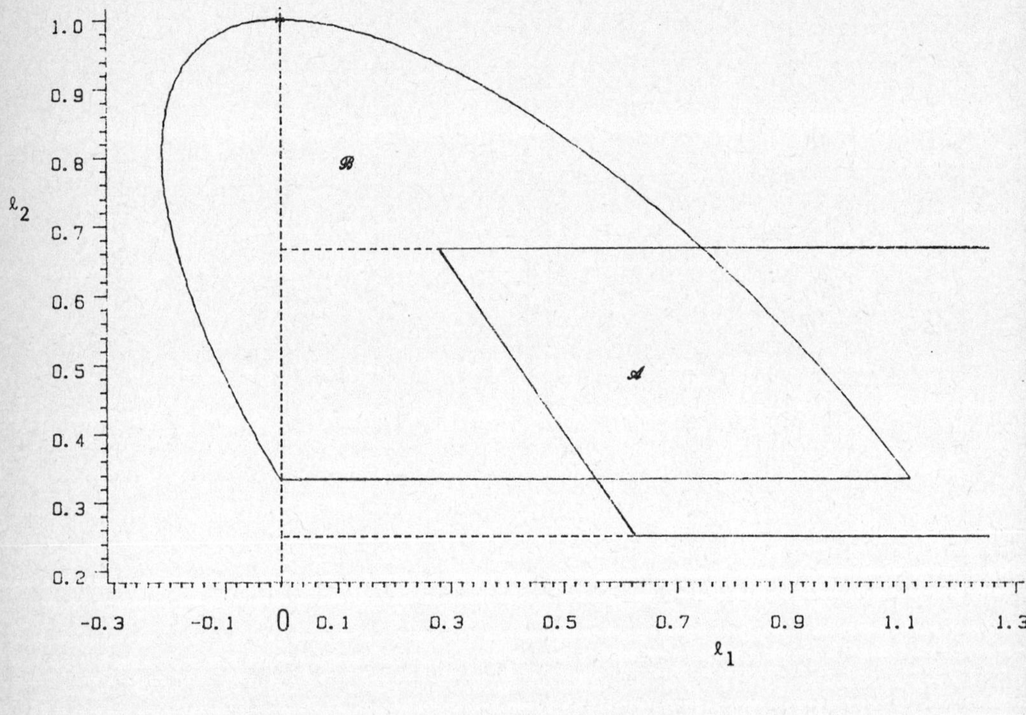

Legend. Solid lines refer to Ω_+; dashed lines show extension to Ω.

* - (p_1,p_2)

$\mathcal{A}$ - admissible

$\mathcal{B}$ - as good as (p_1,p_2)

Figure 4

Admissible IQEs of $p_1\gamma_1 + p_2\gamma_2$ and IQEs as good as (p_1,p_2)

$p_1 = -1,\ p_2 = 1$

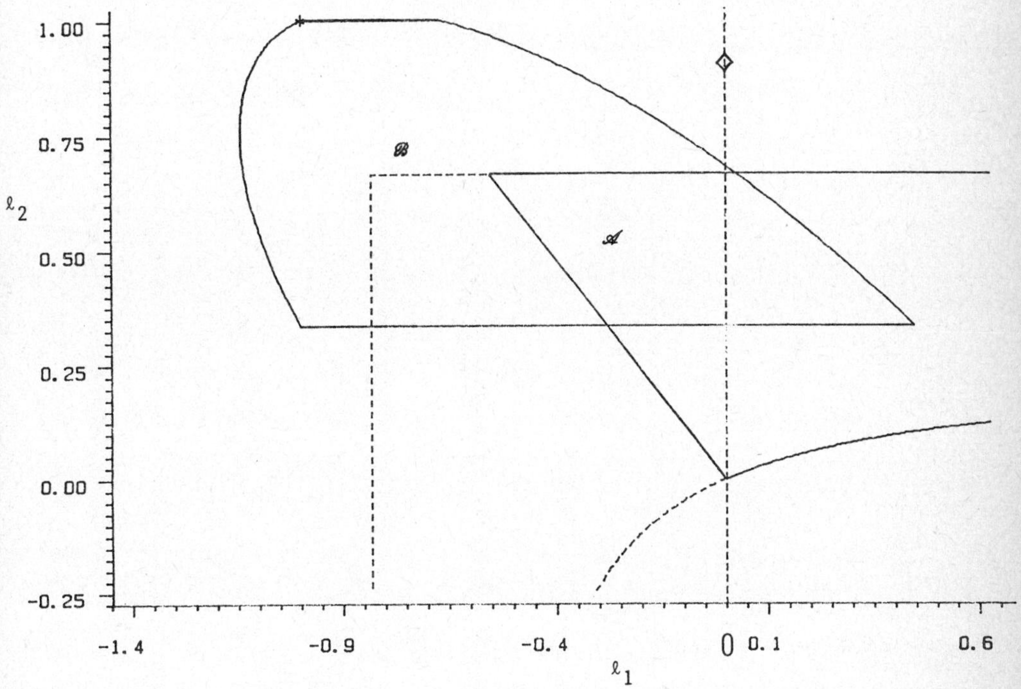

Legend. Solid lines refer to Ω_+; dashed lines show extension to Ω.

$*$ - (p_1,p_2)

◇ - the minimum bias nonnegative IQE

𝒜 - admissible

ℬ - as good as (p_1,p_2)

In Ω_+, the admissible IQEs are the same set up to the line segment joining $((1+c_1)^{-1},0)$ and $(c_2/(c_1+c_2+c_1c_2),\ c_1/(c_1+c_2+c_1c_2))$. With respect to Ω the admissible IQEs form an unbounded convex set, while with respect Ω_+ the set is bounded. In neither case is the set of admissible IQEs closed. The IQEs which are as good as best unbiased IQE are those with $\ell_2 = 0$ and $(1-c_1)/(1+c_1) \leq \ell_1 \leq 1$. Of these, only one is admissible, and it is the scalar multiple $(1+c_1)^{-1}$ of the best unbiased IQE, which is the Hodges-Lehmann (1951) estimator. Figure 3 shows admissible and "better" IQEs of $\gamma_2 = \sigma_e^2 + n\sigma_b^2$, $p_1 = 0$, $p_2 = 1$. With respect to Ω, the picture would be similar to Figure 2, so only relations with respect to Ω_+ are shown in Figure 3. Here the set of admissible IQEs is closed and convex but not bounded. No IQE with $\ell_1 = 0$ is admissible; in particular, $\ell_1 = 0$, $\ell_2 = (1+c_2)^{-1}$ is not admissible. However, some such scalar multiples of the best

Figure 5

Admissible IQEs of $p_1\gamma_1 + p_2\gamma_2$ and IQEs as good as (p_1,p_2)

$p_1 = 1$, $p_2 = 1$

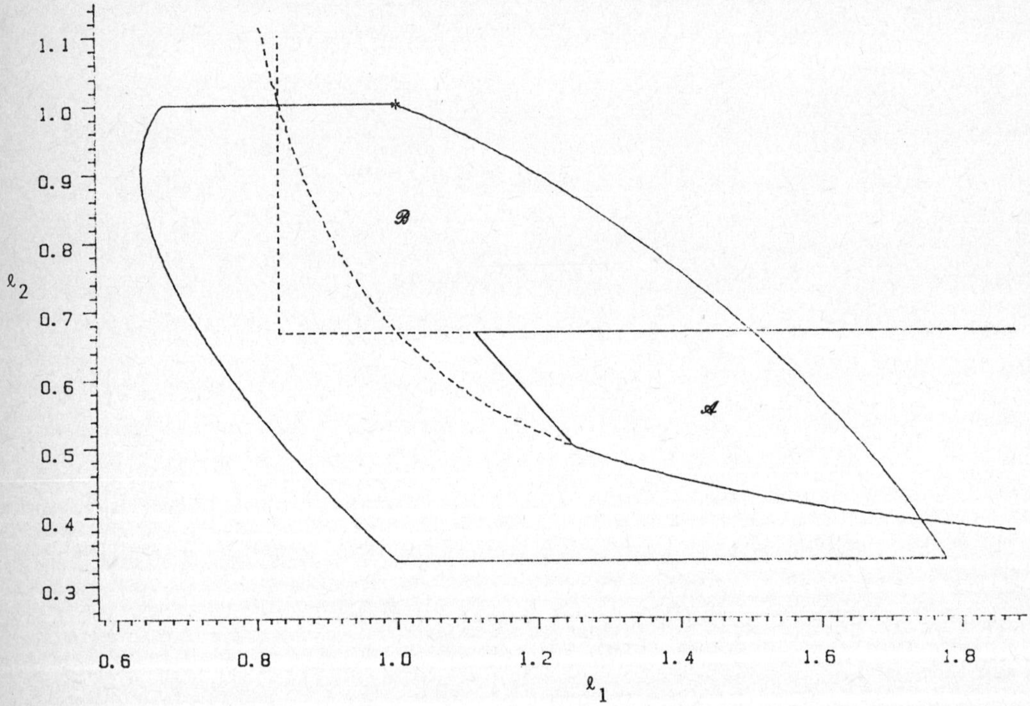

Legend. Solid lines refer to Ω_+; dashed lines show extension to Ω. * - (p_1,p_2); $\mathscr{A}$ -admissible; $\mathscr{B}$- as good as (p_1,p_2).

unbiased IQE ($\ell_1 = 0$, $\ell_2 = 1$) are better than $\ell_1 = 0$, $\ell_2 = 1$, and some estimators with $\ell_1 < 0$ are better than the best unbiased IQE.

In the remaining two cases both p_1 and p_2 are nonzero. With respect to Ω, the admissible estimators form the set of points between an "L" with vertex at $(p_1/(1+c_1), p_2/(1+c_2))$ and the $\ell_*(k_1)$ curve (3.12). If $p_1p_2 < 0$, the $\ell_*(k_1)$ curve does not intersect the L; if $p_1p_2 > 0$, it intersects each leg of the L once.

Figure 4 illustrates the case $p_1 = -1$, $p_2 = 1$, where $\gamma_2 - \gamma_1 = n\sigma_b^2$ is being estimated. Only relations with respect to Ω_+ are shown. The function being estimated is nonnegative but the best unbiased IQE is not. The set of admissible IQEs is closed but it is neither bounded nor convex. No scalar multiple of the best unbiased IQE is admissible. There are some nonnegative IQEs which are admissible and better than the best unbiased IQE. The minimum bias nonnegative IQE is not admissible and not better than the best unbiased IQE. The trivial IQE $\ell_1 = \ell_2 = 0$ is admissible among IQEs but it is not as good as the best unbiased IQE.

When both p_1 and p_2 are positive, the set of admissible IQEs with respect to Ω (shown by light lines in Figure 5) appears in three segments. The two segments along the legs of the L are individually closed and convex, while the segment inside the vertex of the L is closed but not convex. For $p_1 = p_2 = 1$, Figure 5 shows the IQEs which are admissible with respect to Ω_+ and the IQEs which are as good as the best unbiased IQE. With respect to Ω_+, the admissible IQEs form a closed convex set in Figure 5. (With $p_1 > 0$, $p_2 > 0$, it may be shown that this property holds if $c_1p_1(1+c_2)/c_2p_2 \leq 1$.) All the admissible IQEs are nonnegative, as are all the IQEs which are as good as the best unbiased IQE.

6. More on nonnegativity

In examining Figure 4 (for estimating $n\sigma_b^2$) we see that there exist nonnegative IQEs which are admissible, some of which are better than the best unbiased IQE. Throughout this discussion, admissibility has been defined with relation to the class of estimators $\{\ell'Q:\ell\in R^2\}$. Suppose instead that we impose the nonnegativity constraint first so that the class of estimators is $\{\ell'Q, \ell_1 \geq 0, \ell_2 \geq 0\}$.

First let us find the estimators which are admissible among this restricted class. Clearly all the estimators which are admissible in the full class ($\ell \varepsilon R^2$) and nonnegative are admissible in the restricted class. Given $\ell \geq 0$ which is not admissible in the full class, ℓ is admissible in the restricted class only if no nonnegative estimator is better than ℓ. However, the set of estimators in R^2 as good as ℓ is closed and convex. Thus if $\ell > 0$ (i.e., $\ell_1 > 0$ and $\ell_2 > 0$) and ℓ is not admissible in the full class then ℓ is not admissible in the restricted class. In the present model it may be demonstrated that if ℓ is on the boundary of the restricted class (i.e., $\ell_1 = 0$ or $\ell_2 = 0$) and not admissible in the full class then there is a better nonnegative estimator and hence ℓ is not admissible in the restricted class. Thus the set of admissible estimators in the restricted class consists of the estimators which are nonnegative and admissible in the full class.

Previous papers have described the admissible estimators by way of best estimators when the class of estimators forms an affine set. It is interesting to note that in this example the same connection holds when the class of estimator is not an affine set. To see that the admissible estimators in the restricted class are those which are best among the restricted class, it is necessary to note that, for $\gamma \varepsilon \Omega_+$ for which the "unrestricted" best estimator has $\ell_1 < 0$, the "restricted" best estimator has $\ell_1 = 0$ and $0 \leq \ell_2 \leq 1/(1+c_2)$. However, the example discussed here has such special characteristics that it is not clear to what extent, if any, these results might be generalized.

References

[1] Cohen, A. (1966). All admissible linear estimates of the mean vector. Annals of Mathematical Statistics, 37, 458-463.

[2] Hartung, J. (1981). Nonnegative minimum bias invariant estimation in variance component models. Annals of Statistics, 9, 278-292.

[3] Hodges, J.L. and Lehmann, E.L. (1951). Some applications of the Cramèr-Rao inequality. Proceedings of the Second Berkeley Symposium on Mathematical Statistics and Probability, 1.

[4] Klotz, J.H., Milton, R.C. and Zacks, S. (1969). Mean square efficiency of estimators of variance components. Journal of the American Statistical Association, 64 , 1383-1402.

[5] LaMotte, L.R. (1976). Invariant quadratic estimators in the random, one-way ANOVA model. Biometrics, 32,793-804.

[6] LaMotte, L.R. (1979). On admissible invariant quadratic estimators of variance components. Variance Components and Animal Breeding: Proceedings of a Conference in Honor of C.R.Henderson, Cornell University, 181-188.

[7] LaMotte, L.R. (1980). Some results on biased linear estimation applied to variance component estimation. Mathematical Statistics and Probability Theory: Proceedings, Sixth International Conference, Conference, Wisła, Poland, Springer, 266-274.

[8] LaMotte, L.R. (1982). Admissibility in linear estimation. Annals of Statistics, 10, 245-255.

[9] Olsen, A., Seely, J. and Birkes, D.(1976). Invariant quadratic estimation for two variance components. Annals of Statistics, 4, 878-890.

[10] Perlman, M.D. (1972). Reduced mean square error estimation for several parameters. Sankhya B, 89-92.

[11] Pukelsheim, F. (1976). Estimating variance components in linear models. Journal of Multivariate Analysis 6, 626-629.

[12] Pukelsheim, F. (1981). On the existence of unbiased nonnegative estimates of variance-covariance components. Annals of Statistics, 9, 293-299.

[13] Rao, C.R.(1976). Estimation of parameters in a linear model. Annals of Statistics, 4, 1023-1037.

[14] Rockafellar, R.T.(1970). Convex Analysis. Princeton University Press.

ON INFERENCE IN A GENERAL LINEAR MODEL WITH AN INCORRECT DISPERSION MATRIX

T.Mathew
Indian Statistical Institute
New Delhi, India

Summary: The problem of best linear unbiased estimation and testing of hypothesis in a general linear model with an incorrectly specified dispersion matrix is investigated. Previously known results regarding the equality of best linear unbiased estimators and of test statistics under two linear models which differ in their dispersion matrices follow as special cases.

1. Introduction

Consider the Gauss-Markov model $(Y,X\beta,\sigma^2V)$, where Y is a random R^n-vector, X is a known $n\times m$ matrix, V is an $n\times n$ nonnegative definite (nnd) matrix, β is an R^m-vector of unkown parameters and σ^2 is an unknown positive parameter. A problem that has received considerable attention is the characterization of nnd matrices V such that the best linear unbiased estimators (BLUE's) computed under a given model $(Y,X\beta,\sigma^2V_1)$ continue to be corresponding BLUE's under the alternative model $(Y,X\beta,\sigma^2V)$ also (see Rao (1967, 1968, 1971), Zyskind (1967, 1969) Watson (1967), Mitra and Moore (1973, 1976), Haberman (1975), Baksalary and Kala (1978, 1980) and Mathew (1983)). Here V_1 is a given nnd matrix. In particular when $V_1 = I$ we get the conditions under which the simple least squares estimator (SLSE) of $X\beta$ is its BLUE under $(Y,X\beta,\sigma^2V)$. When these conditions are satisfied, the SLSE and BLUE of $X\beta$ coincide with probability one, i.e. they coincide for all $Y \in \underline{R}(X:V)$, the range space of the matrix $(X:V)$. However, even though the SLSE and BLUE can not coincide for all $Y \in \underline{R}(X:V)$, there may exist a subspace of $\underline{R}(X:V)$ such that the BLUE and SLSE coincide whenever Y is in this subspace. Thus if the observed value of Y lies in this subspace, the computationally simpler SLSE could be used to estimate $X\beta$ under $(Y,X\beta,\sigma^2V)$. This problem has been considered by Krämer (1980) when X is of full rank and V is positive definite (pd). Given such an X and V Krämer has characterized the subspace of R^n such that the SLSE and BLUE of $X\beta$ coincide whenever Y is in this subspace. In the next section this problem is taken up in a more general framework. We consider the models $(Y,X\beta,\sigma^2V_1)$ and $(Y,X\beta,\sigma^2V)$ where the matrices V_1 and V could be singular. Hence BLUE of an estimable vector parametric function $A\beta$ may not have a unique linear representation under $(Y,X\beta,\sigma^2V_1)$ as observed by Mitra and Moore (1973). Thus we are lead to the problem of characterizing the subspace of $\underline{R}(X:V)$ such that whenever Y belongs to this subspace, a given linear representation (some linear representation) every linear representation of BLUE of $A\beta$ under $(Y,X\beta,\sigma^2V_1)$ coincides with its BLUE under $(Y,X\beta,\sigma^2V)$. The conditions under which the BLUE's coincide with probability one follow as a special case. Thus, the results we have obtained generalize the already known results regarding the equality of BLUE's in two models which differ in the dispersion matrices.

Characterization of nnd matrices V for which the likelihood ratio test statistic (LRT Statistic) L_V to test a hypothesis $H_o:A\beta = 0$ under $(Y,X\beta,\sigma^2V)$ coincides with the corresponding LRT statistic L under the model $(Y,X\beta,\sigma^2I)$ has been considered by Khatri (1981) and Mathew and Bhimasankaram (1983a). The latter authors also derived the conditions for $L_V-L \geq 0$ (or $L_V-L \leq 0$). This will guarantee that the rejection (or acceptance) of H_o under $(Y,X\beta,\sigma^2I)$ will imply its rejection (respectively acceptance) under $(Y,X\beta,\sigma^2V)$ also, even though L_V and L do not coincide. In Section 3 we consider the more general problem of obtaining conditions under which $L_V-L_{V_1} \geq 0$ or $L_V-L_{V_1} \leq 0$. The equality of L_V and L_{V_1} has been discussed in Mathew and Bhimasankaram (1983b).

The following notations are used in the paper. For a matrix A, $\underline{R}(A)$, $\underline{N}(A)$, A^- and $r(A)$, respectively, denote the range space, the null space, a g-inverse and the rank of A. $A^\perp$ denotes a matrix of maximum rank satisfying $A'A^\perp = 0$. P_A denotes the expression $A(A'A)^-A'$.

2. BLU estimation with an incorrect dispersion matrix.

Consider the models $(Y,X\beta,\sigma^2V_1)$ and $(Y,X\beta,\sigma^2V)$ and let G_1 and G

denote g-inverses of V_1+XX' and $V+XX'$, respectively. If $A\beta$ is an estimable vector parametric function, its BLUE under $(Y,X\beta,\sigma^2V)$ has linear representation $A(X'GX)^-X'GY$. Let $X_o = X(I-A^-A)$. Let $Z_o = X_o^\perp$ and $Z = X^\perp$. The following lemma is given in Mathew and Bhimasankaram (1983a).

Lemma 1. Given matrices A and X such that $R(A') \subset R(X')$, let $X_o = X(I-A^-A)$ and $Z_o = X_o^\perp$. Then $\underline{R}(A') = \underline{R}(X'Z_o)$.

We now prove the following result.

Theorem 1. Let G_1 be a given g-inverse of V_1+XX' and let W_1 be such that $\underline{R}(W_1) = \underline{N}(X'G_1)$. Then, under the model $(Y,X\beta,\sigma^2V)$, $A(X'G_1X)^-X'G_1Y = A(X'GX)^-X'GY$ iff $Y \in \underline{R}(X) \oplus \underline{R}(VZ) \cap \underline{R}(X_o:W_1)$.

Proof. It is easy to show that $\underline{R}(X:V) = \underline{R}(X) \oplus \underline{R}(VZ)$. Hence, since $Y \in \underline{R}(X:V)$ with probability one, we can write $Y = X\lambda_1+VZ\lambda_2$ for some vectors λ_1 and λ_2. Then $A(X'G_1X)^-X'G_1Y = A(X'GX)^-X'GY$ iff

$$Z_o'X(X'G_1X)^-X'G_1(X\lambda_1+VZ\lambda_2) = Z_o'X(X'GX)^-X'G(X\lambda_1+VZ\lambda_2)$$

(using Lemma 1) or, equivalently,

$$Z_o'X(X'G_1X)^-X'G_1VZ\lambda_2 = 0 \qquad (1)$$

using the fact that $X'GVZ = 0$. Since $Z_o = X_o^\perp$, we see that (1) holds iff $X(X'G_1X)^-X'G_1(VZ\lambda_2-X_o\lambda_3) = 0$ for some λ_3, or equivalently, $VZ\lambda_2 - X_o\lambda_3 \in \underline{N}(X'G_1) = \underline{R}(W_1)$. Hence $VZ\lambda_2 \in \underline{R}(X_o:W_1)$ and the proof of Theorem 1 is complete.

Corollary 1. Consider the set up described in Theorem 1.

(a) The linear representation $A(X'G_1X)^-X'G_1Y$ of BLUE of $A\beta$ under $(Y,X\beta,\sigma^2V_1)$ continues to be its BLUE under $(Y,X\beta,\sigma^2V)$ iff $\underline{R}(VZ) \subset \underline{R}(X_o:W_1)$ or, equivalently, $V = XD_1X' + W_1D_2W_1' + X_oD_3W_1' + W_1D_3'X_o'$, where D_1, D_2, D_3 are arbitrary matrices subject to the condition that V is nnd.

(b) The linear representation $X(X'G_1X)^-X'G_1Y$ of BLUE of $X\beta$ under $(Y,X\beta,\sigma^2V_1)$ continues to be its BLUE under $(Y,X\beta,\sigma^2V)$ iff $\underline{R}(VZ) \subset \underline{R}(W_1)$ or, equivalently, $V = XD_1X'+W_1D_2W_1'$, where D_1 and D_2 are arbitrary matrices subject to the condition that V is nnd.

Proof. From Theorem 1, we see that $A(X'G_1X)^-X'G_1Y = A(X'GX)^-X'GY$ for all $Y \in \underline{R}(X:V) = \underline{R}(X:VZ)$ iff $\underline{R}(VZ) \subset \underline{R}(X_o:W_1)$. Since $\underline{R}(X) \oplus \underline{R}(W_1) = R^n$ (Lemma 2.2 in Mitra and Moore (1973)) we can write $V = XD_1X' + W_1D_2W_1' + XD_4W_1' + W_1D_4'X'$ for some D_1, D_2, D_4. Hence $\underline{R}(VZ) \subset \underline{R}(X_o:W_1) \Leftrightarrow W_1D_2W_1'Z + XD_4W_1'Z = X_oK_1 + W_1K_2$ for some K_1 and K_2 or, equivalently, $XD_4W_1'Z = X_oK_1$ (using $\underline{R}(X) \cap \underline{R}(W_1) = \{0\}$). Since $\underline{R}(G_1'X) \oplus \underline{R}(Z) = R^n$, (Lemma 2.1 in Mitra and Moore (1973)) and $X'G_1W_1 = 0$, we see that the above condition is equivalent to $XD_4W_1' = X_oD_3W_1'$ for some D_3. This proves part (a).

Part (b) follows from part (a), once it is observed that $X_o = 0$ when $A = X$.

From Theorem 1 it is also clear that some linear representation of BLUE of $A\beta$ under $(Y, X\beta, \sigma^2V_1)$ coincides with $A(X'GX)^-X'GY$ iff Y belongs to the set $U_{G_1}\{\underline{R}(X) \oplus \underline{R}(X_o:W_1)\}$ (union taken over all g-inverses G_1 of V_1+XX').

Theorem 2. Under the model $(Y, X\beta, \sigma^2V)$, $A(X'G_1X)^-X'G_1Y = A(X'GX)^-X'GY$ for all g-inverses G_1 of V_1+XX' iff $Y \in \underline{R}(X) \oplus \underline{R}(VZ) \cap \underline{R}(X_o:V_1Z)$.

Proof. In view of Theorem 1 we need a representation of the subspace $\underline{R}(W_1^*) = \cap \underline{R}(W_1)$, where the intersection is taken over all $G_1 = (V_1+XX')^-$. But $X'G_1 1 = 0$ for every G_1 or, equivalently, $X'G_1(X\lambda_1 + V_1Z\lambda_2) = 0$ holds if and only if $X'G_1X\lambda_1 = 0$. This simplifies to $X\lambda_1 = 0$, thus showing that a choice of W_1^* is V_1Z, which concludes the proof.

Corollary 2. Consider the set-up described in Theorem 2.

(a) Every linear representation of BLUE of $A\beta$ under $(Y, X\beta, \sigma^2V_1)$ continues to be its BLUE under $(Y, X\beta, \sigma^2V)$ iff $\underline{R}(VZ) \subset \underline{R}(X_o:V_1Z)$ or, equivalently,

$$V = XD_1X' + V_1ZD_2Z'V_1 + X_oD_3Z'V_1 + V_1ZD_3'X_o', \qquad (2)$$

where D_1, D_2, D_3 are arbitrary matrices subject to the condition that V is nnd.

(b) Every linear representation of BLUE of $X\beta$ under $(Y, X\beta, \sigma^2V_1)$ continues to be its BLUE under $(Y, X\beta, \sigma^2V)$ iff $\underline{R}(VZ) \subset \underline{R}(V_1Z)$ or, equivalently, $V = XD_1X' + V_1ZD_2Z'V_1$, where D_1 and D_2 are arbitrary matrices subject to the condition that V is nnd.

Proof. Follows from Theorem 2.

Remark. Part (a) of Corollary 1 and Corollary 2 are given in Mathew and Bhimasankaram (1983b). Part (b) of Corollary 1 is given in Mitra and Moore (1973). Part (b) of Corollary 2 is given in Rao (1968, 1971). If X is of full rank, V is nonsingular and $V_1 = I$, then Theorem 2 reduces to the result obtained by Krämer (1980).

3. The likelihood ratio test

Consider the model $(Y, X\beta, \sigma^2 V)$ and the hypothesis $H_o: A\beta = 0$, where $A\beta$ is estimable. We assume that Y has a multivariate normal distribution. As in the previous section let $X_o = X(I-A^-A)$. Let $Z_o = X_o^\perp$ and $Z = X^\perp$ further satisfy the conditions $Z_o'Z_o = I$ and $Z'Z = I$ (identity matrices of appropriate order). Then $I - P_{X_o} = Z_oZ_o'$ and $I - P_X = ZZ'$. Since $\underline{R}(X_o) \subset \underline{R}(X)$, we have $\underline{R}(Z) \subset \underline{R}(Z_o)$. It has been observed in Mathew and Bhimasankaram (1983b, p.304) that the hypothesis H_o is consistent with the model only if $Y \in \underline{R}(X_o:V)$. If this condition is violated, H_o stands rejected. This condition was first given in Rao and Mitra (1971a, p.300) in an alternative form. When $Y \in \underline{R}(X_o:V)$, the LRT statistic for testing H_o under $(Y, X\beta, \sigma^2 V)$ as given in Mathew and Bhimasankaram (1983b) is

$$L_V = \delta\left[\frac{Y'(I-P_{X_o})((I-P_{X_o})V(I-P_{X_o}))^-(I-P_{X_o})Y}{Y'(I-P_X)((I-P_X)V(I-P_X))^-(I-P_X)Y} - 1\right] = \delta\left[\frac{Y'Z_o(Z_o'VZ_o)^-Z_o'Y}{Y'Z(Z'VZ)^-Z'Y} - 1\right],$$

where $\delta = \dfrac{r(VZ)}{r(VZ_o) - r(VZ)}$. It should be noted that since $Y \in \underline{R}(X_o:V)$, L_V doesn't depend on the choice of the g-inverses in the above expressions. We shall be interested in obtaining conditions under which $L_V - L_{V_1} \geq 0$ or $L_V - L_{V_1} \leq 0$ with probability one (i.e. for all $Y \in \underline{R}(X_o:V)$), where V_1 is a given nnd matrix. We shall do so under the assumption that the BLUE of $A\beta$ under $(Y, X\beta, \sigma^2 V_1)$ irrespective of its linear representation is also BLUE under $(Y, X\beta, \sigma^2 V)$. Such an assumption necessarily holds if we want the equality of L_V and L_{V_1} (See Lemma 3.1 in Mathew and Bhimasankaram (1983b)). Hence, matrices V under consideration have the representation (2). L_V and L_{V_1} are defined for $Y \in \underline{R}(X_o:V)$ and $Y \in \underline{R}(X_o:V_1)$, respectively. Hence the comparison of L_V and L_{V_1} under the model $(Y, X\beta, \sigma^2 V)$ is meaningful only if $\underline{R}(X_o:V) \subset \underline{R}(X_o:V_1)$ or, equivalently, if $\underline{R}(V) \subset \underline{R}(X_o:V_1)$. The condition $L_V - L_{V_1} \geq 0$ (or ≤ 0) will guarantee that the rejection (or acceptance) of H_o under $(Y, X\beta, \sigma^2 V_1)$ implies its rejection (respectively, acceptance) under $(Y, X\beta, \sigma^2 V)$ also, provided the associated F-distributions (under H_o) have the same degrees of freedom. This condition is met iff $r(VZ) = r(V_1 Z)$ and $r(VZ_o) = r(V_1 Z_o)$ (since the F-distribution associated with L_V under $(Y, X\beta, \sigma^2 V)$ has degrees of freedom $(r(VZ_o) - r(VZ),\ r(VZ))$). Thus the class of nnd matrices V under consideration are of the form (2) further satisfying the conditions $\underline{R}(V) \subset \underline{R}(X_o:V_1)$, $r(VZ) = r(V_1 Z)$ and $r(VZ_o) = r(V_1 Z_o)$. If V_1 and V are pd and if V has the representation (2), then all these conditions are satisfied.

Let $r(VZ) = r(V_1 Z) = r$ and $r(VZ_o) = r(V_1 Z_o) = r+s$. Let λ_{oi}, $i = 1,2,\ldots, r+s$, denote the nonzero proper eigenvalues of $(I-P_{X_o})V(I-P_{X_o})$ with respect to $(I-P_{X_o})V_1(I-P_{X_o})$ (See Rao and Mitra (1971b), Section 6.3) and let λ_i, $i = 1,2,\ldots,r$, denote the nonzero proper eigenvalues of $(I-P_X)V(I-P_X)$ with respect to $(I-P_X)V_1(I-P_X)$ (from the assumptions on V it follows that the number of nonzero λ_{oi}'s and λ_i's are, respectively, r+s and r).

Theorem 3. Let V_1 be a given nnd matrix and let V have the representation (2). Assume further that V satisfies $\underline{R}(V) \subset \underline{R}(X_o:V_1)$, $r(VZ_o) = r(V_1 Z_o)$ and $r(VZ) = r(V_1 Z)$. Let λ_{oi} and λ_i be as defined above. Then, under the model $(Y, X\beta, \sigma^2 V)$

(i) $L_V - L_{V_1} \geq 0$ with probability one iff $\lambda_i \geq \lambda_{oi}$, $i=1,\ldots,r$ and $\lambda_i \geq \lambda_{oj}$, $i=1,\ldots,r$, $j=r+1,\ldots,s$;

(ii) $L_V - L_{V_1} \leq 0$ with probability one iff $\lambda_i \leq \lambda_{oi}$, $i=1,\ldots,r$ and $\lambda_i \leq \lambda_{oj}$, $i=1,\ldots,r$, $j=r+1,\ldots,s$.

Proof. Under the given assumptions, we have

$$L_V - L_{V_1} = \delta\left[\frac{Y'Z_o(Z_o'VZ_o)^- Z_o'Y}{Y'Z(Z'VZ)^- Z'Y} - \frac{Y'Z_o(Z_o'V_1 Z_o)^- Z_o'Y}{Y'Z(Z'V_1 Z)^- Z'Y}\right],$$

where $Y \in \underline{R}(X_o:V)$. Write $V = CC'$ and $Z_o'Y = Z_o'C\Theta$.

Then $L_V - L_{V_1} \geq 0$ iff for each Θ

$$\frac{\theta'C'Z_o(Z_o'VZ_o)^-Z_o'C\theta}{\theta'C'Z(Z'VZ)^-Z'C\theta} - \frac{\theta'C'Z_o(Z_o'V_1Z_o)^-Z_o'C\theta}{\theta'C'Z(Z'V_1Z)^-Z'C\theta} \geq 0 . \qquad (3)$$

As in the proof of Theorem 3.2 in Mathew and Bhimasankaram (1983b) it can be shown that when V admits the representation (2) the matrices $C'Z_o(Z_o'VZ_o)^-Z_o'C$, $C'Z(Z'VZ)^-Z'C$, $C'Z_o(Z_o'V_1Z_o)^-Z_o'C$ and $C'Z(Z'V_1Z)^-Z'C$ commute pairwise and hence can be reduced to diagonal forms using the same orthogonal matrix P. Let the corresponding diagonal matrices be $\text{diag}(I_{r+s},0)$, $\text{diag}(I_r,0)$, $\text{diag}(\Lambda_o,0)$ and $\text{diag}(\Lambda,0)$, respectively, where $\Lambda_o = \text{diag}(\lambda_{o1},\ldots,\lambda_{or+s})$ and $\Lambda = \text{diag}(\lambda_1,\ldots,\lambda_r)$. Writing $P'\theta = t = (t_1,\ldots,t_n)'$ we get that (3) holds iff for all t_i

$$\frac{\sum_1^{r+s} t_i^2}{\sum_1^{r} t_i^2} - \frac{\sum_1^{r+s} \lambda_{oi} t_i^2}{\sum_1^{r} \lambda_i t_i^2} \geq 0$$

or, equivalently, $\lambda_i \geq \lambda_{oj}$ for $i = 1,\ldots,r$ and $\lambda_i \geq \lambda_{oj}$ for $i = 1,\ldots,r$, $j = r+1,\ldots,s$.

The proof is complete once it is observed that the nonzero eigenvalues of $C'Z_o(Z_o'V_1Z_o)^-Z_o'C$ are the nonzero proper eigenvalues of $(I-P_{X_o})V(I-P_{X_o})$ with respect to $(I-P_{X_o})V_1(I-P_{X_o})$.

Corollary 3: Let V admit the representation (2) with $V_1 = I$ and let $r(VZ_o) = r(Z_o)$ and $r(VZ) = r(Z)$. Let λ_{oi}, $i = 1,\ldots,r+s$, and λ_i, $i = 1,\ldots,r$, respectively, denote the nonzero eigenvalues of $(I-P_{X_o})V(I-P_{X_o})$ and $(I-P_X)V(I-P_X)$. Then, under the model $(Y,X\beta,\sigma^2V)$, $L_V - L_I \geq 0$ with probability one iff $\lambda_i \geq \lambda_{oi}$ for $i=1,\ldots,r$, and $\lambda_i \geq \lambda_{oj}$ for $i=1,\ldots,r$, $j=r+1,\ldots,s$ and $L_V - L_I \leq 0$ with probability one iff $\lambda_i \leq \lambda_{oi}$ for $i=1,\ldots,r$ and $\lambda_i \leq \lambda_{oj}$ for $i=1,\ldots r, j=r+1,\ldots,s$.

Corollary 4. Let V_1 be a given nnd matrix and let V satisfy $\underline{R}(V) \subset \underline{R}(X_o:V_1)$, $r(VZ_o) = r(V_1Z_o)$ and $r(VZ) = r(V_1Z)$. Then, under the model $(Y,X\beta,\sigma^2V)$, $L_V = L_{V_1}$ with probability one iff $(I-P_{X_o})V(I-P_{X_o}) = \lambda(I-P_{X_o})V_1(I-P_{X_o})$ for same $\lambda > 0$.

Proof. The corollary follows from Theorem 3 once it is observed that the equality of L_V and L_{V_1} necessarily implies that V has the representation (2) (see Lemma 3.1 in Mathew and Bhimasankaram (1983b)).

In general, for applying Theorem 3 one has to compute λ_{oi} and λ_i. However there are many interesting situations where Theorem 3 could by applied without computing λ_{oi} and λ_i. A few such examples are discussed below.

Example 1. Consider the dispersion matrix $\sigma^2 V_\rho$, where $V_\rho = (1-\rho)I_n + \rho 1_n 1_n'$ with $1_n = (1,1,\dots,1)'$ and $-\frac{1}{n-1} < \rho < 1$. Let $V_1 = I$. Assume that $1_n \in \underline{R}(X)$. Then it can be verified that the BLUE of $X\beta$ under $(Y, X\beta, \sigma^2 I)$ (which is the simple least squares estimator) continues to be its BLUE under $(Y, X\beta, \sigma^2 V_\rho)$. Hence V_ρ has the representation (2) with $V_1 = I$ and V_ρ also satisfies the specifications to apply Theorem 3 or Corollary 3. Let L_ρ and L_o denote the LRT statistics for testing $H_o: A\beta = 0$ under $(Y, X\beta, \sigma^2 V_\rho)$ and $(Y, X\beta, \sigma^2 I)$, respectively. Ghosh and Sinha (1980) proved that for each fixed ρ, $L_\rho = L_o$ iff $1_n \in \underline{R}(X_o)$ (see also Theorem 2.3 in Mathew and Bhimasankaram (1983a)). If $1_n \notin \underline{R}(X_o)$, one can apply Corollary 3. The nonzero eigenvalues of $(I-P_{X_o})V_\rho(I-P_{X_o})$ are the eigenvalues of $Z_o' V_\rho Z_o$. It can be shown that the eigenvalues of $Z_o' V_\rho Z_o$ are $1-\rho$ (of multiplicity $r+s-1$) and $1-\rho+\rho 1_n' Z_o Z_o' 1_n$ of multiplicity one. Since $1_n \in \underline{R}(X)$, $Z' 1_n = 0$ and the nonzero eigenvalues of $(I-P_X)V_\rho(I-P_X)$ are all equal to $i-\rho$. Applying Corollary 3 we get, $L_\rho - L_o \geq 0$ iff $\rho \leq 0$ and $L_\rho - L_o \leq 0$ iff $\rho \geq 0$. Thus when $1_n \in \underline{R}(X)$, but $1_n \notin \underline{R}(X_o)$ the rejection (or acceptance) of H_o under $(Y, X\beta, \sigma^2 I)$ implies its rejection (respectively, acceptance) under $(Y, X\beta, \sigma^2 V_\rho)$ also if $\rho \leq 0$ (or $\rho \geq 0$).

Example 2. Consider the dispersion matrix $\sigma^2 V_c$, where $V_c = \alpha I_n + c 1_n' + 1_n c'$ with $\alpha > 0$. It can be shown that the eigenvalues of V_c are $\alpha + 1_n' c + (nc'c)^{\frac{1}{2}}$, $\alpha + 1_n' c - (nc'c)^{\frac{1}{2}}$ and α (of multiplicity $n-2$). The vectors c under consideration are those which make the above eigenvalues positive, so that V_c is pd. Baldessari (1966) studied the validity of χ^2 and F-tests for dependent normal data and came up with a model having the above dispersion matrix. If L_c and L_o, respectively, denote the LRT statistics for testing $H_o: A\beta = 0$

under $(Y,X\beta,\sigma^2V_c)$ and $(Y,X\beta,\sigma^2I)$, then applying Corollary 4 (with $V_1=I$) we see that $L_c=L_o$ iff $Z_o'(\alpha I+1_nc'+c1_n')Z_o=\lambda I$ for some $\lambda>0$ or, equivalently,

$$Z_o'(1_nc'+c1_n')Z_o=(\lambda-\alpha)I. \tag{4}$$

For $n>2$, (4) leads to a contradiction if $\lambda-\alpha\neq 0$, since $(\lambda-\alpha)I$ in such a case has rank greater than 2, whereas the matrix on the left hand side of (4) has rank less than or equal to 2. If $n=2$, writing $a=Z_o'1_n$ and $b=Z_o'c$ (4) becomes $ab'+ba'=(\lambda-\alpha)I$. Here a and b are 2×1 vectors and it can be easily shown that the above relation leads to a contradiction if $\lambda-\alpha\neq 0$. Thus $\lambda-\alpha=0$ in either case and we see that $L_c=L_o$ iff $Z_o'(1_nc'+c1_n')Z_o=0\Leftrightarrow Z_o'1_nc'Z_o=0\Leftrightarrow Z_o'1_n=0$ or $Z_o'c=0$ or both are zeros $\Leftrightarrow 1_n$ or c or both belong to $\underline{R}(X_o)$. This result has also been obtained by Sinha and Mukhopadhyay (1980) by a different approach.

Our next attempt is to apply Corollary 3 when L_c and L_o are different, i.e. when neither 1_n nor c belong to $R(X_o)$. We shall however assume that both 1_n and c are in $\underline{R}(X)$. In such a case the BLUE of $X\beta$ under $(Y,X\beta,\sigma^2I)$ is its BLUE under $(Y,X\beta,\sigma^2V_c)$ also. The nonzero eigenvalues of $(I-P_{X_o})V_c(I-P_{X_o})$ are the same as those of $Z_o'V_cZ_o=\alpha I+Z_o'(1_nc'+c1_n')Z_o$. Also $Z_o'(1_nc'+c1_n')Z_o$ and $(c:1_n)'Z_oZ_o'(1_n:c)$ have the same nonzero eigenvalues. Direct computation gives the eigenvalues of $(c:1_n)'Z_oZ_o'(1_n:c)$ as $1_n'Z_oZ_o'c\pm((1_n'Z_oZ_o'1_n)(c'Z_oZ_o'c))^{\frac{1}{2}}$. Hence the nonzero eigenvalues of $(I-P_{X_o})V_c(I-P_{X_o})$ are $\alpha+1_n'X_oZ_o'c\pm((1_n'Z_oZ_o'1_n)(c'Z_oZ_o'c))^{\frac{1}{2}}$ and α. Since both 1_n and c are in $\underline{R}(X)$, $Z'1_n=Z'c=0$ and hence the nonzero eigenvalues of $(I-P_X)V(I-P_X)$ are all equal to α. Applying Corollary 3, we see that $L_c-L_o\geq 0$ with probability one iff $1_n'Z_oZ_o'c\pm((1_n'Z_oZ_o'1_n)(c'Z_oZ_o'c))^{\frac{1}{2}}\leq 0\Leftrightarrow Z_o'c=\lambda Z_o'1_n$ for some $\lambda\leq 0$ or, equivalently, $(I-P_{X_o})c=\lambda(I-P_{X_o})1_n$ for some $\lambda\leq 0$. Similarly $L_c-L_o\leq 0$ with probability one iff $(I-P_{X_o})c=\lambda(I-P_{X_o})1_n$ for some $\lambda\geq 0$.

Acknowledgement. The author is grateful to a referee for suggestions which simplified the proof of Theorem 2 and improved the presentation of the paper.

References

[1] Baksalary, J.K. and Kala, R.(1978): A bound for the euclidean norm of the difference between the least squares and the best linear unbiased estimators. Ann.Statist. 6, 1390-1393.

[2] Baksalary, J.K. and Kala, R. (1980): A new bound for the euclidean norm of the difference between the least squares and the best linear unbiased estimators. Ann.Statist.8, 679-681.

[3] Baldessari, B. (1966): Analysis of variance of dependent data. Statistica, XXVI, 895-903.

[4] Ghosh, M. and Sinha, B.K.(1980): On the robustness of least squares procedures in regression models. J.Multivariate Analvsis 10, 332- 342.

[5] Haberman, S.J. (1975): How much do Gauss-Markov and least squares estimators differ. A co-ordinate free approach. Ann. Statist. 3, 982-990.

[6] Khatri, C.G. (1981): Study of F-test under dependent model. Sankhya. Ser.A 43, 107-110.

[7] Krämer, W.(1980): A note on the equality of ordinary least squares and Gauss-Markov estimates in the general linear model. Sankhya, Ser. A. 42, 130-131.

[8] Mathew, T. (1983): Linear estimation with an incorrect dispersion matrix in linear models with a common linear part. J. Amer. Statist. Assoc. 78, 468-471.

[9] Mathew, T. and Bhimasankaram, P. (1983a): On the robustness of the LRT with respect to specification errors in a linear model. Sankhya. Ser. A. 45, 212-225.

[10] Mathew, T. and Bhimasankaram, P. (1983b): On the robustness of LRT in singular linear models. Sankhya, Ser.A. 45, 301-312.

[11] Mitra, S.K. and Moore, B.J. (1973): Gauss-Markov estimation with an incorrect dispersion matrix. Sankhya. ser. A. 35, 139-152.

[12] Mitra, S.K. and Moore, B.J.(1976): Linear estimation in general linear models, Essays in Probability and Statistics. (S.Ikeda et. al.editors), Shinko Tsusho Co., Tokyo, 189-201.

[13] Rao, C.R. (1967): Least squares theory using an estimated dispersion matrix and its application to measurament of signals. Proc.Fifth

Berkely Symposium on Math. Stat. and Prob., Vol.1, 355-372

[14] Rao, C.R.(1968): A note on a previous lemma in the theory of least squares and some further results. Sankhya, Ser.A. 30, 245-252.

[15] Rao, C.R.(1971): Unified theory of linear estimation. Sankhya, Ser. A., 33, 371-394.

[16] Rao, C.R. and Mitra, S.K. (1971a); Further contributions to the theory of generalized inverse of matrices and its applications. Sankhya, Ser.A. 33, 289-300.

[17] Rao, C.R. and Mitra, S.K,(1971b): Generalized Inverse of Matrices and its Applications, John Wiley.

[18] Sinha, B.K. and Mukhopadhyay, B.B.(1980): On a characterization of regression models under a specified covariance structure for the robustness of the usual LRT. Tech. Report No. 26/80, Stat-Math Division, Indian Statistical Institute, Calcutta.

[19] Watson, G.S.(1967): Linear least square regression. Ann.Math. Statist. 38, 1679-1699.

[20] Zyskind, G.(1967): On canonical forms, nonnegative covariance matrices and best and simple least squares estimators in linear models. Ann.Math. Statist. 38, 1092-1110.

[21] Zyskind, G.(1969): Parametric augmentations and error structures under which certain simple least squares and analysis of variance procedures are also best. J.Amer.Statisti. Assoc. 64 1353-1368.

A SPLIT-PLOT DESIGN WITH WHOLEPLOT TREATMENTS IN AN INCOMPLETE BLOCK DESIGN

S. Mejza
Department of Mathematical and Statistical Methods
Academy of Agriculture
Poznań, Poland

Abstract: An incomplete split-plot design, where levels of one factor (say A) are applied to the wholeplots and levels of the other (say B) to subplots, and where the number of wholeplots in each block may be less than the number of levels of factor A, is considered. The m levels of factor A are arranged in a proper incomplete block design. The h levels of factor B are arranged in a randomized complete block design within each level of factor A, by considering the wholeplots as blocks.

1. Introduction

In the paper a split-plot type experiment is considered. Usually in such experiment two factors are involved. Let one of the factors be denoted by A, with m different levels, $A_1,\dots,A_m$ and let the second factor be denoted by B with h different levels, $B_1,\dots,B_h$. Let the A_t, $t=1,\dots,m$, be called the wholeplot treatments and the B_g, $g=1,\dots,h$ be called the subplot treatments.

A traditional split-plot design is such that the levels of factor A are applied to plots, called wholeplots, of a randomized complete block design, and the levels of factor B are applied to plots, called subplots, of another randomized complete block design, different for

KEY WORDS: block design, incomplete split-plot, basic contrasts, analysis of variance.
This work was partially supported by the Polish Academy of Sciences, Grant No. MR.I.-2/2.

each level of A, with the wholeplots considered as blocks. There may arise situations where the number of wholeplots on each block is fewer than the number of levels of A. Then as an appropriate design for the levels of A an incomplete block design can be used. A design obtained in this way will be called a split-plot design with wholeplot treatments in an incomplete block design. In the present paper this type of design is examined.

The analysis is based on the estimators of a set of estimable contrasts called basic contrasts. From the BLUE's of basic contrasts the whole analysis of variance is obtainable. It is worth noting that the levels of A as well as of B can be equally or unequally replicated.

Incompleteness in split-plot design has been considered in several papers. Finney (1946a) and (1946b) described examples of experiments involving a balanced or an unbalanced system of split-plot confounding. Robinson (1967) considered incomplete split-plot designs in which m levels of factor A are applied to k ($\leq m$) wholeplots and h levels of factor B are arranged in a BIB designs within each level of factor A, considering the wholeplots as blocks of the BIB design. In this design the number of subplots in each wholeplot is restricted to say l ($\leq h$). Mejza and Mejza (1984) generalized the Robinson's result to the case in which levels of factor B are arranged in any proper incomplete block design.

Bhargava and Shah (1975) considered, among other things, a split-plot design in which wholeplot treatments are arranged in a binary incomplete block design. In that paper analyses of several models, different from that of the present paper, were considered.

2. Linear model estimation

In statistical textbooks and research papers there are many approaches to the linear model of a split-plot design (see, e.g. Federer (1975)). In the experiment considered here it is assumed that there exists additivity between treatment combinations A_tB_g and the experimental units. Moreover, it is assumed that there exist some correlations, as it will be defined in (2.3). For tests of hypotheses normal distributions of random components are assumend, as usually.

Split-plot design belongs to a general class of designs called block designs. Hence, some results from the theory of block designs are adopted here. In particular the treatment combinations A_tB_g, t=1,...,m, g=1,...,h, can be considered as treatments of a block design. In this sense there are v=mh treatments.

Furthermore, by the effect of the i-th treatment (or a corresponding treatment combination) we mean

$$\tau_i=\alpha_t+\beta_g+(\alpha\beta)_{tg} \tag{2.1}$$

with i=(t-1)h+g, t=1,2,...,m, g=1,2,...,h, where α_t is the effect of the t-th level of A, β_g is the effect of the g-th level of B and $(\alpha\beta)_{tg}$ is the effect of the interaction of the t-th level of A and the g-th level of B.

Let b denote the number of blocks of size $k(\leq m)$ with respect to wholeplots and of size kh with respect to subplots, as only the case of wholeplots containing complete sets of subplot treatments are considered here.

The observed yield y_{jtg} of the g-th level of B within the t-th level of A in the j-th block may be written as

$$y_{jtg}=\mu+\rho_j+\alpha_t+\eta_{jt}+\beta_g+(\alpha\beta)_{tg}+e_{jtg}+e^*_{jtg}, \tag{2.2}$$

where μ denotes general parameter, ρ_j an effect of the j-th block, η_{jt} a random variable concerning the t-th level of A (a wholeplot error) within the j-th block, e_{jtg} a random unit error (a subplot error), e^*_{jtg} a random technical error.

It is assumed that

$$E(\eta_{jt})=0, \quad E(e_{jtg})=0, \quad E(e^*_{jtg})=0,$$

$$\mathrm{Cov}(\eta_{jt}, \eta_{j't'}) = \begin{cases} \sigma_1^2 & , \text{ if } j=j', t=t', \\ \rho_1 \sigma_1^2, & \text{ if } j=j', t\neq t', \\ 0 & , \text{ if } j\neq j', \end{cases}$$

$$\mathrm{Cov}(e_{jtg} e_{j't'g'}) = \begin{cases} \sigma_e^2 & , \text{ if } j=j', t=t', g=g', \\ \rho_e \sigma_e^2, & \text{ if } j=j', t=t', g\neq g', \\ 0 & , \text{ if } j\neq j' \text{ or } t\neq t', \end{cases} \tag{2.3}$$

$$\mathrm{Cov}(e^*_{jtg}, e^*_{j't'g'}) = \begin{cases} \sigma^2 & , \text{ if } j=j', t=t', g=g', \\ 0 & \text{otherwise;} \end{cases}$$

also it is assumed that the three sets of random components $\{\eta_{jt}\}$, $\{e_{jtg}\}$, $\{e^*_{jtg}\}$ are distributed independently.

These assumptions can be considered as resulting from the randomization procedures involved.

In the matrix notation model (2.2) can be written as

$$\underline{y} = \underline{1}_n \mu + \underline{D}'\underline{\rho} + \underline{\Delta}'\underline{\tau} + \underline{G}'\underline{\eta} + \underline{e}_1 + \underline{e}, \tag{2.4}$$

where $\underline{1}_n$ is an $n\times 1$ vector of ones, $n=bkh$, $\underline{D}'$ is an $n\times b$ design matrix for blocks, $\underline{\rho}$ is a $b\times 1$ vector of block effects, $\underline{\Delta}'$ is an $n\times v$ design matrix for treatments, $\underline{\tau}$ is a $v\times 1$ vector of treatment effects (2.1), $\underline{G}'$ is an $n\times bk$ design matrix for wholeplots, $\underline{\eta}$ is a $bk\times 1$ vector of wholeplot errors, $\underline{e}_1$ is an $n\times 1$ vector of subunit errors, $\underline{e}$ is an $n\times 1$ vector of technical errors.

Under the lexicographical order of observations, adopted here, assumptions (2.3) can in matrix notation be written as

$$E(\underline{\eta}) = \underline{0}, \quad E(\underline{e}_1) = \underline{0}, \quad E(\underline{e}) = \underline{0},$$

$$E(\underline{\eta}\,\underline{\eta}') = \sigma_1^2 \underline{I}_b \otimes [(1-\rho_1)\underline{I}_k + \rho_1 \underline{1}_k \underline{1}_k'] = \underline{V}_\eta \tag{2.5}$$

$$E(\underline{e}_1 \underline{e}_1') = \sigma_e^2 \underline{I}_b \otimes \underline{I}_k \otimes [(1-\rho_e)\underline{I}_h + \rho_e \underline{1}_h \underline{1}_h'] = \underline{V}_e$$

$$E(\underline{e}\,\underline{e}') = \sigma^2 \underline{I}_n,$$

where $\otimes$ denotes the Kronecker's product of matrices, while $\underline{I}_b$, $\underline{I}_k$, $\underline{I}_h$ and $\underline{I}_n$ denote the identity matrices of indicated order.

Under (2.5) and under the assumption of independence of $\{\eta_{jt}\}$, $\{e_{jtg}\}$ and $\{e^*_{jtg}\}$ the covariance matrix of $\underline{y}$ becomes

$$\underline{V} = \mathrm{Cov}(\underline{y}) = \underline{G}'\underline{V}_\eta\underline{G} + \underline{V}_e + \sigma^2\underline{I}_n. \tag{2.6}$$

If, in addition of these assumptions, normality is assumed, i.e. that $\underline{y} \sim N(\underline{1}_n\mu + \underline{D}'\underline{\rho} + \underline{\Delta}'\underline{\tau}, \underline{V})$, (2.4) is called a normal model.

In the analysis of data based on model (2.4) several matrices are helpful.

Note that $\underline{G}' = \underline{I}_b \otimes \underline{I}_k \otimes \underline{1}_h$ and $\underline{D}' = \underline{G}'(\underline{I}_b \otimes \underline{1}_k) = \underline{I}_b \otimes \underline{1}_k \otimes \underline{1}_h$, $\underline{G}'\underline{G} = \underline{I}_b \otimes \underline{I}_k \otimes \underline{1}_h\underline{1}_h'$, $\underline{D}'\underline{D} = \underline{I}_b \otimes \underline{1}_k\underline{1}_k' \otimes \underline{1}_h\underline{1}_h'$. Let

$$\underline{P}_e = \underline{I}_n - h^{-1}\underline{G}'\underline{G}, \quad \underline{P} = (hk)^{-1}\underline{D}'\underline{D}, \quad \underline{P}_1 = \underline{I}_n - \underline{P}_e - \underline{P}.$$

It is easy to verify that

$$\begin{aligned} &\underline{P}\,\underline{P} = \underline{P},\ \underline{P}_e\underline{P}_e = \underline{P}_e,\ \underline{P}_1\underline{P}_1 = \underline{P}_1,\ \underline{P}\,\underline{P}_e = \underline{O},\ \underline{P}_1\underline{P} = \underline{O}, \\ &\underline{P}_1\underline{P}_e = \underline{O},\ \ \underline{P}_e\underline{1}_n = \underline{O},\ \ \underline{P}_e\underline{D}' = \underline{O} \ \text{ and } \ \underline{P}_1\underline{D}' = \underline{O}. \end{aligned} \tag{2.7}$$

The matrix $\underline{V}$ can then be written as

$$\underline{V} = \gamma_1\underline{P}_1 + \gamma_e\underline{P}_e + \gamma\underline{P}, \tag{2.8}$$

where

$$\gamma_1 = \sigma^2 + h\sigma_1^2(1-\rho_1) + \sigma_e^2[1+(h-1)\rho_e],$$

$$\gamma_e = \sigma^2 + \sigma_e^2(1-\rho_e),$$

while

$$\gamma = \sigma^2 + h\sigma_1^2[1+(k-1)\rho_1] + \sigma_e^2[1+(h-1)\rho_e].$$

In the analysis based on model (2.4) the generalized least squares method is to be used. For this it is necessary to find the matrix $\underline{V}^{-1}$. It follows from (2.7) and (2.8) that

$$\underline{V}^{-1} = \gamma_1^{-1}\underline{P}_1 + \gamma_e^{-1}\underline{P}_e + \gamma^{-1}\underline{P}. \tag{2.9}$$

The (reduced) normal equation for $\underline{\tau}$ obtained after eliminating μ^o and $\underline{\beta}^o$ is then of the form

$$[h^{-1}\gamma_1^{-1}(\underline{\tilde{N}}\,\underline{\tilde{N}}'-k^{-1}\underline{N}\,\underline{N}')+\gamma_e^{-1}(\underline{r}^{\delta}-h^{-1}\underline{\tilde{N}}\,\underline{\tilde{N}}')]\underline{\tau}^o=\gamma_1^{-1}h^{-1}(\underline{\tilde{N}}\,\underline{\tilde{B}}-k^{-1}\underline{N}\,\underline{B})+ \gamma_e^{-1}(\underline{T}-h^{-1}\underline{\tilde{N}}\,\underline{\tilde{B}}), \tag{2.10}$$

shortly written as $\underline{C}_o\underline{\tau}^o = \underline{Q}_o$ (with obvious definitions of $\underline{C}_o$ and $\underline{Q}_o$), where $\underline{N} = \underline{\Delta}\,\underline{D}'$ is the treatment - block incidence matrix, $\underline{\tilde{N}} = \underline{\Delta}\,\underline{G}'$ is the treatment-wholeplot incidence matrix $\underline{T} = \underline{\Delta}\,\underline{y}$, $\underline{\tilde{B}} = \underline{G}\,\underline{y}$, $\underline{B} = \underline{D}\,\underline{y}$ and $\underline{r} = \underline{N}\,\underline{1}_b$, and where the symbol $\underline{x}^{p\delta}$ (with proper p) represents a diagonal matrix with the diagonal elements corresponding to these of $\underline{x}$ (respectively) but each raised to the power p.

The notation above can be simplified for the case considered in this paper, i.e. when the wholeplots are complete blocks for subplot. The following relations can then be applied $\underline{\Delta}' = \underline{\Delta}_1'\otimes\underline{I}_h$, $\underline{D}' = \underline{D}_1'\otimes\underline{1}_h$ and hence

$$\underline{N} = \underline{\Delta}_1\underline{D}_1'\otimes\underline{1}_h = \underline{N}_1\otimes\underline{1}_h',\quad \underline{\tilde{N}} = \underline{\Delta}_1\otimes\underline{1}_h',\quad \underline{r} = \underline{r}_1\otimes\underline{1}_h. \tag{2.11}$$

Also note that in general

$$\underline{Q}_o = h^{-1}\gamma_1^{-1}(\underline{\tilde{N}\tilde{B}}-k^{-1}\underline{NB})+\gamma_e^{-1}(\underline{T}-h^{-1}\underline{\tilde{N}\tilde{B}}) = (\gamma_1^{-1}\underline{\Delta P}_1 + \gamma_e^{-1}\underline{\Delta P}_e)\underline{y},$$

$$\underline{C}_o = h^{-1}\gamma_1^{-1}(\underline{\tilde{N}\tilde{N}}'-k^{-1}\underline{NN}')+\gamma_e^{-1}(\underline{r}^{\delta}-h^{-1}\underline{\tilde{N}\tilde{N}}') = \gamma_1^{-1}\underline{\Delta P}_1\underline{\Delta}'+\gamma_e^{-1}\underline{\Delta P}_e\underline{\Delta}',$$

while using (2.11)

$$\underline{Q}_o = h^{-1}\gamma_1^{-1}[(\underline{\Delta}_1-k^{-1}\underline{N}_1)\otimes\underline{1}_h\underline{1}_h']\underline{y} + \gamma_e^{-1}\underline{\Delta}_1\otimes(\underline{I}_h-h^{-1}\underline{1}_h\underline{1}_h')\underline{y}, \tag{2.12}$$

$$\underline{C}_o = h^{-1}\gamma_1^{-1}(\underline{r}_1^{\delta}-k^{-1}\underline{N}_1\underline{N}_1')\otimes\underline{1}_h\underline{1}_h'+\gamma_e^{-1}\underline{r}_1^{\delta}\otimes(\underline{I}_h-h^{-1}\underline{1}_h\underline{1}_h'). \tag{2.13}$$

Now we want to find which linear parametric functions $\underline{c}'\underline{\tau}$ are estimable in model (2.4), if no restrictions are imposed on the parameters. Note that from $\underline{C}_o\underline{1}_v = \underline{O}$ and from the estimability criterion $\underline{c}'\underline{C}_o^-\underline{C}_o = \underline{c}'$ (see Rao and Mitra (1971), §2.3) it follows that if a linear function is estimable it must be a contrast. Hence, it is sufficient to examine which contrasts are estimable.

The properties of the design considered can be derived from patterns of the matrix $\underline{C}_1 = \underline{r}_1^{\delta} - k^{-1}\underline{N}_1\underline{N}_1'$ which is the $\underline{C}$-matrix of the (reduced) normal equation for the wholeplot treatment effects.

Let $d(<m)$ denote the rank of $\underline{C}_1$. We consider here a connected design only, i.e. the case of $d = m-1$. If the design is disconnected, then some formulae in this paper need certain modifications. Let ε_t be an eigenvalue and let $\underline{s}_{1t}(\neq \underline{0})$ be a corresponding eigenvector of the matrix $\underline{C}_1$ with respect to $\underline{r}_1^{\delta}$, i.e. let $\underline{C}_1\underline{s}_{1t} = \varepsilon_t \underline{r}_1^{\delta}\underline{s}_{1t}$, for $t=1,\dots,m$. The eigenvectors $\underline{s}_{1t}$ can be chosen to be $\underline{r}_1^{\delta}$-orthonormal in pairs, i.e. so that they satisfy the equalities $\underline{s}_{1t}'\underline{r}_1^{\delta}\underline{s}_{1t'} = \delta_{tt'}$. Since $\underline{C}_1\underline{1}_m = \underline{0}$ the last (say) eigenvector $\underline{s}_{1m}$ can be taken equal to $(1/\sqrt{n})\underline{1}_m$. Hence, $\underline{s}_{1t}'\underline{r}_1 = 0$, for $t=1,\dots,m-1$. It can be shown that $\varepsilon_t \in (0,1]$ for any $t \leq m-1$, while $\varepsilon_m = 0$.

Let $\underline{q}_1,\dots,\underline{q}_{h-1}$ be a set of normalized vectors orthogonal to the vector $\underline{q}_h = (1/\sqrt{h})\underline{1}_h$.

Moreover, let

$$\underline{s}_i = \underline{s}_{1t} \otimes \underline{q}_g \quad \text{and} \quad \underline{c}_i = \underline{r}_1^{\delta}\underline{s}_{1t} \otimes \underline{q}_g \tag{2.14}$$

for $i=(t-1)h+g$, with $t=1,2,\dots,m$, $g = 1,2,\dots,h$.

It is easy to verify that $\underline{s}_i'\underline{r}^{\delta}\underline{s}_{i'} = \underline{c}_i'\underline{r}^{-\delta}\underline{c}_{i'} = 0$ or 1, depending on whether $i \neq i'$ or $i=i'$, respectively, $\underline{c}_i'\underline{1}_v = 0$ for $i=1,2,\dots,v-1$. The contrasts $\underline{c}_i'\underline{\tau}$, called $r^{-\delta}$-basic contrasts (see Pearce et al. (1974) play a special role in the analysis of model (2.4).

Let K_1 be the set of indices i such that $i=(t-1)h+g$ for $t=1,..,m$, $g=1,\dots,h-1$ and let K_2 be the set of indices i such that $i=ht$, where $t=1,\dots,m-1$.

Note that the vectors $\underline{s}_i$ are eigenvectors of the matrix $\underline{C}_o$ with respect to the matrix $\underline{r}^{\delta}$, corresponding, respectively, to the eigenvalue $\lambda_i = \gamma_e^{-1}$ for $i=(t-1)h+g$, where $t=1,\dots,m$ and $g=1,\dots,h-1$, and to the eigenvalues

$$\lambda_i = \gamma_1^{-1}\varepsilon_i, \quad \text{for} \quad i=ht,\ t=1,\ldots,m-1.$$

Hence the matrix $\underline{C}_o$ has the (spectral decomposition) representation

$$\underline{C}_o = \sum_{i=1}^{v-1} \lambda_i \underline{r}^{\delta} \underline{s}_i \underline{s}_i' \underline{r}^{\delta} = \sum_{i=1}^{v-1} \lambda_i \underline{c}_i \underline{c}_i'. \tag{2.15}$$

This can be used in solving the equation (2.10). It follows that

$$\underline{\tau}^o = \underline{C}_o^- \underline{Q}_o, \tag{2.16}$$

where $\underline{C}_o^- = \sum_{i=1}^{v-1} \lambda_i^{-1} \underline{s}_i \underline{s}_i'$.

Note, however, that $\underline{c}_i' \underline{C}_o^- \underline{C}_o = \underline{c}_i'$ for any $i < mh-1$ and hence all $r^{-\delta}$ basic contrasts are estimable. From the above it follows that all treatment contrasts are estimable. Their best linear unbiased estimates (BLUE's) are provided by the following

<u>Theorem 2.1.</u> In model (2.4)
(i) the BLUE of an $r^{-\delta}$-basic contrast $\underline{c}_i'\underline{\tau}$ is given by

$$\widehat{\underline{c}_i'\underline{\tau}} = \begin{cases} \underline{s}_i' \underline{\Delta}_1 \otimes (\underline{I}_h - h^{-1} \underline{1}_h \underline{1}_h') \underline{y} = \underline{s}_i' \underline{\Delta} \underline{P}_e \underline{y}, & i \in K_1, \\ \varepsilon_i^{-1} \underline{s}_i' [(\underline{\Delta}_1 - k^{-1} \underline{N}_1) \otimes \underline{1}_h \underline{1}_h'] \underline{y} = \varepsilon_i^{-1} \underline{s}_i' \underline{\Delta} \underline{P}_1 \underline{y}, & i \in K_2, \end{cases} \tag{2.17}$$

and

(ii) the variance of $\widehat{\underline{c}_i'\underline{\tau}}$ is

$$\mathrm{Var}(\widehat{\underline{c}_i'\underline{\tau}}) = \begin{cases} \gamma_e, & i \in K_1, \\ \gamma_1 \varepsilon_i^{-1}, & i \in K_2. \end{cases} \tag{2.18}$$

<u>Proof</u>. From the estimability of $\underline{c}_i'\underline{\tau}$ it follows that $\widehat{\underline{c}_i'\underline{\tau}} = \underline{c}_i'\underline{\tau}^o$, and hence the form (2.17) results from (2.16), leading to (2.18) on account of (2.8).

If model (2.4) is a normal model, i.e. $\underline{\eta}, \underline{e}_1, \underline{e}$ have a joint normal distribution, the following theorem is true.

<u>Theorem 2.2</u>. If model (2.4) is a normal model then all estimators

$\widehat{\underline{c}_i'\underline{\tau}}$ are distributed independently according to the formulae

$\widehat{\underline{c}_i'\underline{\tau}} \sim N(\underline{c}_i'\underline{\tau}, \gamma_e)$ for $i \in K_1$ and $\widehat{\underline{c}_i'\underline{\tau}} \sim N(\underline{c}_i'\underline{\tau}, \varepsilon_1^{-1}\gamma_1)$ for $i \in K_2$.

To test a hypothesis of the type $H_{oi}: \underline{c}_i'\underline{\tau} = 0$ it is necessary to find unbiased estimators of γ_e and γ_1 that are distributed independently of $\widehat{\underline{c}_i'\underline{\tau}}$. Such estimators are given in the following theorems.

<u>Theorem 2.3</u>. For model (2.4) the sum of squares $SS_e = \underline{y}'\underline{H}\underline{y}$, where $\underline{H} = \underline{P}_e[\underline{I}_n - \underline{\Delta}'(\underline{\Delta}\underline{P}_e\underline{\Delta}')^-\underline{\Delta}]\underline{P}_e$ is an unbiased estimator of $\nu_e\gamma_e$, where $\nu_e = (bk-m)(h-1)$.

<u>Proof</u>. The known result that $E(\underline{y}'\underline{H}\underline{y}) = tr(\underline{H}\underline{V}) + \underline{\mu}'\underline{H}\underline{\mu}$ can be used here. Note that $\underline{H}\underline{H} = \underline{H}$, $\underline{H}\underline{1}_n = \underline{0}$, $\underline{H}\,\underline{\Delta}' = \underline{0}$, $\underline{H}\underline{D}' = \underline{0}'$ and $\underline{H}\underline{V} = \gamma_e\underline{H}$. From this we get in our case $\underline{\mu}'\underline{H}\underline{\mu} = \underline{0}$ and $tr(\underline{H}\underline{V}) = tr(\gamma_e\underline{H}) = \gamma_e tr(\underline{H}) =$
$\gamma_e[n-bk-tr(\underline{P}_e\underline{\Delta}'(\underline{\Delta}\underline{P}_e\underline{\Delta}')^-\underline{\Delta}\underline{P}_e)] = \gamma_e[n-bk-tr((\underline{\Delta}\underline{P}_e\underline{\Delta}')(\underline{\Delta}\underline{P}_e\underline{\Delta}')^-)] =$

$\gamma_e[n-bk-r(\underline{\Delta}\underline{P}_e\underline{\Delta}')] = \gamma_e[n-bk-m(h-1)]$.

Now let $\underline{W} = \underline{I}_{bk} - (kh)^{-1}\underline{G}\underline{D}'\underline{D}\underline{G}'$. Note that $\underline{W}\underline{W} = \underline{W}$, $\underline{W}\underline{G}\underline{D}' = \underline{0}$ and that $\underline{W}\underline{1} = \underline{0}$. Moreover, let

$$\underline{H}_o = \underline{G}'\underline{W}[\underline{I}_{kb} - \underline{\tilde{N}}'(\underline{\tilde{N}}\underline{W}\underline{\tilde{N}}')^-\underline{\tilde{N}}]\underline{W}\underline{G}\,, \tag{2.19}$$

and note that $\underline{H}_o\underline{H}_o = h\underline{H}_o$, from which $h^{-1}\underline{H}_o$ is idempotent and that $\underline{\Delta}\underline{H}_o = \underline{0}$. The above properties are used in the proof of

<u>Theorem 2.4</u>. In model (2.4) the sum of squares

$$SS_E = \underline{y}'\underline{H}_o\underline{y}, \tag{2.20}$$

with $\underline{H}_o$ defined in (2.19), is an unbiased estimator of $\gamma_1\nu_E$, where $\nu_E = bk-b-t+1$.

<u>Proof</u>. Note that

$$E(\underline{y}'\underline{H}_o\underline{y}) = tr(\underline{H}_o\underline{V}) = tr(\underline{H}_o(\gamma_1\underline{P}_1 + \gamma_e\underline{P}_e + \gamma\underline{P})) = \gamma_1 tr(\underline{H}_o)$$

$$tr(\underline{H}_o) = tr(\underline{G}'\underline{W}\underline{G}) - tr(\underline{G}'\underline{W}\underline{\tilde{N}}'(\underline{\tilde{N}}\underline{W}\underline{\tilde{N}}')^-\underline{\tilde{N}}\underline{W}\underline{G}) =$$

$= bkh-bh-tr(\tilde{\underline{N}}\ \underline{W}\ \underline{G}\ \underline{G}'\ \underline{W}\ \tilde{\underline{N}}'(\tilde{\underline{N}}\ \underline{W}\ \tilde{\underline{N}}')^{-}) = bkh-bh-h\ r(\tilde{\underline{N}}\ \underline{W}\ \tilde{\underline{N}}') = h(bk-b-t+1).$

Hence $E(\underline{y}'\underline{H}_o\underline{y}) = h\gamma_1(bk-b-t+1).$

3. Test of hypotheses

Tests of hypotheses are considered here under the assumptions that model (2.4) is a normal model. The following theorems, applicable to the BLUE's from Theorem 2.1 and to the sums of squares from Theorems 2.3 and 2.4 give justification for the tests involved in the analysis.

Theorem 3.1. If model (2.4) is a normal model, then $\widehat{\underline{c}_i'\underline{\tau}}$, $i \in K_1$, and SS_e are distributed independently.

Proof. The proof consists in noting that $\underline{\Delta P}_e(\gamma_1\underline{P}_1+\gamma_e\underline{P}_e+\gamma\underline{P})\underline{H} = \gamma_e\underline{\Delta H} = \underline{O}$, which satisfies a relevant necessary and sufficient condition (see, e.g., Searle (1971), p. 59).

Theorem 3.2. If model (2.4) is a normal model, then SS_e is distributed as $\gamma_e\chi^2$ with ν_e degrees of freedom.

Proof. From the proof of Theorem 2.3, $\underline{\mu}'\underline{H\mu} = 0$ and $\underline{HV} = \gamma_e\underline{H}$. Hence $\gamma_e^{-1}\underline{HV}$ is idempotent, and this completes the proof (see, Searle (1971), p.57).

From Theorems 3.1 and 3.2. we get

Theorem 3.3. The ratio $F_i = [(\widehat{\underline{c}_i'\underline{\tau}})^2\nu_e]/SS_e$ is distributed as F with 1 and ν_e degrees of freedom, provided the hypothesis $H_{oi}: \underline{c}_i'\underline{\tau} = 0$ for $i \in K_1$, is true.

Theorem 3.4. If model (2.4) is a normal model, then $\widehat{\underline{c}_i'\underline{\tau}}$, $i \in K_2$, and SS_E are distributed independently.

Proof. The proof consists in showing that

$$\underline{\Delta}\ \underline{P}_1(\gamma_1\underline{P}_1 + \gamma_e\underline{P}_e + \gamma\underline{P})\underline{H}_o = \gamma_1\underline{\Delta}\ \underline{P}_1\underline{H}_o.$$

For this note that

$\underline{P}_1\underline{G}'\underline{W} = \underline{G}'\underline{W}$, from which $\underline{P}_1\underline{H}_o = \underline{H}_o$, giving

$\underline{\Delta}\underline{P}_1\underline{H}_o = \underline{O}$, as $\underline{\Delta}\underline{H}_o = \underline{O}$.

Theorem 3.5. If model (2.4) is a normal model, then SS_E is distributed as $h\gamma_1\chi^2$ with ν_E degrees of freedom.

Proof. Note that $\underline{H}_o\underline{V} = \gamma_1\underline{H}_o\underline{P}_1 = \gamma_1\underline{H}_o$, from which the matrix $h^{-1}\gamma_1^{-1}\underline{H}_o\underline{V}$ is idempotent, and the theorem follows.

From Theorems 3.4 and 3.5 we get

Theorem 3.6. If model (2.4) is a normal model, then the ratio $F_{1i}=[\varepsilon_i(\widehat{\underline{c}_i'\tau})^2\nu_E]/SS_E$ is distributed as F with 1 and ν_E degrees of freedom provided the hypothesis $H_{1oi}:\underline{c}_i'\underline{\tau} = 0$, for $i \in K_2$, is true.

From the structure of vectors $\underline{c}_i$ note that contrasts $\underline{c}_i'\underline{\tau}$, for $i \in K_2$ concern the levels of factor A, these for $i=(m-1)h+g$, with $g=1,\dots,h$ concern the levels of factor B, and the remaining contrasts concern the interaction $A \times B$. Utilizing all of them the whole analysis of variance can be performed.

Let us consider, as a special case the traditional split-splot design (a complete case), for which $k=m$ and $\underline{r}=r\underline{1}_v$. Then the matrix $\underline{C}_1$ equal to $r\,\underline{I}_m - m^{-1}\underline{1}_m\underline{1}_m'$, has only one nonzero eigenvalue, $\varepsilon=1$ with multiplicity $m-1$.

In fact, in an incomplete split-plot design considered in this paper, the eigenvalue ε_i can be meant as an efficiency factor of the design for the corresponding contrast between levels of factor A (the wholeplot treatments). Only that contrast is partially confounded with blocks for which $\varepsilon_i < 1$. Contrasts between subplot treatments and interaction contrasts are in our case estimated with full efficiency, i.e. with the same efficiency as in complete split-plot design.

Finally it should be noticed that if in model (2.2) the natural restrictions $\sum_t \alpha_t = \sum_g \beta_g = \sum_t (\alpha\beta)_{tg} = \sum_g (\alpha\beta)_{tg} = 0$ are assumed then it is possible to base the analysis on two independent models, one containing subplot effects and wholeplot - subplot interactions, while the other containing the block and wholeplot treatment effects. The analysis could then be presented in a more straightforward way, but all the results concerning inferences on the contrasts would remain the same.

Acknowledgement

The author would like to thank Professor T.Caliński and a referee for many helpful comments on an earlier draft of this paper.

References

[1] R.P.Bhargava and K.R.Shah (1975). Analysis of some mixed-models for block and split-plot designs. Ann.Inst. Statist.Math. 27 365-375.

[2] W.T.Federer (1975). The misunderstood split-plot. In: R.P.Gupta, editor, Applied Statistics, North-Holland, Amsterdam, 9-39 .

[3] D.J.Finney (1946a). Recent developments in the design of field experiments I. Split-plot confounding. J.Agric. Sci., 36 56-62.

[4] D.J.Finney,(1946b). Recent developments in the design of field experiments II. Unbalanced split-plot confounding. J.Agric. Sci., 36 63-68.

[5] I.Mejza and S.Mejza (1984). Incomplete split-plot designs. Statistics and Probability Letters 2 327-332.

[6] S.C.Pearce, T.Caliński and T.F. de C.Marshall.(1974). The basic contrasts of an experimental design with special reference to the analysis of data. Biometrika 61 449-460.

[7] C.R.Rao and S.K.Mitra (1971). Generalized inverse of matrices and its applications, Wiley, New York.

[8] J.Robinson (1967). Incomplete split-plot designs. Biometrics 23 793-802.

[9] S.R.Searle, (1971). Linear Models, Wiley, New York.

SENSITIVITY OF LINEAR MODELS WITH RESPECT TO THE COVARIANCE MATRIX

E.Neuwirth
Wien, Austria

Summary. We consider the Gauss-Markov estimator (GME) in a linear model and the influence of using a wrong covariance matrix $V+\alpha R$ instead of the correct matrix V. For this purpose we give conditions that the derivative of the GME with respect to α vanishes. In many cases it turns out that derivatives only can be zero in directions where the GME is also constant for some finite interval.

1. Statement of the problem, results and applications

We consider a classical linear model where the random vector y is distributed according to $E(y) = X\beta$ and $\text{cov}(y) = V$. Then the classical Gauss-Markov-estimator (GME) of β has the following form $\tilde{\beta}(y) = \tilde{\beta}_V(y) = (X'V^{-1}X)^{-1}X'V^{-1}y$. For the rest of the paper we assume V to be regular and X having linearly independent column vectors.

We will study the influences of wrongly specifying V, i.e. taking another matrix W and using

$$\tilde{\beta}_W(y) = (X'W^{-1}X)^{-1}X'W^{-1}y.$$

It is well known (see e.g. [1],[3],[5],[6]) that $\tilde{\beta}_V$ and $\tilde{\beta}_W$ are identical for all y under the following conditions: Let $X^\perp$ be a matrix consisting of a maximal linear independent set of column vectors orthogonal to all column vectors of X. Then $\tilde{\beta}_V = \tilde{\beta}_W$ iff $VX^\perp$ and $WX^\perp$ span the same space. (The statement in this form is true only in case of V and W regular. There are also conditions for the singular case (see[5]), but in the following we only consider regular V and W.)

We will study now the case $W(\alpha) = V+\alpha R$ and differentiate $\tilde{\beta}(\alpha,y)=K(\alpha)y$, where

$$K(\alpha) = (X'W^{-1}(\alpha)X)^{-1}X'W^{-1}(\alpha),$$

with respect to α and find conditions that this derivative is 0 for some y or for all y and compare these conditions with the classical noninfinitesimal results.

<u>Theorem 1</u>: Let $W(\alpha)$ be regular on an interval. Then

$$\frac{dK}{d\alpha} = -K \frac{dW}{d\alpha} W^{-1}(I-P),$$

where P is the projection onto span (X) with respect to the inner product given by W^{-1} and therefore parallel to $\mathrm{span}(WX^{\perp})$, while K gives the coordinates of any projected vector with respect to the base consisting of the column vectors of X, and we have $P = XK$.

Therefore, $I-P$ is the projection onto the W^{-1}-orthogonal complement of span(X) which is $\mathrm{span}(WX^{\perp})$.

<u>Corollary 1</u>: For $W(\alpha) = V+\alpha R$ we have

$$\frac{dK}{d\alpha} = -KRW^{-1}(I-P)$$

and for $\alpha = 0$ we have

$$\frac{dK}{d\alpha} = -KRV^{-1}(I-P).$$

<u>Theorem 2</u>: Let $W(\alpha) = V+\alpha R$. Then we have $\frac{d\tilde{\beta}(\alpha,y)}{d\alpha} = 0$ for $\alpha=0$ iff one of the following conditions holds:

(i) $RV^{-1}(I-P)y \in \mathrm{span}(VX^{\perp})$,

(ii) $V^{-1}RV^{-1}(I-P)y \in \mathrm{span}(X^{\perp})$,

(iii) $y \in \mathrm{span}(X) \oplus (\mathrm{span}(VX^{\perp}) \cap \{z/RV^{-1}z \in \mathrm{span}(VX^{\perp})\})$,

(iv) $y \in \mathrm{span}(X) \oplus (\{z/V^{-1}z \in \mathrm{span}(X^{\perp})\} \cap \{z/V^{-1}RV^{-1}z \in \mathrm{span}(X^{\perp})\})$.

If we are interested in $A\tilde{\beta}(y)$ instead of $\tilde{\beta}(y)$ we have the following result.

<u>Theorem 3</u>: Let $W(\alpha) = V+\alpha R$. Then $\frac{dA\tilde{\beta}(\alpha,y)}{d\alpha} = 0$ for $\alpha = 0$ iff one of

the following conditions holds:

(i) $y \in \mathrm{span}(X) \oplus (\{z / V^{-1}z \in \mathrm{span}(X^{\perp})\} \cap \{z / V^{-1}RV^{-1}z \in \mathrm{span}(X^{\perp})\})$

$\oplus (\{z / V^{-1}z \in \mathrm{span}(X^{\perp})\} \cap \{z / RV^{-1}z \in \mathrm{span}(X(I-A^{-}A))\})$;

(ii) $y \in \mathrm{span}(X) \oplus (\mathrm{span}(VX^{\perp}) \cap \{z / V^{-1}RV^{-1}z \in \mathrm{span}(VX) \cap \mathrm{span}(V^{-1}X(I-A^{-}A))\})$,

where A^{-} denotes a generalized inverse of A.

Theorem 4: Let $W(\alpha) = V+\alpha R$. Then $\frac{d\tilde{\beta}(\alpha,y)}{d\alpha} = 0$ for $\alpha = 0$ and all $y \in R^n$ iff one of the following conditions holds:

(i) $V^{-1}RV^{-1}y \in \mathrm{span}(X)$ for all $y \in \mathrm{span}(VX^{\perp})$,

(ii) $\mathrm{span}(RX^{\perp}) \subset \mathrm{span}(VX^{\perp})$.

Since $\tilde{\beta}_V(y)$ and $\tilde{\beta}_W(y)$ are identical iff $\mathrm{span}(VX^{\perp}) = \mathrm{span}(WX^{\perp})$, condition (ii) implies $\mathrm{span}((V+\alpha R)X^{\perp}) \subset \mathrm{span}(VX^{\perp})$ and if we consider an interval containing 0 such that $V+\alpha R$ is regular on the whole interval we have

Corollary 2: Within the defined set up $\frac{d\tilde{\beta}(\alpha,y)}{d\alpha} = 0$ for all $y \in R^n$ and $\alpha = 0$ iff $\tilde{\beta}(0,y) = \tilde{\beta}(\alpha,y)$ for α from an interval containing 0 and for all $y \in R^n$.

Loosely speaking this means the following. If we search for directions of deviations of the covariance matrix where the local (infinitesimal) change of the estimate $\tilde{\beta}_{V+\alpha R}(y)$ is 0 for all $y \in R^n$ this is only possible in directions where $\tilde{\beta}_{V+\alpha R}(y)$ is constant on an interval. Conversely, this fact clearly implies zero derivative. So if we want "infinitesimal" insensibility of $\tilde{\beta}_{V+\alpha R}$ for all data points this only is possible in directions with "finite" insensibility, i.e. constancy. This result does not remain true if we want to have zero derivatives only for some data points y. Then it is possible to have zero derivatives for nonconstant $\tilde{\beta}_{V+\alpha R}$ as shown in the following example: Let $n = 3$, $V = I$ and let

$$X = \begin{bmatrix} 1 \\ 1 \\ 1 \end{bmatrix}, \quad R = \begin{bmatrix} 1 & -1 & 0 \\ -1 & 0 & 0 \\ 0 & 0 & 0 \end{bmatrix}, \quad y = \begin{bmatrix} 1 \\ 0 \\ -1 \end{bmatrix}.$$

Then we have

$$\tilde{\beta}_{V+\alpha R}(y) = \frac{\alpha^2}{3+4\alpha-\alpha^2}$$

and the derivative is 0 for $\alpha = 0$ but $\tilde{\beta}_{V+\alpha R}$ is strictly convex in some neighbourhood of 0. One can see a reason for this. Let $z=(I-P)y$. Then the condition for constancy is $z \in \text{span}((V+\alpha R)X^{\perp})$ and this is equivalent to $(V+\alpha R)^{-1} z \in \text{span}(X^{\perp})$. Since $(V+\alpha R)^{-1} = V^{-1}(\sum_{k=0}^{\infty} (-\alpha RV^{-1})^k)$, a sufficient condition for constancy of $\tilde{\beta}(y)$ in direction R is $(RV^{-1})^k z \in \text{span}(VX^{\perp})$ for all $k \geq 1$. When we have

$$RV^{-1} z \in \text{span}(VX^{\perp}) \quad \text{for all} \quad z \in \text{span}(VX^{\perp}),$$

we also have

$$(RV^{-1})^k z \in \text{span}(VX^{\perp}) \quad \text{for all} \quad z \in \text{span}(VX^{\perp}) \quad \text{for all} \quad k \geq 2$$

and so we have

$$(V+\alpha R)^{-1} z \in \text{span}(VX^{\perp}) \quad \text{for all} \quad z \in \text{span}(VX^{\perp}) \quad \text{and} \quad \alpha \text{ small enough.}$$

This implies constancy of $\tilde{\beta}$.

So for constructing a counterexample we need $z \in \text{span}(VX^{\perp})$ such that $RV^{-1} z \in \text{span}(VX^{\perp})$ and $(RV^{-1})^k z \in \text{span}(VX^{\perp})$ for some $k \geq 2$. So we have to solve the following system

$$X'V^{-1} z = 0,$$

$$X'V^{-1} RV^{-1} z = 0,$$

$$X'V^{-1}(RV^{-1})^k z = 0 \quad \text{for some} \quad k \geq 2.$$

We can conclude that in the case where $\text{span}(RX^{\perp})$ is not contained in $\text{span}(VX^{\perp})$ but $z \in \text{span}(VX^{\perp})$ and $RV^{-1} z \in \text{span}(VX^{\perp})$ already imply $(RV^{-1})^k z \in \text{span}(VX^{\perp})$ for $k \geq 2$ a derivative 0 also only is possible in directions R, where $\tilde{\beta}$ already is constant on a finite interval.

A special case is $\text{rank}(R) = 1$ implying $R = aa'$ with $a \in R^n$. Since in this case we have

$$(V+\alpha R)^{-1} = V^{-1} + \frac{\alpha}{1+\alpha \text{tr}(R)} V^{-1} RV^{-1}$$

the condition $\frac{d\tilde{\beta}(y)}{d\alpha} = 0$ implies $\tilde{\beta}_{V+\alpha R}(y) = \tilde{\beta}_V(y)$ for α taken from an interval containing 0. In this case the condition

$$RV^{-1}z \in \text{span}(VX^{\perp}) \quad \text{for} \quad z \in \text{span}(VX^{\perp})$$

is fulfilled if

(a) $a \in \text{span}(VX^{\perp})$

or

(b) $a'V^{-1}z = 0.$

Condition (a) implies $RV^{-1}z \in \text{span}(VX^{\perp})$ for all $z \in \text{span}(VX^{\perp})$ so in this case $\frac{d\tilde{\beta}(y)}{d\alpha} = 0$ for all $y \in R^n$. If on the other hand $a \in \text{span}(X)$ then $a'V^{-1}z = 0$ for all $z \in \text{span}(VX^{\perp})$ and in this case we also have $\frac{d\tilde{\beta}(y)}{d\alpha} = 0$ for all $y \in R^n$. So - roughly speaking - if we add aa' to V, where $a \in \text{span}(X)$ or $a \in \text{span}(VX^{\perp})$, then $\tilde{\beta}(y)$ does not change infinitesimally and finitely in this direction. If we have $R=ab'+ba'$ with $a,b \in R^n$ then $a,b \in \text{span}(X) \cup \text{span}(VX^{\perp})$ is sufficient for $\frac{d\tilde{\beta}(y)}{d\alpha}=0$ for all $y \in R^n$, and then again we have constancy on a finite interval. It is trivial that $\frac{d\tilde{\beta}(y)}{d\alpha} = 0$ for all $y \in \text{span}(X)$ but we furthermore have

<u>Theorem 5</u>: $\frac{d\tilde{\beta}(y)}{d\alpha} \neq 0$ for all $y \notin \text{span}(X)$ and $\alpha = 0$ if $\text{span}(RX^{\perp}) \cap \text{span}(VX^{\perp}) = \{0\}$ and $\text{span}(R) \oplus \text{span}(X) = R^n$.

In this case we also have $\tilde{\beta}_{V+\alpha R}(y) \neq \tilde{\beta}_V(y)$ for all α so once more "infinitesimal" and "finite" results coincide.

Now as a final result we will show a sufficient condition that for some $y \notin \text{span}(X)$

$$\frac{d\tilde{\beta}(y)}{d\alpha} = 0.$$

<u>Theorem 6</u>: Let $\text{rank}(R)+\text{rank}(X) < n$ or $\text{rank}(R) \geq \text{rank}(X)$ and $\text{rank}(X) < \frac{n}{2}$. Then there exists $y \notin \text{span}(X)$ with $\frac{d\tilde{\beta}(Y)}{d\alpha} = 0$ for $\alpha = 0$.

2. Proofs

For proving Theorem 1 we need the following

<u>Lemma 1</u>: Let $W(\alpha)$ be

a matrix-valued function of a real argument α, let $W(\alpha)$ be differentiable with respect to α and let $W(\alpha)$ be regular. Then $W^{-1}(\alpha)$ is differentiable and

$$\frac{dW^{-1}(\alpha)}{d\alpha} = W^{-1}\frac{dW}{d\alpha}W^{-1}.$$

Essentially this result is well known, a proof of a slightly different version can be found in [2].

Proof of Theorem 1: Notice that

$$\frac{dK}{d\alpha} = \frac{d(X'W^{-1}X)^{-1}}{d\alpha} X'W^{-1} + (X'W^{-1}X)^{-1}X'\frac{dW^{-1}}{d\alpha}$$

$$= -(X'W^{-1}X)^{-1}\frac{d(X'W^{-1}X)}{d\alpha}(X'W^{-1}X)^{-1}X'W^{-1} + (X'W^{-1}X)^{-1}X'\frac{dW^{-1}}{d\alpha}$$

$$= -(X'W^{-1}X)^{-1}X'\frac{dW^{-1}}{d\alpha}X(X'W^{-1}X)^{-1}X'W^{-1} + (X'W^{-1}X)^{-1}X'\frac{dW^{-1}}{d\alpha}$$

$$= (X'W^{-1}X)^{-1}X'\frac{dW^{-1}}{d\alpha}(-X(X'W^{-1}X)^{-1}X'W^{-1}+I)$$

$$= -(X'W^{-1}X)^{-1}X'W^{-1}\frac{dW}{d\alpha}W^{-1}(I-X(X'W^{-1}X)^{-1}X'W^{-1}) = K\frac{dW}{d\alpha}W^{-1}(XK-I).$$

Proof of Theorem 2: Clearly,

$$\frac{d\tilde{\beta}(y)}{d\alpha} = 0 \quad \text{iff} \quad KRV^{-1}(I-P)y = 0 \quad \text{iff} \quad RV^{-1}(I-P)y \in \ker(K) = \operatorname{span}(VX^{\perp})$$

$$\text{iff} \quad V^{-1}RV^{-1}(I-P)y \in \operatorname{span}(X^{\perp}).$$

Now every $y \in R^n$ can uniquely be decomposed into $y = x+z$ with $x \in \operatorname{span}(X)$ and $y \in \operatorname{span}(VX^{\perp})$ or, equivalently, $V^{-1}z \in \operatorname{span}(X^{\perp})$.

Therefore: $V^{-1}RV^{-1}(I-P)y \in \operatorname{span}(X^{\perp})$ iff $V^{-1}RV^{-1}z \in \operatorname{span}(X^{\perp})$ and, therefore, we arrive at $y \in X \oplus (\{z/V^{-1}z \in \operatorname{span}(X^{\perp})\} \cap \{z/V^{-1}RV^{-1}z \in \operatorname{span}(X^{\perp})\})$ or the equivalent formulation in the theorem.

Proof of Theorem 3: Clearly,

$$\frac{dA\tilde{\beta}(y)}{d\alpha} = 0 \quad \text{iff} \quad AKRV^{-1}(I-P)y = 0 \quad \text{iff} \quad RV^{-1}(I-P)y \in \ker(AK).$$

Decomposing $y = x+z$ as before we need $v=RV^{-1}z \in \ker(AK)$. Now v can be decomposed in $v=r+s$ with $r \in \text{span}(VX^{\perp})$ and $s \in \text{span}(X)$. Since $Kr=0$ we have $v \in \ker(AK)$ iff $s \in \ker(AK)$. Since $s = Xd$ for some d we need $AKXd = 0$. But by definition of K we have $KX = I$ and therefore $AKXd = Ad$. Now $Ad = 0$ iff $d \in \text{span}(I-A^{-}A)$ and so for $s \in \text{span}(X)$ we have $s \in \ker(AK)$ iff $s \in \text{span}(X(I-A^{-}A))$. Therefore $\ker(AK) = \text{span}(VX^{\perp}) \oplus \text{span}(X(I-A^{-}A))$ and we have $RV^{-1}y \in \ker(AK)$ iff $RV^{-1}z \in \text{span}(VX^{\perp}) \oplus \text{span}(X(I-A^{-}A))$ or, equivalently, $V^{-1}RV^{-1}z \in \text{span}(X^{\perp}) \oplus \text{span}(V^{-1}X(I-A^{-}A))$.

Proof of Theorem 4: Condition (i) is just a reformulation of condition (i) of Theorem 3 and is equivalent to

$$RV^{-1}y \in \text{span}(VX^{\perp}) \quad \text{for all} \quad y \in \text{span}(VX^{\perp})$$

and this is equivalent to

$$Rv \in \text{span}(VX^{\perp}) \quad \text{for all} \quad v \in \text{span}(X^{\perp})$$

or

$$\text{span}(RX^{\perp}) \subset \text{span}(VX^{\perp}).$$

Proof of Theorem 5: Clearly, $y \notin \text{span}(X)$ iff $y = x+z$ with $x \in \text{span}(X)$, $z \neq 0$ and $z \in \text{span}(VX^{\perp})$.

Now $\frac{d\tilde{\beta}(y)}{d\alpha} = \frac{d\tilde{\beta}(z)}{d\alpha}$ and $\frac{d\tilde{\beta}(z)}{d\alpha} = 0$ iff $RV^{-1}z \in \text{span}(VX^{\perp})$.

Also $\text{span}(R) \oplus \text{span}(X) = R^{n}$ imply $Rv \neq 0$ for all $v \in \text{span}(X^{\perp})$, since $\ker(R) = \text{span}(R^{\perp})$ and $\text{span}(X^{\perp}) \cap \text{span}(R^{\perp}) = \{0\}$ imply injectivity of R on $\text{span}(X^{\perp})$. So we have $RV^{-1}z = 0$ for all $z \in \text{span}(VX^{\perp})$ and $\text{span}(RX^{\perp}) \cap \text{span}(VX^{\perp}) = \{0\}$ then imply $RV^{-1}z \notin \text{span}(VX^{\perp})$ for all $z \neq 0$.

Proof of Theorem 6: Let $G = \text{span}(VX^{\perp})$ and $H = \{z / RV^{-1}z \in \text{span}(VX^{\perp})\}$. Then it suffices to show that $\dim(G) + \dim(H) > n$.

Now $\dim(G) = n-\text{rank}(X)$ and $\dim(H) \geq n-\text{rank}(R) + \max(0,\text{rank}(R)-\text{rank}(X))$. The second inequality follows from the fact that

$$\dim(H) \geq \dim(\ker(RV^{-1})) + \dim(\text{span}(VX^{\perp}) \cap \text{span}(R))$$

so $\text{rank}(X) + \text{rank}(R) < n$ implies $\dim(G)+\dim(H) > n$. For $\text{rank}(R) > \text{rank}(X)$ we have $\dim(H) \geq n-\text{rank}(X)$ and then $\text{rank}(X) < \frac{n}{2}$ implies dim(G) + dim(H) > n.

References

[1] W.Kraemer (1980). A note on the equality of ordinary least squares and Gauss-Markov-estimates in the general linear model. Sankhya A 42 130-131.

[2] K.V. Mardia, J.T.Kent and J.M.Bibby.(1979).Multivariate Analysis, Academic Press, New York.

[3] T.Mathew (1983). Linear estimation with an incorrect dispersion matrix in linear models with a common linear part. Journal of the Am.Stat. Ass. 78 468-471.

[4] W.Kruskal (1968). When are Gauss-Markov and least squares estimators identical ? A coordinate-free approach. Annals of Math. Statist 39 70-75.

[5] S.J.Haberman (1975). How much do Gauss-Markov and least squares estimators differ ? A coordinate-free approach. Annals of Statist. 3 982-990.

[6] E.Neuwirth, (1982). Parametric deviations in linear models. Probability and Statistical Inference (ed. by W.Grossmann et al.) Dordrecht 1982, 257-264.

ON A DECOMPOSITION OF THE SINGULAR GAUSS-MARKOV MODEL

K. Nordström
Swedish School of Economics and
Buisness Administration
Helsinki, Finland

1. Summary and notation

It is well-known that singularity of the dispersion matrix in a Gauss-Markov model may have various consequences, which remain obscure in models furnished with a regular dispersion matrix (see e.g. Rao [9] or Zyskind [16]). Roughly speaking, these consequences appear in form of inherent restrictions on the vector of observations or parameters (or both).

In this paper we revied these inherent properties from a geometrical point of view, and, in order to make these properties more explicit, decompose the model using a suitable column space decomposition of Euclidean n-space. More specifically, we begin by stating in Section 2 most of the results that are needed in the deduction in later Sections. In Section 3 we commence by deriving a simple result stating the relation between the column space of the dispersion matrix and the linear support of the unknown distribution of the error vector of the model (Theorem 3.1). Thenattention is turned to the peculiarities arising from a singular dispersion matrix, leading to Theorem 3.2, which provides the affine subspaces to which the vector of observations and parameters are confined by the singularity of the dispersion matrix. These results are then compared with results obtained by Rao [9]. In Section 4 we derive a decomposition of the model, which makes the results of Section 3 rather evident. The decomposition, which is somewhat different from the one given by Alalouf [1], is obtained combining column space decompositions given by Stein [15] (independently by Rao [10]) and Baksalary and Kala [2]. In Section 5 the adopted decomposition is

Key words: Singular Gauss-Markov models, projections, column space decompositions, affine subspaces.

utilized in determining the possible structures of model, resulting from different cases of relative location of the column spaces of the design and dispersion matrix. To emphasize the apparent complexity of singularity, a division of the concept of singularity into weak and strong singularity is proposed. In order to maintain the logical pattern of the development, detailed matrix representations of the projectors and vectors appearing in Section 4 have been deferred to a short Appendix.

Given a real matrix A, the symbols A', A^+, $r(A)$, $\mathscr{C}(A)$ and $\mathscr{N}(A)$ denote transpose, Moore-Penrose inverse, rank, column space and null space,respectively, of A. A^- and A_ℓ^- stand for generic elements of the class of g-inverses and least squares g-inverses of A, respectively, whereas $(A:B)$, $\mathscr{A}(a, \mathscr{C}(A))$ and $\mathscr{C}^\perp(A)$ denote a partitioned matrix with A and B as submatrices, the affine subspace with translating vector a and parallel to $\mathscr{C}(A)$ and the orthogonal complement of $\mathscr{C}(A)$, respectively. Orthogonality in this contex, and throughout the paper, means orthogonality with respect to the Euclidean (canonical) inner product. Given a subset S of a linear space, sp(S) denotes the span (i.e. linear extension)of S. Finally, given linear subspaces $\mathscr{U}$ and $\mathscr{V}$, $\mathscr{U} + \mathscr{V}$ stands for their sum, the sing + being replaced by $\oplus$, if $\mathscr{U}$ and $\mathscr{V}$ are disjoint (i.e. $\mathscr{U} \cap \mathscr{V} = \{0\}$), and by the sign $\boxplus$, if $\mathscr{U}$ and $\mathscr{V}$ are further known to be orthogonal.

2. Preliminaries

In this paper we will consider the singular Gauss-Markov model

$$\{y : y = X\beta + e, \beta \in \mathbb{R}^m\}, \tag{2.1}$$

where X is a known $n\times m$ matrix of rank $r \leq m$, $\beta \in \mathbb{R}^m$ is considered fixed but unknown in a classical spirit, and e is a random error vector with expectation $\mathscr{E}(e) = 0$ and dispersion matrix $\mathscr{D}(e) = V$. The matrix V is assumed to be known at least up to a positive scalar multiplier and singular with rank $k < n$. In the sequel the model (2.1) will be denoted by the triplet $(y, X\beta, V)_s$, the subscript s indicating that V is assumed to be singular.
To make the treatment reasonably self-contained, the rest of this section is devoted to results that are used extensively in later sections.

Lemma 2.1 (e.g. [Halmos 6, p.19]): Let A,B and C be matrices with columns in $\mathbb{R}^n$. Then

$$\mathscr{C}(A) \cap \{ \mathscr{C}(B) + [\mathscr{C}(A) \cap \mathscr{C}(C)] \} = \mathscr{C}(A) \cap \mathscr{C}(B) + \mathscr{C}(A) \cap \mathscr{C}(C). \quad (2.2)$$

Corollary 2.2: Let A,B and C be such that $\mathbb{R}^n = \mathscr{C}(B) \oplus \mathscr{C}(C)$ and $\mathscr{C}(C) \subset \mathscr{C}(A)$. Then

$$\mathscr{C}(A) = \mathscr{C}(A) \cap \mathscr{C}(B) \oplus \mathscr{C}(A) \cap \mathscr{C}(C). \quad (2.3)$$

Lemma 2.3 (Stein[15, Lemma IV.7], Rao [10, Lemma 2.1]): Let A be a matrix with columns in $\mathbb{R}^n$, Π the matrix of the orthogonal projector onto $\mathscr{C}^{\perp}(A)$ and B a n.n.d. (non-negative definite) matrix of order n. Then

$$\mathscr{C}(A : B) = \mathscr{C}(A) \oplus \mathscr{C}(B\Pi). \quad (2.4)$$

Lemma 2.4 (Baksalary and Kala [2, Lemma 1]): Let A and B be matrices with columns in $\mathbb{R}^n$ and Π the matrix of the orthogonal projector onto $\mathscr{C}^{\perp}(A)$. Then

$$\mathbb{R}^n = \mathscr{C}(A) \boxplus \mathscr{C}(\Pi B) \boxplus \mathscr{C}^{\perp}(A) \cap \mathscr{C}^{\perp}(B). \quad (2.5)$$

Note that (2.4) implies

$$\mathbb{R}^n = [\mathscr{C}(A) \oplus \mathscr{C}(B\Pi)] \boxplus \mathscr{C}^{\perp}(A) \cap \mathscr{C}^{\perp}(B), \quad (2.6)$$

which is analogous to (2.5), but depends crucially on B being n.n.d.

Lemma 2.5 (Ben-Israel [4, Theorem 1]): Let $\mathscr{A}(a, \mathscr{C}(A))$ and $\mathscr{A}(b, \mathscr{C}(B))$ be affine subspaces of $\mathbb{R}^n$. Then

$$\mathscr{A}(a, \mathscr{C}(A)) \cap \mathscr{A}(b, \mathscr{C}(B)) \neq \emptyset \Leftrightarrow (b-a) \in \mathscr{C}(A : B), \quad (2.7)$$

in which case

$$\mathscr{A}(a, \mathscr{C}(A)) \cap \mathscr{A}(b, \mathscr{C}(B)) = \mathscr{A}(a+AA_{\ell}^{-}(AA_{\ell}^{-} + BB_{\ell}^{-})^{+}(b-a), \mathscr{C}(A) \cap \mathscr{C}(B)). \quad (2.8)$$

The following results are well-known and simple, but rather fundamental in the theory of the model $(y, X\beta, V)_s$.

Lemma 2.6: In the model $(y,X\beta,V)_s$ the error vector e satisfies

$$e \in \mathcal{C}(V) \quad (a.s.). \tag{2.9}$$

Proof. The linear form $a'e$ vanishes (a.s.) for all a $\mathcal{N}(V)$. Q.E.D.

Corollary 2.7: In the model $(y,X\beta,V)_s$

$$y \in \mathcal{C}(X:V) \quad (a.s.). \tag{2.10}$$

3. Geometry of the singular model

(i) Dispersion matrix and linear support

To clarify the implications of a singular dispersion matrix, we begin by deriving a simple result stating the relation between $\mathcal{C}(V)$ and the linear support $sp(S_\mu)$ of the unknown distribution μ of e. Consider the support of μ

$$S_\mu = \{a \in \mathbb{R}^n : \mu(n(a)) > 0 \text{ for all neighbourhoods } n(a) \text{ of } a\}, \tag{3.1}$$

where the neighbourhoods are relative to the usual topology of $\mathbb{R}^n$. S_μ is a closed set, and may be characterized as the smallest closed set in $\mathbb{R}^n$ having μ-measure 1, i.e.

$$S_\mu = \cap\{F \subset \mathbb{R}^n : \mu(F) = 1 \text{ and } F \text{ is closed in } \mathbb{R}^n\} \tag{3.2}$$

[3, p.90]. Obviously S_μ need not be a linear space, and hence we consider $sp(S_\mu)$, the linear support of μ. Since $(2.9) \Leftrightarrow \mu(\mathcal{C}(V)) = 1$, and $\mathcal{C}(V)$ is closed in $\mathbb{R}^n$, (3.2) implies $S_\mu \subset \mathcal{C}(V)$. Hence we have

Theorem 3.1: With $\mathcal{C}(V)$ and $sp(S_\mu)$ defined as above, the following relation holds

$$sp(S_\mu) \subset \mathcal{C}(V). \tag{3.3}$$

Remark 1: Stein gives an erroneous proof of the converse inclusion, and claims that strict inclusion may hold [15, p.100-103]. In view of Theorem 3.1, this is impossible.

(ii) Restrictions inherent in the model

As is well-known , (2.9) and (2.10) have a number of consequences, which have to be taken into account in procedures of estimation and tests of hypotheses. Denoting $P = I - VV_{\ell}^{-}$ and $Q = I - XX_{\ell}^{-}$, we observe that (2.9) implies

$$Py = PX\beta \text{ (a.s.)}, \tag{3.4}$$

or more explicitly

$$\begin{aligned} P_1y &= P_1X\beta \quad \text{(a.s.)} \\ P_2y &= 0 \quad \text{(a.s.)}, \end{aligned} \tag{3.5}$$

where $P_1 = PX(PX)_{\ell}^{-}$ and $P_2 = 2Q(Q+P)^{-}P$ are the matrices of the orthogonal projectors corresponding to the decomposition

$$\mathcal{N}(V) = \mathcal{C}(PX) \boxplus \mathcal{C}^{\perp}(X) \cap \mathcal{N}(V), \tag{3.6}$$

obtainable from (2.5), choosing $A = V$, $B = X$ and $\Pi = P$. (The expression for P_2 follows from a result on parallel sums of matrices [13, Theorem 10.1.8].) As pointed out by Seely and Zyskind [14, p.693], the set

$$\mathcal{B} = \{\beta \in \mathbb{R}^m : Py = PX\beta\} \tag{3.7}$$

cannot be considered a (restricted) parameter space in the usual sense, since Py is unknown and $\mathcal{B}$ hence unspecifiable (i.e. equal to $\mathbb{R}^m$) prior to an observation of y. Denoting an observed value of y by $\underline{y}$ (and the corresponding realization of e by $\underline{e}$), the model $(y, X\beta, V)_S$ has,since the initial papers by Goldman and Zelen [5] and Mitra and Rao [8], often been converted to a model of smaller dimension and regular dispersion matrix, but with restrictions on β given by

$$\underline{\mathcal{B}} = \{\beta \in \mathbb{R}^m : P\underline{y} = PX\beta\}. \tag{3.8}$$

This reformulation is, however, logically unclear, as noted by Rao [12, p.1354], since it entails $Py = P\underline{y}$, i.e. rules out sets of measure zero. To avoid this reformulation, Harville [7] and Rao [12] have considered a class of estimators where the stochastic relationship (3.4) is exploited in a direct manner.

In view of the preceding, it is of some theoretical interest to determine the different cases that result in violation of

$$P\underline{y} = PX\beta . \tag{3.9}$$

Assume $\underline{e} \notin \mathscr{C}(V)$ is such that $P_2\underline{e} \neq 0$. Irrespective of whether $P_1\underline{e}$ equals zero or not, we have $P\underline{y} \neq PX\beta$, violating (3.9). Moreover, $P_2\underline{e} \neq 0$ implies violation of

$$\underline{y} \in \mathscr{C}(X : V), \tag{3.10}$$

which means that the model $(y, X\beta, V)_s$ is not consistent in the sense of Rao [9, p.278]. If, however, $\underline{e} \notin \mathscr{C}(V)$ is such that $P_1\underline{e} \neq 0$ and $P_2\underline{e} = 0$, the conclusion is $P\underline{y} \neq PX\beta$ and $\underline{y} \in \mathscr{C}(X : V)$, i.e. the observation $\underline{y}$ violates (3.9) but satisfies the condition of consistency. The last case is interesting, confirming in a detailed manner that consistency of the model $(y, X\beta, V)_s$ is not sufficient for $\underline{y}$ to satisfy (3.9), for which we must require

$$\underline{e} \in \mathscr{C}(V) \tag{3.11}$$

to hold. (3.11) can thus be viewed as expressing a form of stronger consistency of the model $(y, X\beta, V)_s$, but is, of course, unsuitable as a practical criterion of consistency, since $\underline{e}$ is, in general, only partly observable.

The preceding considerations provide a good starting point for a geometrical interpretation of the consequences of a singular dispersion matrix, which is the topic of the rest of this section. As pointed out by Rao [9, p. 278], (2.10) is the only statement concerning y that can be made prior to an observation. In particular, $\mathscr{B} = \mathbb{R}^m$, allowing β to be arbitrary in $\mathbb{R}^m$. Now assume $\underline{y}$ is recorded, and assume further that $\underline{y}$ satisfies (3.9). Then we have

<u>Theorem</u> 3.2: In the model $(y, X\beta, V)_s$ with observation $\underline{y}$ satisfying (3.9), the following relations hold

$$y \in \mathscr{A}(c, \mathscr{C}(V)) \quad \text{(a.s.)}, \tag{3.12}$$

$$X\beta \in \mathscr{C}(X) \cap \mathscr{A}(c, \mathscr{C}(V)), \tag{3.13}$$

$$X\beta \in \mathscr{A}(XX_\ell^-(XX_\ell^- + VV_\ell^-)^+c, \mathscr{C}(X) \cap \mathscr{C}(V)), \tag{3.13'}$$

$$\beta \in \mathscr{A}((PX)^{-}c,\ \overleftarrow{X}\{\mathscr{C}(V)\}), \tag{3.14}$$

$$\beta \in \mathscr{A}(X^{-}c,\ \mathscr{N}(X)) + \overleftarrow{X}\{\mathscr{C}(V)\}, \tag{3.14'}$$

where $c = P\underline{y}(=PX\beta)$ and $\overleftarrow{X}\{\mathscr{C}(V)\}$ is the inverse image of $\mathscr{C}(V)$ in the linear transformation with matrix X.

<u>Proof</u>. (3.12) follows from (3.4) observing that $\mathscr{N}(P) = \mathscr{C}(V)$. (3.13) follows from $c \in \mathscr{C}(X:V)$ and (3.9) by virtue of (2.7), and using (2.8), (3.13) takes the form (3.13').
(3.14) follows from (3.9) noting that $\mathscr{N}(PX) = \overleftarrow{X}\{\mathscr{C}(V)\}$, whereas (3.13) yields $\beta \in \overleftarrow{X}\{\mathscr{A}(c,\mathscr{C}(V))\} = \overleftarrow{X}\{c\} + \overleftarrow{X}\{\mathscr{C}(V)\} = \mathscr{A}(X^{-}c,\mathscr{N}(X)) + \overleftarrow{X}\{\mathscr{C}(V)\}$, i.e. (3.14'). Q.E.D.

An interesting question in this context is the relationship between the results obtained by Rao [9, Lemmas 2.3 and 2.4], and the ones above.
Denoting $N = Pc^{\perp}$, where $c^{\perp} = \{a \in \mathbb{R}^n : a \perp c\}$, the results given by Rao are as follows:

$$y \in \mathscr{C}^{\perp}(N) \quad \text{(a.s.)} \tag{3.15}$$

$$X\beta \in \mathscr{C}^{\perp}(N), \tag{3.16}$$

[9, Lemma 2.3], and

$$y \in \mathscr{C}(XS : V) \quad \text{(a.s.)}, \tag{3.17}$$

where $S = (X'N)^{\perp}$, i.e. a matrix of maximum rank such that $N'XS = 0$ [9, Lemma 2.4]. Since $\mathscr{C}(XS:V) = \mathscr{C}^{\perp}(N)$, we may concentrate on (3.15) and (3.16). Now

$$\mathscr{C}(c^{\perp}) = \{a \in \mathscr{C}(PX) : a \perp c\} \boxplus \mathscr{C}^{\perp}(X) \cap \mathscr{N}(V) \boxplus \mathscr{C}(V) \Rightarrow$$

$$\mathscr{C}(N) = P\,\mathscr{C}(c^{\perp}) = \{a \in \mathscr{C}(PX) : a \perp c\} \boxplus \mathscr{C}^{\perp}(X) \cap \mathscr{N}(V) \Rightarrow$$

$$\mathscr{C}^{\perp}(N) = [\,\mathscr{C}(c) \boxplus \mathscr{C}^{\perp}(X) \cap \mathscr{N}(V) \boxplus \mathscr{C}(V)\,] \cap \mathscr{C}(X:V) = \mathscr{C}(c) \boxplus \mathscr{C}(V),$$

where the last equality follows by applying (2.2). Thus (3.15) can be written as

$$y \in \mathscr{C}(c) \boxplus \mathscr{C}(V) \quad \text{(a.s.)}, \tag{3.18}$$

and comparing (3.18) with (3.12) it is seen that (3.15) provides the smallest linear subspace of $\mathbb{R}^n$ containing the affine subspace of (3.12). Similarly it is seen that (3.16) provides the smallest linear subspaces of $\mathbb{R}^n$ and $\mathbb{R}^m$ that contain the affine subspaces of (3.13') and (3.14), respectively.

From the preceding considerations it is obvious that $X\beta$ can be viewed as being the sum of a component that is essentially known as $\underline{y}$ is known, and one that remains unaffected by the knowledge of $\underline{y}$. To make this property explicit, the rest of this paper is devoted to obtaining a suitable decomposition of the model $(y, X\beta, V)_s$, and to the possible implications of such a decomposition.

4. Decomposition of the model

To obtain the desired decomposition, we proceed as follows. Choose $A = X$, $B = V$ and $\Pi = Q$ in (2.4), yielding

$$\mathscr{C}(X : V) = \mathscr{C}(X) \oplus \mathscr{C}(VQ), \tag{4.1}$$

which is the fundamental decomposition applied by Rao [10]. Now choose $A = Q$, $B = P$ and $\Pi = XX_\ell^-$ in (2.5), to obtain the orthogonal direct sum

$$\mathscr{C}(X) = \mathscr{C}(X) \cap \mathscr{C}(V) \boxplus \mathscr{C}(XX_\ell^- P), \tag{4.2}$$

which, combined with (4.1), and choosing $\mathscr{C}^\perp(X : V)$ as the complementary space of $\mathscr{C}(X : V)$ in $\mathbb{R}^n$, establishes

$$\mathbb{R}^n = \{[\mathscr{C}(X) \cap \mathscr{C}(V) \boxplus \mathscr{C}(XX_\ell^- P)] \oplus \mathscr{C}(VQ)\} \boxplus \mathscr{C}^\perp(X) \cap \mathscr{N}(V). \tag{4.3}$$

According to (4.3), y may be expressed as the sum of four unique components $y_1, \dots, y_4$:

$$\begin{aligned}
y_1 &= \Pi_{\mathscr{C}(X) \cap \mathscr{C}(V)} [P_{\mathscr{C}(X) | \mathscr{C}(VQ)} (\Pi_{\mathscr{C}(X:V)} y)] \\
y_2 &= \Pi_{\mathscr{C}(XX_\ell^- p)} [P_{\mathscr{C}(X) | \mathscr{C}(VQ)} (\Pi_{\mathscr{C}(X:V)} y)] \\
y_3 &= P_{\mathscr{C}(VQ) | \mathscr{C}(X)} (\Pi_{\mathscr{C}(X:V)} y) \\
y_4 &= \Pi_{\mathscr{C}^\perp(X:V)} y,
\end{aligned} \tag{4.4}$$

where $\Pi_{\mathscr{U}}$ denotes the orthogonal projector onto $\mathscr{U}$ and $P_{\mathscr{V}|\mathscr{W}}$ denotes a member of the class of generalized projectors, $\{P_{\mathscr{V}|\mathscr{W}}\}$, with $\mathscr{V}$ as projection and $\mathscr{W}$ as directionspace, respectively [10, Definition 1]. (Matrix representations of the projectors and vectors in (4.4) and (4.5) are given in the Appendix). Substituting $X\beta + e$ for y in (4.4) yields

$$\begin{aligned} y_1 &= X\beta_1 + e_1 \\ y_2 &= X\beta_2 + e_2 \\ y_3 &= e_3 \\ y_4 &= e_4 \end{aligned} \tag{4.5}$$

from which it is seen that only y_1 and y_2 provide information on $X\beta$. Moreover, by virtue of (2.2) we have

$$\mathscr{C}(V) = \mathscr{C}(X) \cap \mathscr{C}(V) \oplus \mathscr{C}(VQ), \tag{4.6}$$

and combining (4.6) with (2.9) it is seen that e_2 and e_4 vanish (a.s.), implying that y_1 is stochastic and y_2 deterministic (i.e. essentially non-stochastic). Hence $X\beta_2$ is the part of $X\beta$ that becomes known once $\underline{y}$ is known. Observing that composite operators of the form $P_{\mathscr{C}(X)|\mathscr{C}(VQ)} \circ \Pi_{\mathscr{C}(X:V)}$ constitute a subclass of $\{P_{\mathscr{C}(X)|\mathscr{C}(VQ)}\}$, it is evident that the only difference between Rao's decomposition and the one given in (4.4), is the further decomposition of $P_{\mathscr{C}(X)|\mathscr{C}(VQ)}(\Pi_{\mathscr{C}(X:V)}y)$ into two orthogonal components. Thus $y_1 + y_2$ is the BLUE of $X\beta$, y_3 the vector of residuals providing the MINQUE of the scalar multiplier of the dispersion matrix, and y_4 a component that equals zero, provided the model is consistent. For consistent models the expressions in (4.4) simplify, as $\Pi_{\mathscr{C}(X:V)}y$ can be replaced by y.

<u>Remark</u> 2: The particular choice of $\mathscr{C}^{\perp}(X:V)$ as complementary space of $\mathscr{C}(X:V)$ in (4.3) implies the use of the Euclidean metric in $\mathbb{R}^n \setminus \mathscr{C}(X:V)$.

It is well-known that the direct sum in (4.1) is non-orthogonal, unless $\mathscr{C}(X)$ is V-invariant (e.g. Zyskind [16, p. 654]), but one may ask whether (4.6) is an orthogonal direct sum, i.e. whether $e_1 + e_3$ is an orthogonal decomposition of the effective error vector. Choosing $A = P$, $B = Q$ and $\Pi = VV_\ell^-$ in (2.5) implies

$$\mathscr{C}(V) = \mathscr{C}(X) \cap \mathscr{C}(V) \boxplus \mathscr{C}(VV_\ell^- Q), \tag{4.7}$$

and hence the question reduces to the coincidence of $\mathscr{C}(VQ)$ with $\mathscr{C}(VV_\ell^- Q)$. For this we state

<u>Theorem</u> 4.1; $\mathscr{C}(VQ) = \mathscr{C}(VV_\ell^- Q) \Leftrightarrow \mathscr{C}(VQ)$ is V-invariant.

<u>Proof</u>. Since $\mathscr{C}\{V[I - VQ(VQ)_\ell^-]\} = \mathscr{C}(X) \cap \mathscr{C}(V)$ [8, Lemma 2], we have

$$\mathscr{C}(VQ) = \mathscr{C}(VV_\ell^- Q) \Leftrightarrow (VQ)'\, V[I - VQ(VQ)_\ell^-] = 0 \Leftrightarrow$$

$\mathscr{C}^\perp(VQ)$ is V-invariant $\Leftrightarrow \mathscr{C}(VQ)$ is V-invariant, by symmetry of V.

Q.E.D.

5. Singularity and structure of the model

In this section we examine in some detail the possible structures of model, resulting from different cases of relative location of $\mathscr{C}(X)$ and $\mathscr{C}(V)$ in $\mathbb{R}^n$. Before turning to these structures, we note that the dimensions of the linear spaces corresponding to the different cases to be treated, are as follows:

	$\mathscr{C}(X) \cap \mathscr{C}(V)$	$\mathscr{C}(XX_\ell^- P)$	$\mathscr{C}(VQ)$	$\mathscr{C}^\perp(X) \cap \mathscr{N}(V)$
A. General case	s	r-s	k-s	n-[k+(r-s)]
B. Special cases				
(i) $\mathscr{C}(X) = \mathscr{C}(V)$	r=k	0	0	n-r=n-k
(ii) $\mathscr{C}(X) \subsetneqq \mathscr{C}(V)$	r	0	k-r	n-k
(iii) $\mathscr{C}(X) \supsetneqq \mathscr{C}(V)$	k	r-k	0	n-r
(iv) $\mathscr{C}(X) = \mathscr{N}(V)$	0	r=n-k	k	0
(v) $\mathscr{C}(X) \subsetneqq \mathscr{N}(V)$	0	r	k	n-k-r
(vi) $\mathscr{C}(X) \supsetneqq \mathscr{N}(V)$	r-(n-k)	n-k	n-r	0

<u>Remark</u> 3: In the cases (i), (iii), (iv) and (vi), (4.2) becomes

$$\mathscr{C}(X) = \mathscr{C}(X) \cap \mathscr{C}(V) \boxplus \mathscr{C}(X) \cap \mathscr{N}(V), \tag{4.2'}$$

by virtue of (2.3).

A. General case.
In the general case, when all the linear spaces in (4.3) are non-trivial, the structure of the model $(y,X\beta,V)_s$ corresponding to the decomposition (4.3) is given by

$$y = X\beta_1 + X\beta_2 + e_1 + e_2 + e_3 + e_4 \tag{5.1}$$

or

$$y = X\beta_1 + X\beta_2 + e_1 + e_3 \quad \text{(a.s.)}, \tag{5.2}$$

since e_2 and e_4 vanish (a.s.).

In what follows, we only state the structure corresponding to (5.2), omitting the repetitious (a.s.).

B. Special cases

(i) $\mathcal{C}(X) = \mathcal{C}(V)$

Since $\mathcal{C}(XX_\ell^- P) = \mathcal{C}(VQ) = \{0\}$, we have the structure

$$y = X\beta_1 + e_1, \tag{5.3}$$

showing that the model lacks a deterministic component providing sure information on $X\beta$, and is non-residual.

(ii) $\mathcal{C}(X) \underset{\neq}{\subset} \mathcal{C}(V)$

In this case $\mathcal{C}(XX_\ell^- P) = \{0\}$, yielding

$$y = X\beta_1 + e_1 + e_3. \tag{5.4}$$

Hence this case differs from (i) in that the model may have a non-zero vector of residuals.

(iii) $\mathcal{C}(X) \underset{\neq}{\supset} \mathcal{C}(V)$

Since $\mathcal{C}(VQ) = \{0\}$, we have

$$y = X\beta_1 + X\beta_2 + e_1, \tag{5.5}$$

from which it is seen that the model has a deterministic component providing sure information on $X\beta$, and is non-residual.

(iv) $\mathscr{C}(X) = \mathscr{N}(V)$

This case is somewhat trivial, since $\mathscr{C}(X) \cap \mathscr{C}(V) = \mathscr{C}^{\perp}(X:V) = \{0\}$ implies the structure

$$y = X\beta_2 + e_3. \tag{5.6}$$

As $X\beta_2$ is known (in practice) after observing $\underline{y}, \underline{y}$ can actually be resolved into two unique components. Hence there is no estimation problem in standard sense. In addition the model is always consistent, a possibility noted earlier by Rao [9, p. 282].

(v) $\mathscr{C}(X) \overset{\subset}{\neq} \mathscr{N}(V)$

Since $\mathscr{C}(X) \cap \mathscr{C}(V) = \{0\}$, we have the structure

$$y = X\beta_2 + e_3. \tag{5.7}$$

This case differs from (iv) in that $\mathscr{C}^{\perp}(X:V)$ is non-trivial, allowing for possible inconsistency of the model.

(vi) $\mathscr{C}(X) \overset{\supset}{\neq} \mathscr{N}(V)$

Here $\mathscr{C}^{\perp}(X;V) = \{0\}$, resulting in

$$y = X\beta_1 + X\beta_2 + e_1 + e_3. \tag{5.8}$$

Hence the model is always consistent, has both a stochastic and a deterministic component providing information on $X\beta$ and may have a non-zero vector of residuals.

From the preceding it is apparent that the consequences of singularity on the structure of the model are exceedingly varying. Therefore it is felt that a division of the concept of singularity would be clarifying. At least from the point of view of applications, the most suitable criterion of division appears to be the restriction-inducing property of the model. Thus the following definition is proposed.

Definition 1: The model $(y,X\beta,V)_s$ is strongly (weakly) singular iff it possesses a non-zero (zero) deterministic component providing information on $X\beta$.

Gathering together the results of this section, we obtain the following result.

Theorem 5.1: The model $(y,X\beta,V)_s$ is

(i) weakly singular $\Leftrightarrow$ $\mathscr{C}(X) \subset \mathscr{C}(V)$

(ii) strongly singular $\Leftrightarrow$ $\mathscr{C}(X) \not\subset \mathscr{C}(V)$

(iii) non-residual $\Leftrightarrow$ $\mathscr{C}(X) \supset \mathscr{C}(V)$

(iv) always consistent $\Leftrightarrow$ $\mathscr{C}(X) \supset \mathscr{N}(V)$.

The results (i) and (ii) of Theorem 5.1 have been used in constructing normal equations suitable for the general Gauss-Markov model, and are hence well-known (e.g. [16, p. 658]).

Concluding remarks: It should be stressed that singular dispersion matrices arise naturally in several important classes of statistical problems, even excluding the standard case of representing a restricted non-singular model as an augmented singular model. Completely randomized experiments in experimental design and complete systems of linear demand functions in econometrics are prominent examples. The methods of dealing with a singular dispersion matrix in econometrics, appear, in general, rather involved and ad hoc, and, unfortunately, the elegant theory developed by Rao and others is hardly ever referred to.

Acknowledgement: The author would like to thank Prof. H. Niemi for helpful comments on an earlier draft of the paper.

Appendix

In this Appendix we give matrix representations of projectors and vectors appearing in (4.4) and (4.5).
To begin with, $\Pi_{\mathscr{C}(X)\cap\mathscr{C}(V)}$ may be given the matrix representation

$$2XX_{\ell}^{-}(XX_{\ell}^{-} + VV_{\ell}^{-})^{-}VV_{\ell}^{-}\ , \tag{A.1}$$

by virtue of a result on parallel sums of matrices [13, Theorem 10.1.8]. Further the class of composite operators of the form $P_{\mathscr{C}(X)|\mathscr{C}(VQ)} \circ \Pi_{\mathscr{C}(X:V)}$, may be given the convenient matrix representation

$$X(X'T^{-}X)^{-}X'T_{\ell}^{-}, \tag{A.2}$$

where $T = V + XUX'$ and U is any symmetric matrix such that $r(T) = r(X:V)$ (see [2, Theorem 2] or [11, p. 1203]). Observing that composite operators of the form $P_{\mathscr{C}(VQ)|\mathscr{C}(X)} \circ \Pi_{\mathscr{C}(X:V)}$ can be obtained from the preceding, using the result

$$P_{\mathscr{C}(X)|\mathscr{C}(VQ)} + P_{\mathscr{C}(VQ)|\mathscr{C}(X)} = I \quad \text{in} \quad \mathscr{C}(X:V) \tag{A.3}$$

[10, Lemma 2.10], the vectors of (4.4) and (4.5) may be given the following (not necessarily computationally optimal) matrix representations

$$\begin{aligned}
y_1 &= 2XX_{\ell}^{-}(XX_{\ell}^{-} + VV_{\ell}^{-})^{-}VV_{\ell}^{-}X(X'T^{-}X)^{-}X'T_{\ell}^{-}y \\
y_2 &= XX_{\ell}^{-}P(XX_{\ell}^{-}P)_{\ell}^{-}X(X'T^{-}X)^{-}X'T_{\ell}^{-}y \\
y_3 &= [I - 2Q(Q + P)^{-}P - X(X'T^{-}X)^{-}X'T_{\ell}^{-}]y \\
y_4 &= 2Q(Q + P)^{-}Py
\end{aligned} \tag{4.4'}$$

and

$$\begin{aligned}
y_1 &= 2XX_{\ell}^{-}(XX_{\ell}^{-} + VV_{\ell}^{-})^{-}VV_{\ell}^{-}[X\beta + X(X'T^{-}X)^{-}X'T_{\ell}^{-}e] \\
y_2 &= XX_{\ell}^{-}\ P(XX_{\ell}^{-}P)_{\ell}^{-}[X\beta + X(X'T^{-}X)^{-}X'T_{\ell}^{-}e] \\
y_3 &= [I - 2Q(Q + P)^{-}P - X(X'T^{-}X)^{-}X'T_{\ell}^{-}]e \\
y_4 &= 2Q(Q + P)^{-}Pe.
\end{aligned} \tag{4.5'}$$

REFERENCES

[1] Alalouf, I.S. (1978): An explicit treatment of the general linear model with singular covariance matrix. Sankhyā, Ser. B, 40, 65-73.

[2] Baksalary, J.K. and Kala, R. (1978): Relationships between some representations of the best linear unbiased estimator in the general Gauss-Markoff model. SIAM J. Appl. Math. 35, 515-520.

[3] Barndorff-Nielsen, O. (1978): Information and Exponential Families in Statistical Theory. New York: Wiley.

[4] Ben-Israel, A. (1967): On the geometry of subspaces in Euclidean n-spaces. SIAM J. Appl. Math. 15, 1184-1198.

[5] Goldman, A. J. and Zelen, M. (1964): Weak generalized inverses and minimum variance linear unbiased estimation. J. Res. Nat. Bur. Standards Sect. B, 68, 151-172.

[6] Halmos, P.R. (1974): Finite-dimensional Vector Spaces, 2nd ed. New York: Springer-Verlag.

[7] Harville, D.A. (1981): Unbiased and minimum-variance unbiased estimation of estimable functions for fixed linear models with arbitrary covariance structure. Ann. Statist. 9, 633-637.

[8] Mitra, S.K. and Rao, C.R. (1968): Some results in estimation and tests of linear hypotheses under the Gauss-Markoff model. Sankhyā, Ser. A, 30, 281-290.

[9] Rao, C.R. (1973): Representations of best linear unbiased estimators in the Gauss-Markoff model with a singular dispersion matrix. J. Multiv. Anal. 3, 276-292.

[10] Rao, C.R. (1974): Projectors, generalized inverses and the BLUE's. J. Roy. Stat. Soc., Ser. B, 36, 442-448.

[11] Rao, C.R. (1978): Choice of best linear estimators in the Gauss--Markoff model with a singular dispersion matrix. Commun. Statist. A 7 (13), 1199-1208.

[12] Rao, C.R. (1979): Estimation of parameters in the singular Gauss--Markoff model. Commun. Statist. A 8 (14), 1353-1358.

[13] Rao, C.R. and Mitra, S.K. (1971): Generalized Inverse of Matrices and its Applications. New York: Wiley.

[14] Seely, J. and Zyskind, G. (1971): Linear spaces and minimum variance unbiased estimation. Ann. Math. Statist. 42, 691-703.

[15] Stein, R.A. (1972): Linear model estimation, projection operators, and conditional inverses. Unpublished Ph. D. Thesis, Iowa State University Library, Ames, Iowa.

[16] Zyskind, G. (1975): Error structures, projections and conditional inverses in linear model theory. In: A Survey of Statistical Design and Linear Models (Shrivastava, U.N., ed.). Amsterdam: North-Holland.

RIDGE TYPE M-ESTIMATORS

H. Nyquist
Department of Statistics
Umea University,
Sweden

Abstract. In this paper we introduce a new class of estimators, ridge type M-estimators, designed for analyzing linear regression models when regressor variables are multicollinear and residual distributions display long tails. The estimators are defined as weighted maximum likelihood type (M-) estimators when additional information about the parameters is given. An example shows that conclusions based on a ridge type M-estimator can be rather different from conclusions based on the M-estimator and from the ordinary ridge regression estimator.

1. Introduction

A serious problem that can occur when analyzing linear regression models is the presence of multicollinearity among the regressor variables, causing highly unstable least squares estimates of the regression parameters. Therefore, considerable attention has been given to alternative estimation techniques for non-orthogonal problems. One such technique is the ridge regression estimator, originally introduced by Hoerl and Kennard (1970 a) and followed by a large number of papers in the statistical literature, see Hoerl and Kennard (1981) for a recent bibliographical survey.

Another serious problem that often occurs in applications of regression models, in its nature completely different from the multicollinearity problem, is that of residual term distributions characterized by long

KEY WORDS: multicollinearity, long tailed residual distributions, ridge regression, maximum likelihood type estimators.

Research reported in this paper has been supported by the Swedish Research Council for the Humanities and the Social Sciences.

tails. It is now recognized that outliers which are the direct result of non-normal, longtailed residual distributions do have an unsatisfactory large influence on the least squares estimator. In order to "robustify" estimators several procedures have been proposed. A class of such estimators introduced by Huber (1964,1973) is the class of maximum likelihood type (M-) estimators.

When multicollinearity among regressor variables in a regression model is present it is easily shown that M-estimates, like least squares estimates, possess large asymptotic variances. The purpose of this paper is to define a ridge type M-estimator which is robust against fat-tailed residual distributions and, as well, multicollinearity. Thus, we desire an estimator of the $m\times 1$ parameter vector β in the linear regression model

$$y = X\beta + u, \tag{1}$$

where y is a $n\times 1$ vector of observations on a dependent variable, X is a $n\times m$ matrix of nonstochastic regressors, and u is a $n\times 1$ vector of unknown residuals. The residual variance is denoted by σ^2.

2. The ordinary ridge regression estimator

The ordinary ridge regression (ORR) estimator is defined by

$$b(k) = (X'X + kI)^{-1}X'y, \quad k \geq 0, \tag{2}$$

and indexed by a scalar parameter k. Note that $b(0)$ is the least squares estimator and that $b(k)$ is a biased estimator of β if $k > 0$ is a fixed constant. Provided that $\lim n^{-1} X'X$ exists and is a non--singular matrix, $b(k)$ is consistent. Letting k tend to infinity we find that $b(k)$ approaches zero in the limit. We can therefore consider the ORR estimator as a compromise between the least squares estimator $b(0)$ which is unbiased, but unstable (if multicollinearity is present) and $\lim_{k\to\infty} b(k) = 0$ which is biased, but stable. One of the more interesting results for the ORR estimator is the existence of a positive k such that

$$\text{tr MSE } [b(k)] < \text{tr MSE } [b(O)], \tag{3}$$

see Hoerl and Kennard (1970 a) and the generalization in Theobald (1974).

Several interpretations of the ORR estimator have been proposed. One possible interpretation is that additional information about the parameters is provided by the linear expression

$$O = \beta + \delta, \tag{4}$$

where δ is an mx1 vector of independent disturbances whith expectation zero and variance τ^2. Assuming u and δ to be independent, the weighted least squares estimator of β has the form (2) when (1) and (4) are combined, with $k = \sigma^2/\tau^2$. The additional information can be based on a prior distribution of β in the usual Bayesian sense (Lindley and Smith, 1972), on theoretical restrictions on the parameters (Smith, 1976), or on experience from other sets of observations (Anderson and Battiste, 1975).

In the original articles by Hoerl and Kennard (1970 a,b) k is not interpreted as containing any a priori information. It is used in the ridge trace (the plot of the components of $b(k)$ and corresponding estimates of σ^2 versus k) to illustrate the effect of the multicollinearity on the estimation of β. The optimal value of k minimizing MSE is a function of the unknown β and σ^2. Many attempts to utilize information in data have therefore been made when selecting k, implying k to be stochastic. However, when k is stochastic there is no guarantee that ORR dominates least squares in the MSE sense described by (3). The problem of selecting a "good" value of k is unsolved but often discussed in the literature. In this paper we leave that question and consider only the ridge trace and generalizations of it to analyze the effects of multicollinearity.

3. M-estimators

It is well known that the least squares estimator of β in the linear regression model (1) is sensitive to departures from normality in the residual distribution. In particular, the least squares estimator possesses intractable properties when the residual distribution has long tails. As alternatives to the least squares estimator members of the class of maximum likelihood type (M-) estimators have been proposed, see Huber (1973). For a given real valued criterion function $\rho(\cdot)$ the corresponding M-estimator is defined by minimizing

$$\sum_{i=1}^{n} \rho(y_i - x_i\beta),$$

where y_i and x_i, $i = 1, \ldots, n$ are the components and the rows of y and X, respectively. Note that the particular choice $\rho(u) = u^2$ provides the least squares estimator and the choice $\rho(u) = -\log f(u)$ with $f(u)$ the probability density function of the residual distribution provides the maximum likelihood estimator.

Unfortunately, such M-estimators are not generally scale invariant. Therefore to obtain a scale invariant version we instead minimize

$$\sum_{i=1}^{n} \rho((y_i - x_i\beta)/s), \tag{5}$$

where s is a robust estimate of scale.

Assuming $\rho(\bullet)$ to be differentiable with derivate $\psi(\cdot)$, a necessary condition for a minimum of (5) is that the estimate $\hat{\beta}$ satisfies the equations

$$\sum_{i=1}^{n} x_{ij}\, \psi((y_i - x_i\hat{\beta})/s) = 0, \quad j = 1, \ldots, m. \tag{6}$$

Under mild regularity conditions on ψ and the residual distribution Huber (1973) showed that $\hat{\beta}$ is consistent and that $\sqrt{n}(\hat{\beta} - \beta)$ is asymptotically normally distributed with mean zero and variance $\omega^2 Q^{-1}$, where

$$\omega^2 = \frac{E[s^2\psi^2(u/s)]}{E^2[\psi'(u/s)]},$$

and

$$Q = \lim_{n\to\infty} n^{-1}X'X.$$

Note that the asymptotic variance depends only on the residual distribution through the constant ω^2. Note also that the asymptotic distribution of M-estimators is of the same form as the distribution of the least squares estimator. In particular, when regressor variables are multicollinear the matrix Q may be ill-conditioned implying relatively large asymptotic variances. Hence, we expect M-estimators to perform asymptotically similarly to the least squares estimator in the presense of multicollinearity.

If the residuals are independent and identically distributed up to a scale constant there exists a diagonal $n \times n$ matrix D with strictly positive diagonal elements, such that the transformed residuals $u^*=Du$ are independent and identically distributed. Thus, the transformed model

$$y^* = X^*\beta + u^*, \tag{7}$$

with $y^* = Dy$ and $X^* = DX$, satisfies the assumptions for M-estimation. We define a weighted M-estimator of β when an M-estimator is applied to (7).

4. Ridge type M-estimators

In order to define a ridge type M-estimator we generalize the interpretation of the ridge estimator considered as a weighted least squares estimator when additional information about the regression parameters is provided. Combining the model (1) with the additional information (4) with weights equal to one for the n observations and equal to the square root of the ridge parameter, $\sqrt{k}$, for the m additional equations the ridge type M-estimator is defined by minimizing

$$\sum_{i=1}^{n} \rho((y_i - x_i\beta)/s) + \sum_{i=1}^{m} \rho(\sqrt{k}\beta_i/s) \tag{8}$$

Assuming a differentiable criterion function we obtain the system of equations

$$\sum_{i=1}^{n} x_{ij} \psi((y_i - x_i\hat{\beta}(k))/s) + \sqrt{k}\,\psi(\sqrt{k}\,\hat{\beta}_j(k)/s) = 0, \quad j=1, \ldots, m. \tag{9}$$

As a special case the ordinary ridge regression estimator discussed in

Section 2 is obtained for $\rho(u) = u^2$. Further, $\hat{\beta}(0)$ is the M-estimator, and for a finite k, $\hat{\beta}(k)$ is a consistent estimator of β provided that $\lim n^{-1}X'X$ exists and is non-singular. Simple numerical counter examples show that ridge type M-estimators are not generally symmetrically distributed. For criterion functions such that $\rho(u) \to \infty$ when $|u| \to \infty$ (excluding the class of redescending M-estimators) we find that $\hat{\beta}(k)$ tends to zero when k tends to infinity suggesting that the ridge type M-estimator can be considered as a compromise between the consistent but possibly unstable M-estimator $\hat{\beta}(0)$ and the biased but stable $\lim_{k\to\infty} \hat{\beta}(k) = 0$.

In general, the equations (9) are nonlinear and must be solved by some iterative technique. In this study we applied the iteratively reweighted least squares technique (see for example Holland and Welsch (1977)) defined as follows: Given the estimate $\hat{\beta}(k)^{(l)}$ in the l-the iteration, weights $w_i^{(l)}$ are computed as

$$w_i^{(l)} = \psi((y_i - x_i\hat{\beta}(k)^{(l)})/s)/((y_i - x_i\hat{\beta}(k)^{(l)})/s)$$

for the n observations and

$$v_i^{(l)} = \psi(\sqrt{k}\hat{\beta}_i(k)^{(l)}/s)/(\hat{\beta}_i(k)^{(l)}/s)$$

for the m additional equations. A new estimate $\hat{\beta}^{(l+1)}$ is found as the solution to the system of linear equations

$$\sum_{i=1}^{n} w_i^{(l)} x_{ij}(y_i - x_i\hat{\beta}(k)^{(l+1)}) + v_j^{(l)}\hat{\beta}_j(k)^{(l+1)} = 0, \quad j=1,\dots,m,$$

or, expressed in matrix form

$$\hat{\beta}(k)^{(l+1)} = (X'W^{(l)}X + V^{(l)})^{-1}X'W^{(l)}Y,$$

where $W^{(l)}$ and $V^{(l)}$ are the nxn and mxm diagonal matrices with elements $w_i^{(l)}$ and $v_i^{(l)}$, respectively.

As starting values $\hat{\beta}(k)^{(0)}$ we used the least absolute deviations estimator ($\rho(u) = |u|$ in (8)) computed by the Barrodale and Roberts (1972) algorithm. As estimator of s we used the symmetrized interquartile range

$$s = \text{median}\,|(y_i - x_i\hat{\beta}(k)^{(0)}) - \text{median}\,(y_i - x_i\hat{\beta}(k)^{(0)})|/\Phi^{-1}(.75)$$

where $\Phi(\cdot)$ is the standard normal cumulative distribution function. The value $\Phi^{-1}(.75) \approx .6745$ is used because s is then a consistent estimator of the standard deviation of a normally distributed stochastic variable.

5. An application

As an application of ridge type M-estimators we consider a model where energy consumption in a pulp industry is linearly related to production levels in two processes (x_1 and x_3) and to energy contents in water used in the processes (x_2 and x_4). It is reasonable to believe that higher production requires more energy implying β_1 and β_3 to be positive. It is also reasonable to believe that when the water entering the industry is warm the requirements for heating it is small and vice versa implying β_2 and β_4 to be negative.

Table 1. Sample correlations in the example

x_1	x_2	x_3	x_4	y
1				
0.0991	1			
0.8358	0.0551	1		
0.0918	0.9939	0.0946	1	
0.7532	-0.2160	0.8439	-0.1909	1

A data set consisting of 51 weekly observations was recorded. The analysis is based on standardized variables with zero mean and unit variance. Table 1 reports the matrices $X'X$ and $X'y$ in correlation form. Some of the correlations between regressor variables are very high, which is reflected in a determinant of $n^{-1}X'X$ close to zero (here equal to 0.0016) and two relatively small eigenvalues (see Table 2). The sum of the reciprocals of the eigenvalues is $\Sigma(1/\lambda_i) = 391.647$, indicating the asymptotic variance of M-estimators to be more than 95 times higher than it would be in a case of orthogonal regressors.

Table 2. Eigenvalues and eigenvectors corresponding to the sample correlation matrix of regressor variables in the example

Eigenvalue	Eigenvectors			
	x_1	x_2	x_3	x_4
2.1028	0.3867	0.5932	0.3741	0.5989
1.7273	0.5909	-0.3860	0.6011	-0.3747
0.1673	-0.7016	-0.0932	0.6980	0.1093
0.0026	-0.0951	0.7001	0.1073	0.6989

The ORR estimates computed for values of k in the interval $[0,1]$ are shown in Figure 1. It appears that β_1 is overestimated by the least squares estimator. For an increasing k the ridge estimate of β_1 rapidly decreases, is negative in a short interval, and finally stays just above zero. The least squares estimate of β_2 is positive, increases rapidly to a maximum for an increasing k, after which it decreases to small positive values. The third factor shows the least dramatic trace: its importance is overestimated by least squares suggested by a decreasing yet always positive ridge trace. The behaviour of $\hat{\beta}_4(k)$ is in a sense a mirror of $\hat{\beta}_2(k)$: the least squares estimate is negative, for increasing k it rapidly decreases to a minimum, after which it increases and stabilizes at a negative value. However $|\hat{\beta}_4(k)|$ is much larger than $\hat{\beta}_2(k)$. A final conclusion is that the energy consumption seems to be dominated by production and water used in the second process, while the activities in the first process seem to have only negligable effects.

It is reasonable to believe that factors such as quality in raw materials and reliability of different components in the processes, which in our model are included in the residual term, might produce outliers. In order to robustify for this nonnormality we apply the ridge type M--estimator with the criterion function suggested by Huber (1964),

$$\rho(u) = \begin{cases} u^2/2 & \text{if } |u| \leq c_H \\ c_H|u| - c_H^2/2 & \text{if } |u| > c_H . \end{cases}$$

The tuning constant c_H is related to the proportion of outliers among

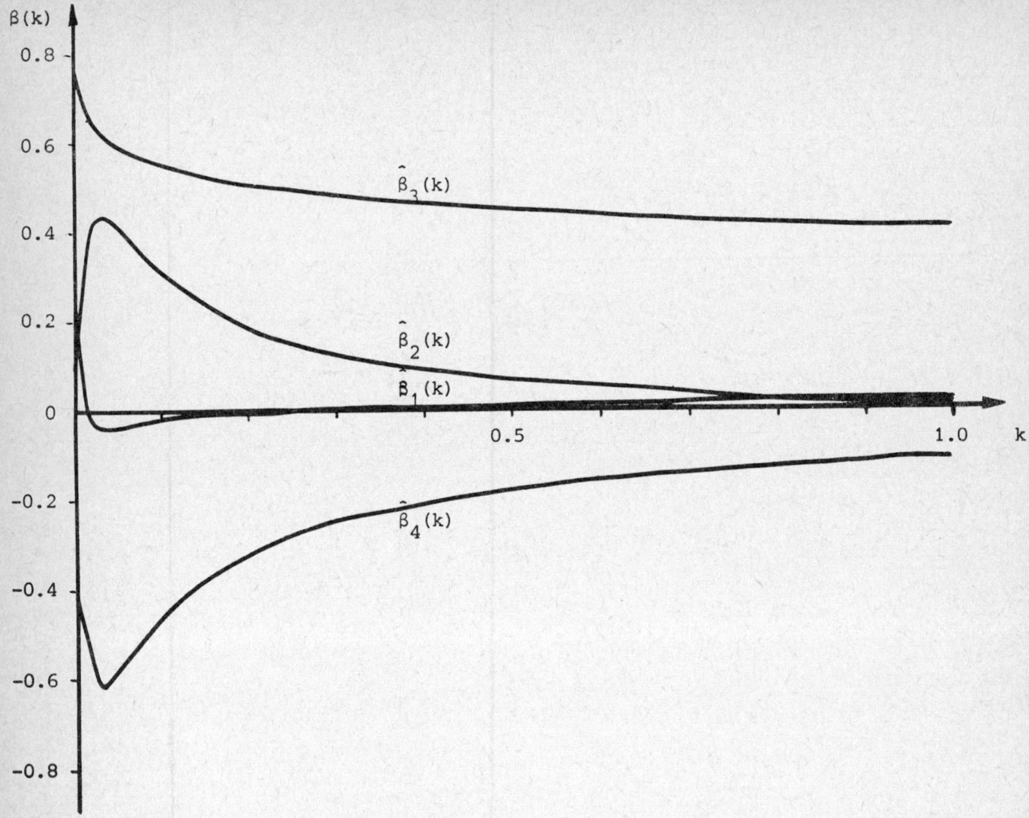

Figure 1. Ridge trace based on least squares estimation.

the residuals. In this study we used $c_H = 1.399$ corresponding to 5% residual outliers.

The ridge trace based on ridge type Huber estimates, shown in Figure 2, is quite different from the ORR ridge trace. Estimates of parameters corresponding to production variables, $\hat{\beta}_1(k)$ and $\hat{\beta}_3(k)$, are both positive and do not change very much for increasing k. The Huber estimate of β_2 is positive but decreases rapidly and stabilizes at a negative value. Again $\hat{\beta}_4(k)$ in a sense mirrors $\hat{\beta}_2(k)$; it is negative and overestimated by the Huber estimator but increase and stabilizes at a negative value for increasing k. The conclusion from this analysis is that production in both processes are positive and important for the total energy consumption. We also conclude that energy contents in water used in both processes has a negative and important effect on total energy consumption. Note that these conclusions agree with the a priori believes.

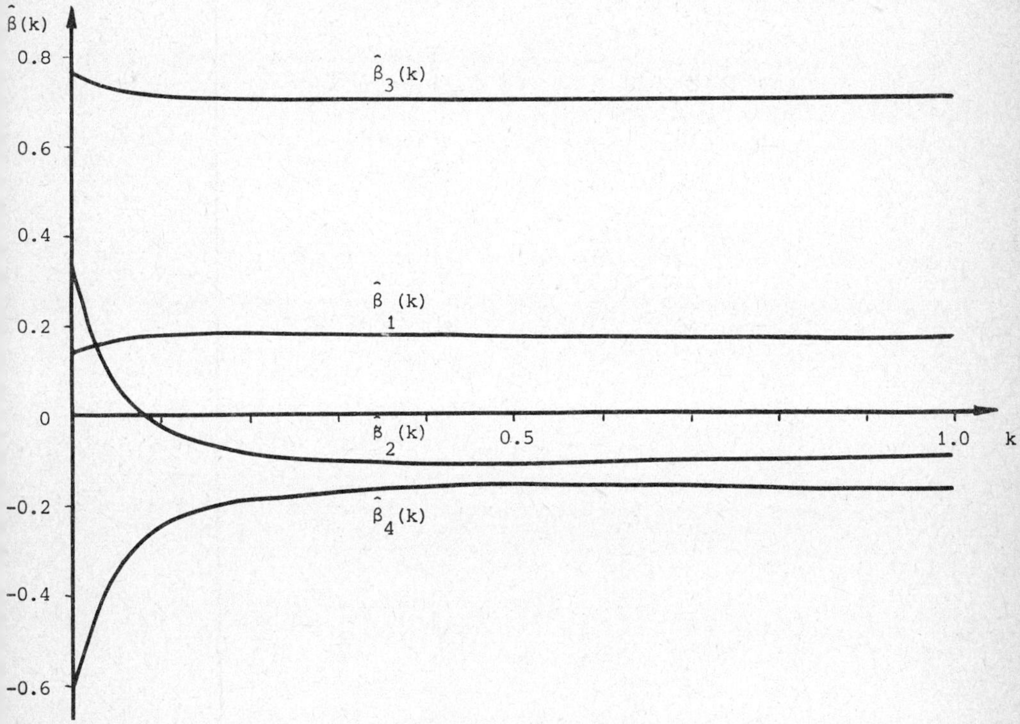

Figure 2. Ridge trace based on Huber estimation.

6. Concluding remarks

The linear regression model has been found useful in a wide variety of practical applications. Usually the method of least squares is used for the parameter estimation. This choice is often based on a more or less implicite assumption about normally distributed residuals or a belief that the performance of the least squares estimator is not severely affected by minor departures from normality. However, it is now recognized that even a small proportion of outliers do have an unusually large effect on least squares estimates and can deteriorate results. As alternatives to the least squares estimator, members of the class of maximum likelihood type (M-) estimators have been proposed and shown to be robust against longtailed residual distributions.

Another problem that can deteriorate results is that of multicollinearity. It is well known that multicollinearities among regressor variables may have severe effects on the least squares estimator. In this paper it is argued that all M-estimators are in an asymptotical sense equally affected by the multicollinearity problem. When multicollinearities are present the ordinary ridge regression (ORR) estimator has become an often used alternative to least squares. The choice of ORR is justified by the fact that it has favourable MSE properties for some values on the ridge parameter.

Writing the ORR estimator as the least squares estimator with additional information about the parameters provided, we find that ORR might be sensitive to outliers in residuals. Thus, there exist no "good" estimation technique for applications in the intersection of the set of models with longtailed residual distributions and the set of models with multicollinear data. In this paper we make use of the interpretation of ORR as an estimator when additional information is provided and apply the alternative criterion functions used in M-estimators to define a ridge type M-estimator.

In a four factor example it appears that conclusions based on a ridge type M-estimator are rather different from conclusions based on the ORR estimator. We also find that conclusions based on the ridge type M-estimator are consistent with our a priori considerations, while conclusions based on the ordinary ridge estimator are not.

One form of the generalized ridge regression (GRR) estimator is given by

$$b_{GRR} = (X'X + MKM')^{-1}X'y ,$$

where $K = \text{diag}(k_1, \ldots, k_m)$ is a diagonal matrix of constants and M is the $m \times m$ matrix of the eigenvectors of $X'X$. If $k_j = k$, $j = 1, \ldots m$, the GRR estimator reduces to the ORR estimator. The GRR estimator can be interpreted as the weighted least squares estimator of β in the model

$$\begin{aligned} y &= x\beta + u \\ 0 &= M\beta + \delta , \end{aligned} \qquad (10)$$

with weights equal to one for the n observations and equal to $\sqrt{k_j}$, $j = 1, \ldots, m$, for the m additional equations. Using this interpretation we can define a generalized ridge type M-estimator by applying a weighted M-estimator for the estimation of β in (10). Of course, an even more general form of ridge type M-estimators is obtained by letting the matrix M in (10) be an arbitrary $m \times m$ matrix with orthonormal columns.

The main conclusion to be drawn from our analysis is that ridge type M-estimators seem to provide a fruitful approach when analyzing linear regression models with multicollinear regressor variables and longtailed residual distributions. However, since much of the detailed mechanical behavior of the estimators is still unknown, we look forward to research in this new branch of estimation theory.

References

ANDERSON, R.L., and BATTISTE, E.L. (1975), The Use of Prior Information in Linear Regression Analysis, Communications in Statistics, 4, 497-517.

BARRODALE, I., and ROBERTS, F.D.K. (1972), Solution of an Overdetermined System of Equations in the L_1-norm, Mathematics Department Report No. 69, University of Victoria.

HOERL, A.E., and KENNARD, R.W. (1970 a), Biased Estimation for Nonorthogonal Problems, Technometrics, 12, 55-67.

- (1970 b), Ridge Regression: Applications to Nonorthogonal Problems Technometrics, 12, 69-82.

- (1981), Ridge Regression 1980: Advances, Algorithms and Applications, Americal Journal of Mathematical and Management Sciences, 1, 5-83.

HOLLAND, P.W., and WELSCH, R.E. (1977), Robust Regression Using Iteratively Reweighted Least-Squares, Communications in Statistics - Theory and Methods, A 6, 813-827.

HUBER, P.J. (1964), Robust Estimation of a Location Parameter , Annals of Mathematical Statistics, 35, 73-101.

- (1973), Robust Regression: Asymptotics, Conjectures and Monte Carlo", Annals of Statistics, 1, 799-821.

LINDLEY, D.V., and SMITH, A.F.M. (1972), Bayes Estimates for the Linear Model , Journal of the Royal Statistical Society, Series B, 34, 1-41.

SMITH, V.K. (1976), A Note on Ridge Regression , Decision Sciences, 7, 562-566.

THEOBALD, C.M. (1974), Generalizations of Mean Square Error Applied to Ridge Regression, Journal of the Royal Statistical Society, Series B, 36, 103-106.

MAJORIZATION AND APPROXIMATE MAJORIZATION FOR FAMILIES OF MEASURES, APPLICATIONS TO LOCAL COMPARISON OF EXPERIMENTS AND THE THEORY OF MAJORIZATION OF VECTORS IN R^n (SCHUR CONVEXITY)

E. Torgersen
Institute of Mathematics
University of Oslo, Oslo, Norway

The general theory of comparison and of approximate comparison of experiments generalizes, as shown by this author in 1969, without too many complications, to a theory of comparison which is applicable to general families of measures. After having introduced a suitable notion of differentiability for statistical experiment it is then shown that local comparison of statistical experiments may be reduced to comparisons of certain finite and quite natural sets of measures. Thus we have at our hand a possible foundation for a non asymptotic theory for local comparison and, in particular, for local sufficiency. Another natural application of this general notion of comparison is to the theory of majorization of vectors in R^n (Schur convexity).

Content

MAJORIZATION AND APPROXIMATE MAJORIZATION. COMPARISON OF FAMILIES OF MEASURES

1. Introduction

One of the main contributions of the theory of comparison of experiments is to show that natural and apparently different criteria for comparison actually are equivalent. Thus we have equivalent criteria in terms of over all comparison of risk functions, in terms of Bayes risk for a fixed prior distribution, in terms of sub linear functionals, in terms of performance functions of decision rules and in terms of randomizations

The underlying framework of these results is the decision theory of Abraham Wald. The first results, of Blackwell and others, were concerned with criteria for the ordering "being more informative than". Later on LeCam in his 1964 paper on sufficiency and approximate sufficiency introduced the concept of deficiency of one experiment w.r. to another. As experiments usually are not ordered and since these deficiencies always exist this extended considerably the applicability of the theory. In 1970 this author considered deficiencies for k-decision problems combining Blackwell's idea of comparison for k-decision problems with LeCam's deficiencies.

The theory has found important applications to asymptotic and non-asymptotic statistical theory. Besides LeCam's works the reader might consult Millar's recent survey on asymptotic minimax theory. Expositions of the theory of statistical experiments may be found in Heyer (1973), LeCam (1974) and in Torgersen (1976). A recent contribution to non-asymptotic theory is due to Lehmann (1983).

The idea that the underlying distributions which constitute a statistical experiment should be proper probability distributions is of course essential for the interpretations of most results from the theory of comparison of experiments. In particular the interpretation of the risk as expected loss depends on this assumption. If, however, we look over the various arguments which are used we see that this assumption is often not needed or may be avoided by a suitable reformulation.

Even the formal expression of the risk of a decision procedure as the value of the corresponding bilinear functional for the underlying distribution and the loss function remains well defined within a much more general set up.

We shall now see that such a generalization, or extension, to "non proper experiments actually yields interesting results on "proper experiments". Another benefit is a unified approach to several results concerning systems of inequalities which are frequently encountered in mathematical statistics. The theory we shall formulate here may be considered as a generalization of the theory of majorization as it is described in a book by Marshall and Olkin (published in 1979), and in later generalizations due to Dahl (1983) and due to Karlin (1983).

Let us at this point briefly indicate by examples, some directions of applications.

Example 1.1: Comparison of modified risks

Assume that all losses are bounded by 1 and let a, b and ε be given functions of the unknown parameter θ.

When are we entitled to claim for all decision problems (of a certain type) and given experiments $\mathcal{E}$ and $\mathcal{F}$ that to each risk function s obtainable in $\mathcal{F}$ there corresponds a risk function r obtainable in $\mathcal{E}$ such that $a(\theta)r(\theta) \leq b(\theta)s(\theta)+\varepsilon(\theta)$ for all θ?

If $a(\theta) \underset{\theta}{\equiv} 1$ and $b(\theta) \underset{\theta}{\equiv} 1$ then this is just a description of the situation where $\mathcal{E}$ is ε-deficient w.r. to $\mathcal{F}$ according to LeCam's definition of deficiency.

For general a,b and ε this problem is easily described, at least theoretically, in terms of general families of measures. In the next section we will see that one possible answer may be obtained by a straightforward generalization of LeCam's randomization (Markov kernel) criterion.

Example 1.2: Local (nonasymptotic) comparison of experiments

Consider experiments having the same subset Θ of R^m as their parameter set. Let θ^o be an interior point of Θ. If $\mathcal{E} = (P_\theta : \theta \in \Theta)$ is an experiment then, assuming differentiability, we may consider the measure P_{θ^o} along with the partial derivatives $[\partial P_\theta/\partial\theta_i]_{\theta=\theta^o}$, $i = 1,\ldots,m$. This family of measures is not an experiment since the total masses of the measures $[\partial P_\theta/\partial\theta_i]_{\theta=\theta^o}$ are all zero.

If θ is close to θ^o then the measure P_θ may be approximated by $P_{\theta^o} + \sum_{i=1}^{m} (\theta-\theta^o)_i [\partial P_\theta/\partial \theta_i]_{\theta=\theta^o}$. Thus it is not surprising that if θ is close to θ^o and if quantities of smaller order than the distance from θ to θ^o are considered negligible then the local behavior of $\mathcal{E}$ around θ^o may be completely described in terms of this family of measures. It is interesting that, as we will see in Section 3, the local comparison of experiments within small neighbourhoods of θ^o may be expressed quite naturally and simply in terms of such families of measures.

Example 1.3: Comparison of measures. Dilations

System of inequalities of the following type have received attention in various connections.

Let μ and ν be measures on a measurable space $(\chi, \mathcal{A})$. Consider also a convex set H of integrable functions which contains the constants and which is closed under the formation of maxima of finite sub sets. What conditions on (H,μ,ν) ensure that the inequality $\int h d\mu \geq \int h d\nu$ holds for every function h in H?

The ordering of experiments is a case closely related with μ and ν being conical measures of experiments.

We may also consider, as we will, systems of inequalities of the form $\int h d\mu \geq \int h d\nu - \varepsilon_h$ for some given function $h \to \varepsilon_h$ on H. Thus we are led to a problem of comparing the families $(h\mu : h \in H)$ and $(h\nu : h \in H)$ of measures.

In Section 4 we will see how the general theory of comparison of families of measures provides a method of attacking such problems.

Another related and most interesting discussion of such problems may be found in Karlin(1983).

Example 1.4: Distributions with given marginals

The fundamental papers by LeCam (1964) and by Strassen (1965) both show that there is a close relationship between the theory of comparison of experiments, on the one hand, and various existence problems for joint distributions with given marginals, on the other. This relationship is also apparent from earlier works of Blackwell (1953) and Boll (1955).

Thus, the dilation criterion for ordering of experiments is, as shown by Strassen (1965), related to the problem of whether or not a sequence of distributions might be the sequence of distributions of a martingale. Strassen considered in his paper the problem of deciding whether one distribution P is the convolution factor of another distribution Q. Boll (1955) has shown that this is tantamount to the problem of deciding whether the translation experiment defined by P is at least as informative as the translation experiment defined by Q. The statistical significance of Strassen's criterion in terms of minimax risk was, using the results in LeCam's paper, clarified in a later paper by Torgersen (1972c), which also generalized these results to the case of ε-deficiency.

Among other results which bring out the connection between these fields we will here only mention the various criteria for stochastic orderings of distributions on a general partially ordered set. The basic criterion in terms of probabilities of "monotone" events follows directly from Strassen's paper. It was shown later, by Torgersen (1982), that this criterion is also the criterion for "being more informative" for experiments associated with sampling plans defined by these distributions. More generally it was shown that the case of ε-deficiency corresponded nicely with one distribution being up to an amount of ε stochastically larger than the other distribution.

In Section 5 we will show that this and other results may be deduced from general principles for comparison of families of measures.

Example 1.5: Majorization

The theory of comparison of measure families may be considered as a natural generalization of most of the various theories of majorization. In particular a functional of experiments which is monotonically increasing (decreasing) for the ordering on the basis of "being more informative" may be considered as Schur convex (concave). Thus Fisher information and statistical distance may be considered Schur convex while the Hellinger transform is Schur concave.

In order to make this clearer let us comment on the well known particular case of majorization of vectors in R^n. Two important sources of information for this case is the book by Hardy, Littlewood and Polya (1934) and the book by Marshall and Olkin (1979). Following the notation used in the last book we will write $(x_{[1]}, x_{[2]},\dots,x_{[n]})$ for the vector in R^n obtained from the vector $(x_1,\dots,x_n)$ by arranging the coordinates of x in decreasing order.

Let $p=(p_1,\dots,p_n)$ and $q=(q_1,\dots,q_n)$ be two vectors in R^n such that $p_1+\dots+p_n = q_1+\dots+q_n$. Then the following conditions are known to be equivalent:

(i) $p_{[1]}+\dots+p_{[j]} \geq q_{[1]}+\dots+q_{[j]}$, $j=1,\dots,n$;

(ii) $\sum_j \phi(p_j) \geq \sum_j \phi(q_j)$ when the function ϕ is convex on R;

(iii) $\sum_j (p_j-c)^+ \geq \sum_j (q_j-c)^+$, $c \in R$;

(iv) $pM = q$ for a doubly stochastic $n\times n$ matrix M.

Now assume that p and q are probability vectors. Define probability distributions P_0, Q_0, P_1 and Q_1 on $\{1,\dots,n\}$ by putting $P_0(j) = Q_0(j) = 1/n$, $P_1(j)=p_j$ and $Q_1(j) = q_j$, $j=1,\dots,n$. Put $\Theta=\{0,1\}$ and consider the experiments (dichotomies) $\mathcal{E} = (P_0,P_1)$ and $\mathcal{F} = (Q_0,Q_1)$.

Using the Neyman-Pearson lemma to the problem of testing "$\theta = 0$" against "$\theta = 1$" we see that (i) tells us that, for any level of significance, $\mathcal{E}$ yields at least as large power as $\mathcal{F}$. By the results of Blackwell (1951 and 1953) this is equivalent to the condition that $\mathcal{E}$ is at least as informative as $\mathcal{F}$. Multiplying (ii) and (iii) by (-1) we see that condition (ii) is the overall minimum Bayes risk criterion for comparability and that (iii) is the minimum Bayes risk criterion for testing problems. Finally, condition (iv) states that $P_\theta M = Q_\theta$, $\theta = 0,1$, for a Markov kernel M from the set $\{1,2,\dots,n\}$ into itself. By the same papers of Blackwell these conditions are all equivalent. Note, in particular, that the condition that the Markov matrix M is doubly stochastic expresses that the corresponding Markov kernel preserves the uniform distribution, i.e. that $P_0M = Q_0$.

Replacing the distribution $P_0 = Q_0$ with more general distributions we arrive at more general notions of majorization, as e.g. the notion of π-majorization considered in Marshall and Olkin's book.

Now majorization is a partial ordering on all of R^n and not just on the probability simplex. The validity of the arguments (e.g. the Neyman-Pearson lemma) used above does not, however, depend on the condition that the vectors p and q are probability vectors. We will see in Section 6 that most of the mathematics remains valid, after some straightforward modifications, for general families of measures. Thus we may, as in Dahl (1983), consider ε-deficiency as well as multivariate majorization. Multivariate majorization was also considered by Karlin (1983).

As it is amply demonstrated by Marshall and Olkin in the already mentioned book, the concept of majorization in R^n is a very useful tool for establishing inequalities in statistics and elsewhere.

The point we are making is that this concept fits naturally into the extension of the framework of comparison of experiments which we will consider here. Majorization in R^n is, however, a distinguished particular case with several features which appear difficult to deduce from this general theory of comparison of families of measures.

In the forthcoming sections we shall take a closer look at the situations described in the examples of this section. It is, however, necessary that we first become acquinted with some basic principles for comparison of families of measures. These principles are, therefore, the topic of the next section.

2. Equivalent criteria for comparison of families of measures

In this section we will be concerned with general principles for comparing families of measures. It will be assumed that all the considered measures belonging to a given family are real valued and that they are defined on a common measurable space.

We have found it convenient to use the term measure family to denote a family of measures satisfying these two requirements. Thus the measures in a measure family are all bounded, but we do not assume that they are non-negative.

A measure family $\mathscr{E}$ with a measurable space $(\chi,\mathscr{A})$ will be denoted by $\mathscr{E} = (\chi,\mathscr{A};\mu_\theta:\theta \in \Theta)$, or just by $\mathscr{E} = (\mu_\theta:\theta \in \Theta)$, where $(\mu_\theta:\theta \in \Theta)$ is the family of finite measures which constitute $\mathcal{E}$. The set Θ is called the <u>parameter set of $\mathcal{E}$</u>. It may be any set, but we shall assume, unless otherwise stated, that if two measure families $\mathcal{E}$ and $\mathcal{F}$ are compared then their parameter sets are the same.

A measure family of probability measures is called an <u>(statistical) experiment</u> and the measurable space of an experiment is usually called <u>the sample space of the experiment</u>.

Consider now in addition to the parameter set Θ, a set T and a family $L = (L_\theta: \theta \in \Theta)$ of real valued functions on T. If the measure families under consideration are experiments then we may think of T as a set of decisions and of L as a loss function.

Let us agree to use the notation $||g||$ for the supremum norm $\sup_t |g(t)|$ of a real valued function g on T. Risking the slight possibility of a conflict of interpretations we shall use the notation $||\mu||$ to denote the total variation of a measure μ. Finally, we will use the term <u>deficiency function</u> for any non-negative function on the parameter set.

Assuming that the parameter set Θ is finite we shall now list four principles for comparing measure families $\mathcal{E} = (\chi, \mathcal{A}, \mu_\theta: \theta \in \Theta)$ and $\mathcal{F} = (\mathcal{Y}, \mathcal{B}, \nu_\theta: \theta \in \Theta)$. The comparison shall be expressed in terms of a fixed deficiency function ε and a fixed set T.

We precede the statement of each of the listed principles by a "headline" indicating the interpretations of the principles when the measure families are experiments.

It should be noted that the only aspect of the set T which matters here is its cardinality, which we shall denote by k.

In the statement of the third of the principles below and also at several other places we are using the convenient notation $\int h(d\mu_\theta: \theta \in \Theta)$ to denote $\int h(f_\theta: \theta \in \Theta)\, d\mu$ whenever the finite measure μ_θ, for each θ, has density f_θ w.r. to the non negative measure μ and h is a non-negative homogeneous measurable function on R^Θ. It is then easily checked that neither the existence nor the value of this integral depends on the choice of the majorizing measure μ, nor on the specifications of the densities.

If θ is finite then the first four principles mentioned in the introduction may be formulated as follows.

(i) Pointwise comparison of risks

To each family $L=\{L_\theta : \theta \in \Theta\}$ and each Markov kernel σ from $\mathcal{F}$ to T, there corresponds a Markov kernel ρ from $\mathcal{E}$ to T such that

$$\mu_\theta \rho L_\theta \leq \nu_\theta \sigma L_\sigma + \varepsilon_\theta \| L \|_\theta ; \quad \theta \in \Theta.$$

(ii) Comparison of Bayes risks

To each family $\{L_\theta : \theta \in \Theta\}$ and each Markov kernel σ from $\mathcal{F}$ to T there corresponds a Markov kernel ρ from $\mathcal{E}$ to T such that

$$\sum_\theta \mu_\theta \rho L_\theta \leq \Sigma_\theta \nu_\theta \sigma L_\theta + \Sigma_\theta \varepsilon_\theta \| L_\theta \| .$$

(iii) Comparison of maximum Bayes utilities.
The sub linear function criterion

For each function ψ on R^Θ which is a maximum of $k = \#T$ linear functionals we have the following inequality

$$\int \psi (d\mu_\theta : \theta \in \Theta) \geq \int \psi (d\nu_\theta : \theta \in \Theta) - \sum_\theta \varepsilon_\theta [\psi(-e^\theta) \vee \psi(e^\theta)].$$

Here, as elsewhere, $e^\theta = (0,\ldots,\overset{\theta}{1},\ldots,0)$ denotes the θ-th unit vector in R^Θ.

(iv) Comparison of performance functions

To each Markov kernel σ from $\mathcal{F}$ to T there corresponds a Markov kernel ρ from $\mathcal{E}$ to T such that

$$\| \mu_\theta \rho - \nu_\theta \sigma \| \leq \varepsilon_\theta , \quad \theta \in \Theta.$$

Following Blackwell (1951, 1953) and LeCam (1964) and proceeding as in Torgersen (1969) we may now state and prove the basic

Theorem 2.1: Equivalent rules for comparison

If Θ is finite then the criteria (i)-(iv) are all equivalent.

<u>Proof</u>: First choose non negative measures μ and ν so that for each θ, μ_θ has density f_θ w.r. to μ, while ν_θ has density g_θ w.r. to ν.

If $L = (L_\theta : \theta \in \Theta)$ is a family of real valued functions on T then $\min_\rho \sum_\theta \mu_\theta \rho L_\theta = \int [\bigwedge_t \sum_\theta L_\theta(t) f_\theta] d\mu$, and $\min_\sigma \sum_\theta \nu_\theta \sigma L_\theta = \int [\bigwedge_t \sum_\theta L_\theta(t) g_\theta] d\nu$. Multiplying (ii) by (-1) yields $\int [\bigvee_t \sum_\theta U_\theta(t) f_\theta] d\mu \geq \int [\bigvee_t \sum_\theta U_\theta(t) g_\theta] d\nu - \sum_\theta \varepsilon_\theta \| U \|_\theta$, where $U = -L$. Putting $\psi(x) = \bigvee_t \sum_\theta U_\theta(t) x_\theta$, $x \in R^\Theta$, we see that (iii) is just a reformulation of (ii).

Since the implications (iv) $\Rightarrow$ (i) $\Rightarrow$ (ii) are trivial it suffices now to show that (ii) $\Rightarrow$ (iv). Assume that (ii) is satisfied. Let σ be a Markov kernel from $\mathcal{F}$ to T and put $\Pi(L,\rho) = \sum_\theta \mu_\theta \rho L_\theta - \sum_\theta \nu_\theta \sigma L_\theta - \sum_\theta \varepsilon_\theta \| L_\theta \|$, where $L = \{L_\theta : \theta \in \Theta\}$, while ρ is a Markov kernel from $\mathcal{E}$ to T. Then Π is concave-convex on (L,ρ). Furthermore, Π is continuous in ρ, for fixed L, if we topologize $\{\rho\}$ by the notion of convergence which states that a net $\{\rho^{(n)}\}$ of Markov kernels from $\mathcal{E}$ to T converges to a Markov kernel ρ from $\mathcal{E}$ to T if and only if $\int \delta \rho^{(n)}(t|\cdot) d\mu_\theta \to \int \delta \rho(t|\cdot) d\mu_\theta$ for each bounded measurable function δ on the sample space of $\mathcal{E}$. The weak compactness lemma implies that the set of Markov kernels from $\mathcal{E}$ to T is compact for this topology. Using the minimax theory, e.g. Theorem 4.2 in Sion (1958), we see that there is a Markov kernel ρ_0 from $\mathcal{E}$ to T so that $\sup_L \Pi(L,\rho_0) = \inf_\rho \sup_L \Pi(L,\rho) = \sup_L \inf_\rho \Pi(L,\rho)$, and the last quantity is, by (iv), non positive. Thus $\Pi(L,\rho_0) \leq 0$ for each L. Choosing a particular real valued function g on T and a point $\theta^0 \in \Theta$ and then putting $L_\theta = g$ or $= 0$ as $\theta = \theta^0$ or $\theta \neq \theta^0$, we see that $0 \geq \Pi(L,\rho_0) = (\mu_{\theta^0} \rho_0 - \nu_{\theta^0} \sigma)(g) - \varepsilon_{\theta^0} \| g \|$. Varying g and θ^0 we find that $\| \mu_\theta \rho_0 - \nu_\theta \sigma \| \leq \varepsilon_\theta$, $\theta \in \Theta$.

Restricting attention in (iii) to funtions ψ such that ψ or $-\psi$ are projections on the coordinate spaces of R^Θ, we see that the equivalent conditions (i)-(iv) imply

$$|\mu_\theta(\chi) - \nu_\theta(\mathcal{Y})| \leq \varepsilon_\theta, \quad \theta \in \Theta . \qquad (*)$$

This is equivalent to (i)-(iv) when T possesses just one element.

Assuming that the deficiency function ε satisfies (*), which it always does when $\mathscr{E}$ and $\mathscr{F}$ are experiments, we find that the "deficiency term" $\sum_\theta \varepsilon_\theta[\psi(-e^\theta)\vee\psi(e^\theta)]$ in (iii) may be replaced by the linear (in ψ) term $\sum_\theta \varepsilon_\theta[\psi(-e^\theta)+\psi(e^\theta)]/2+\sum_\theta(\nu_\theta(\mathscr{Y})-\mu_\theta(\chi))[\psi(e^\theta)-\psi(-e^\theta)]/2$. We may then even assume that $\psi(-e^\theta)\underset{\theta}{\equiv}\psi(e^\theta)$, and thus get the simple expression $\sum_\theta \varepsilon_\theta\psi(e^\theta)$ for the deficiency term. We may also pass from a general sub linear function ψ on R^Θ to a sub linear function which is monotonically decreasing (increasing), by replacing $\psi(x)$ with $\psi(x)-\sum_\theta x_\theta\psi(e^\theta)$ (with $\psi(x)+\sum_\theta x_\theta\psi(-e^\theta)$).

A substantial reduction is available when Θ is a two point set. In this case the fact that convex polygons may be decomposed as vector sums of triangles and line segments tells us that it suffices to consider 3 point sets T and thus functions ψ which are maxima of at most 3 linear functionals. Actually additional assumptions will often guarantee that it suffices to consider two point sets T, and thus functions ψ which are maxima of at most 2 linear functionals. This is the case whenever $\mathscr{E}$ and $\mathscr{F}$ are experiments (i.e. dichotomies), or if $\mu_\theta,\nu_\theta\geq 0$ and $\varepsilon_\theta=0$ for one of the two points θ in Θ. The latter case covers the usual case of majorization as well as the cases of weak majorization as defined in Marshall and Olkin (1979). In the case of dichotomies a proof may be found in Torgersen (1970) and the other case follows by almost the same arguments. It is, however, not true in general that we may reduce any comparison to a comparison for two point sets T (testing problems) whenever Θ is a two point set.

Let us drop the assumption that Θ is finite. We will say that <u>the measure family $\mathscr{E}=(\mu_\theta:\theta\in\Theta)$ is ε-deficient w.r. to the measure family $\mathscr{F}=(\nu_\theta:\theta\in\Theta)$ for k-point sets</u> if conditions (i)-(iv) are satisfied for the restrictions $\mathscr{E}|\Theta_0$, $\mathscr{F}|\Theta_0$ and $\varepsilon|\Theta_0$ for any non empty finite subset Θ_0 of Θ. If $\mathscr{E}$ and $\mathscr{F}$ are experiments, then the qualification "for k-point sets" may be replaced by the qualification "for k-decision problems".

If $\mathscr{E}$ is ε-deficient w.r. to $\mathscr{F}$ for k-point sets, $k=1,2,\ldots$, then we say that <u>$\mathscr{E}$ is ε-deficient w.r. to $\mathscr{F}$ for ∞-point sets</u> or just say that <u>$\mathscr{E}$ is ε-deficient w.r. to $\mathscr{F}$</u>. Thus the notion of ε-deficiency for k-point sets is defined for all positive integers k

and for $k = \infty$.

It is easily checked that $\mathscr{E}$ is ε'-deficient w.r. to $\mathscr{F}$ for k'-point sets problems whenever $\mathscr{E}$ is ε-deficient w.r. to $\mathscr{F}$ for k-point sets for $\varepsilon \leq \varepsilon'$ and $k \geq k'$.

We will see below, without using any new ideas, that the randomization criterion is also valid for general measure families. The framework of measure theory which we are using here does not quite permit the full elegance of LeCam's formulations. Thus we might have worked with families in abstract L-spaces rather than with families of measures. No generality is however lost since any abstract L-space is, by Theorem 7 in Kakutani (1941), isomorphic to a L-space of bounded measures on some mesurable space.

LeCam's randomization criterion has generalized significantly earlier results in this direction by e.g. Blackwell (1951, 1953) and Bahadur (1954, 1955) the theory of sufficiency. The formulation of this criterion and related results in terms of transitions has to a large extent also clarified the distinction between the essential statistical features behind this result, on the one hand, and various "technical" problems (often quite interesting) which occur within the traditional framework, on the other.

Before we embark on the extension of the randomization criterion, let us introduce two concepts which we shall need in this connection. Firstly we define <u>the L-space of the measure family $\mathscr{E} = (\chi, \mathscr{A}, \mu_\theta : \theta \in \Theta)$</u> as the linear space of finite measure on $\mathscr{E}$ which are dominated by finite measures of the form $\Sigma\{c_\theta|\mu_\theta| : \theta \in \Theta_0\}$, where the c_θ's are non negative and Θ_0 is a countable subset of Θ. We use the notation $\Gamma(\mathscr{E})$ for the Banach space of bounded measurable functions on χ equipped with the supremum norm. According to standard notation $\Gamma^*(\mathscr{E})$ then denotes the adjoint space of the Banach space $\Gamma(\mathscr{E})$. Thus $\Gamma^*(\mathscr{E})$ is the Banach space of bounded linear functionals on $\Gamma(\mathscr{E})$ equipped with the supremum norm on the unit ball of $\Gamma(\mathscr{E})$. This Banach space is isomorphic to the linear space $ba(\chi,\mathscr{A})$ of bounded (real valued) additive set functions on $\mathscr{A}$ provided the latter space is normed by the total variation norm. A isomorphisme from $ba(\chi,\mathscr{A})$ onto $\Gamma(\mathscr{E})^*$ may be defined by assigning to each bounded additive set function κ the linear functional on $\Gamma(\mathscr{E})$ which maps $f \in \Gamma(\mathscr{E})$ into the integral $\int f d\kappa = \lim_{n\to\infty} \sum_k \frac{k}{n}\kappa(\frac{k-1}{n} < f \leq \frac{k}{n})$. The L-space $L(\mathscr{E})$ of $\mathscr{E}$ is a closed subspace of $ba(\chi, \mathscr{A})$ and may therefore be represented as a closed subspace of $\Gamma(\mathscr{E})^*$.

The reader may either check out these facts directly or consult, e.g. Section 5 in Chapter IV in Dunford and Schwartz (1958).

If $\mathscr{E} = (\chi, \mathscr{A}, \mu_\theta : \theta \in \Theta)$ and $\mathscr{F} = (\mathscr{Y}, \mathscr{B}, \nu_\theta : \theta \in \Theta)$, then any Markov kernel M from $(\chi, \mathscr{A})$ to $(\mathscr{Y}, \mathscr{B})$ induces a map $\mu \to \mu M$ from $L(\mathscr{E})$ into $\Gamma(\mathscr{F})^*$, in fact into the space of finite measures on $\mathscr{B}$. This map is linear, non negative, i.e. μM is non negative when μ is non-negative, and it preserves total mass. We will, following LeCam (1964), call any map from $L(\mathscr{E})$ into $\Gamma(\mathscr{F})^*$ having these properties a <u>transition</u>. A very usefull property of the set of all transitions is that it, by Tychonoff's theorem on product topologies (see e.g. Kelley (1955)), is compact for the pointwise topology on $L(\mathscr{E}) \times \Gamma(\mathscr{F})$.

Assume now that $\mathscr{E} = (\chi, \mathscr{A}, \mu_\theta : \theta \in \Theta)$ is ε-deficient w.r. to $\mathscr{F} = (\mathscr{Y}, \mathscr{B}, \nu_\theta : \theta \in \Theta)$. Let N denote the set of all triples $n = (\pi, \eta, F)$, where $\pi = (B_1, \ldots, B_k)$ is a measurable partitioning of $\mathscr{Y}$, $\eta = (y_1, \ldots, y_k)$ with $y_i \in B_i$, $i = 1, \ldots, k$, and where F is a finite subset of Θ. Direct the set N by defining that $n_1 = (\pi_1, \eta_1, F_1) \geq n_2 = (\pi_2, \eta_2, F_2)$ whenever $F_1 \supseteq F_2$ and each set in π_2 is a union of sets in π_1. Let $n = (\pi, \eta, F) \in N$ and consider $\{y_1, \ldots, y_k\}$ as a choice for the set T appearing in the statements of conditions (i)-(iv). Let σ_n denote the Markov kernel (function) from $\mathscr{F}$ to $\{y_1, \ldots, y_k\}$ which maps y into y_i when $y \in B_i$. Then, by (iv), there is a Markov kernel ρ_n from $\mathscr{E}$ to the same space such that $\| \mu_\theta \rho_n - \nu_\theta \sigma_n \| \leq \varepsilon_\theta$ when $\theta \in F$. Put $M_n(B|x) = \Sigma I_B(y_i) \rho_n(y_i|x)$ when $B \in \mathscr{B}$ and $x \in \chi$. Then M_n is a Markov kernel from $(\chi, \mathscr{A})$ to $(\mathscr{Y}, \mathscr{B})$, and thus defines a transition from $L(\mathscr{E})$ to $\Gamma(\mathscr{F})^*$. By compactness there is a transition M which is a point of accumulation for the net of transitions defined by the net $\{M_n\}$ of Markov kernels. It follows then that $\mu_\theta M g - \nu_\theta g \leq \varepsilon_\theta \| g \|$ for all $\theta \in \Theta$ and for all $g \in \Gamma(\mathscr{F})$. Hence $\| \mu_\theta M - \nu_\theta \| \leq \varepsilon_\theta$, $\theta \in \Theta$. This proves

<u>Theorem 2.2</u>: The randomization criterion for comparison of measure families

The measure family $\mathscr{E} = (\mu_\theta : \theta \in \Theta)$ is ε-deficient w.r. to the measure family $\mathscr{F} = (\nu_\theta : \theta \in \Theta)$ if and only if there is a transition M from the L-space of $\mathscr{E}$ to the L-space of bounded additive set functions on $\mathscr{F}$ such that

$$\| \mu_\theta M - \nu_\theta \| \leq \varepsilon_\theta ; \; \theta \in \Theta .$$

Remark 1. Using the same arguments as by LeCam (1964) we may always modify M so that it maps $L(\mathscr{E})$ into $L(\mathscr{F})$.

Remark 2. It follows from Theorem 2.2 that one possible answer to the question raised in Example 1.1 is the following.

"If and only if there is a transition M such that $\| a_\theta P_\theta M - b_\theta Q_\theta \| \leq \varepsilon_\theta$ for all $\theta \in \Theta$". If $a_\theta \underset{\theta}{\equiv} 1$ then this implies that $|b_\theta - 1| \leq \varepsilon_\theta$; $\theta \in \Theta$. If $b_\theta = 1-\varepsilon_\theta$ then the "θ-th" inequality states that $P_\theta M \geq (1-\varepsilon_\theta) Q_\theta$. On the other hand, if $b_\theta = 1+\varepsilon_\theta$ then the θ-th inequality states that $P_\theta M \leq (1+\varepsilon_\theta) Q_\theta$.

If we want to ensure that the transition M appearing in the randomization criterion is representable as a Markov kernel then various regularity conditions may be invoked. We say that a measurable space $(\chi, \mathscr{A})$ is Euclidean if it is Borel isomorphic to a Borel sub set of $[0,1]$. Moreover, we say also that a measure family $\mathscr{E} = (\mu_\theta : \theta \in \Theta)$ is coherent if all bounded linear functionals on $L(\mathscr{E})$ are representable as bounded measurable functions. If $\mathscr{E} = (\mu_\theta : \theta \in \Theta)$ is coherent and if $(\mathscr{Y}, \mathscr{B})$ is Euclidean then any transition from $L(\mathscr{E})$ to the L-space of finite measures on $(\mathscr{Y}, \mathscr{B})$ is representable as a Markov kernel.

If the families $\mathscr{E}$ and $\mathscr{F}$ as well as the deficiency function ε are invariant under the actions of a group and this group possesses an invariant mean then under regularity conditions the kernel M may, (see e.g. LeCam (1964 and 1974) and Torgersen (1972c)), be chosen invariant.

We shall not pursue the general theory of comparison of measure families further here. Most of the results are more or less straightforward generalizations of results known from the theory of comparison of experiments. Thus we may define the deficiency $\delta_k(\mathscr{E}, \mathscr{F})$ of $\mathscr{E}$ w.r. to $\mathscr{F}$ for k-point sets (k-decisions when $\mathscr{E}$ and $\mathscr{F}$ are experiments) as the smallest (it exists) constant ε such that $\mathscr{E}$ is ε-deficient w.r. to $\mathscr{F}$ for k-point sets. Symmetrizing the deficiency we define <u>the deficiency distance $\Delta_k(\mathscr{E},\mathscr{F})$ between the measure families $\mathscr{E} = (\mu_\theta:\theta \in \Theta)$ and $\mathscr{F} = (\nu_\theta:\theta \in \Theta)$ for k-point sets</u> as the largest of the numbers $\delta_k(\mathscr{E},\mathscr{F})$ and $\delta_k(\mathscr{F},\mathscr{E})$.

If $\delta_k(\mathscr{E},\mathscr{F}) = 0$ then we will say that <u>$\mathscr{E}$ majorizes $\mathscr{F}$ for k-point sets</u> and write this $\mathscr{E} \underset{k}{\geq} \mathscr{F}$. Then $\underset{k}{\geq}$ is an ordering for measure families and this ordering extends the notion of one experiment being at least as informative as another for k-decision problems. Similarly we shall say that <u>$\mathscr{E}$ and $\mathscr{F}$ are equivalent for k-point sets</u> if $\mathscr{E} \underset{k}{\geq} \mathscr{F}$ and $\mathscr{F} \underset{k}{\geq} \mathscr{E}$, i.e. if $\Delta_k(\mathscr{E},\mathscr{F}) = 0$.

The qualification "for k-point sets" as well as the subscript k may be omitted when $k = \infty$.

Although most of the results known from the theory of comparison of experiments generalize easily there are a few surprises. Thus equivalence for two point sets does no longer imply full equivalence. Equivalence for 3 point sets does, however, imply full equivalence and this in turn is equivalent to the condition that the vector lattices generated by the measure families are isometrically isomorphic by a (and hence the) correspondence making the θ-th measures correspond to each other for each $\theta \in \Theta$. We may proceed, as in Torgersen (1972 a and b), and generalize the theory of sufficiency to the case of general measure families.

Useful characteristics of statistical experiments are, see LeCam (1974), certain functionals called conical measures. These are essentially the functionals which to a sublinear function ψ and for a given measure family $(\mu_\theta:\theta \in \Theta)$ associate the number $\int\psi(d\mu_\theta:\theta \in \Theta)$ according to the recipe given before our statements of principles (i)-(iv). Most of the basic properties of this characteristic extend without difficulties to measure families.

Let us now return to Examples 1.2-1.5 in the introduction and see if this theory can contribute something in each of these situations.

3. Local (fixed sample size) comparison of statistical experiments

We will assume throughout this section that the parameter set Θ is a subset of R^m for some positive integer m. We will be concerned with local comparison within small neighbourhoods of a given point θ^o belonging to the interior of Θ.

An experiment $\mathcal{E} = (\chi, \mathcal{A}, P_\theta : \theta \in \Theta)$ will be called differentiable (in the first mean) at θ^o if the map $\theta \to P_\theta$ from Θ to the Banach space of finite measures on $\mathcal{A}$ is Fréchet differentiable.

Let $e^i \equiv (0,\ldots,\overset{(i)}{1},\ldots,0)$ denote the i-th unit vector in R^m. Then $\mathcal{E}$ is differentiable at θ^o if and only if the partial derivatives defined as $[\partial P_\theta/\partial\theta_i]_{\theta^o} = \lim_{t\to 0}(P_{\theta^o+te^i} - P_{\theta^o})/t$, $i = 1,\ldots,m$, and the limit $\| P_\theta - P_{\theta^o} - \sum_{i=1}^m [\partial P_\theta/\partial\theta_i]_{\theta^o}(\theta_i - \theta_i^o) \| / \| \theta - \theta^o \| \to 0$ as $\theta \to \theta^o$ exist.

The theory for differentiable experiments which will be presented below is by and large self contained. We have, however, described some results without giving complete proofs. The missing proofs are then, if not otherwise stated, given in Torgersen (1972 a and b).

The notion of differentiability used here is weaker than the usual notion of differentiability in quadratic mean. The latter leads (see LeCam (1974) or Millar (1983)) to some basic results concerning the asymptotic behaviour of replicated experiments. If we merely assume differentiability according to the definition above, then these results need not hold. We will not be concerned with asymptotic theory in this sense here and thus the chosen notion of differentiability appears to be appropriate - although there are many other possibilities for weaker as well as for stronger notions.

Let us return to the notion of differentiability (in the first mean) as defined above. If $\mathcal{E} = (P_\theta : \theta \in \Theta)$ is differentiable at θ^o then the family consisting of the (m+1) measures $(P_{\theta^o}, [\partial P_\theta/\partial\theta_i]_{\theta=\theta^o}$, $i = 1,\ldots,m)$ is called the first order characterization of $\mathcal{E}$ at θ^o. We might have used instead the measure family $(\mu_h : h \in R^m)$, where $\mu_h = P_{\theta^o} + \sum_{i=1}^m h_i [\partial P_\theta/\partial\theta_i]_{\theta=\theta^o}$, $h \in R^m$. If we insist that the local approximation should be an experiment, then we should replace μ_h by $|\mu_h| / \| \mu_h \|$.

We shall find it convenient to write $\dot{P}_{\theta^o,i}$ for the partial derivative of P_θ w.r.to θ_i at θ^o, i.e. $\dot{P}_{\theta^o,i} = [\partial P_\theta/\partial\theta_i]_{\theta^o}$. The first order characterization of $\mathscr{E}$ at θ^o will be denoted as $\dot{\mathscr{E}}_{\theta^o}$. Thus $\dot{\mathscr{E}}_{\theta^o} = (P_{\theta^o}, \dot{P}_{\theta^o,1}, \ldots, \dot{P}_{\theta^o,m})$ when $\mathscr{E} = (P_\theta : \theta \in \Theta)$ is differentiable at $\theta = \theta^o$.

Let $e_i = (0,\ldots,\overset{(i)}{1},\ldots,0)$ denote the i-th unit vector of R^m. If $\mathscr{E} = (P_\theta : \theta \in \Theta)$ is differentiable at θ^o then the experiments $(P_{\theta^o + te^i} : |t| < \varepsilon)$ are all well defined and differentiable at $t = 0$, provided ε is sufficiently small. Furthermore, P_θ is approximable by $\mu_\theta = P_{\theta^o} + \sum_{i=1}^m (P_{\theta^o + (\theta_i - \theta^o_i)e^i} - P_{\theta^o})$ in the sense that $\| P_\theta - \mu_\theta \| / \| \theta - \theta^o \| \to 0$ as $\theta \to \theta^o$.

Conversely, these two conditions guarantee that $\mathscr{E}$ is differentiable at $\theta = \theta^o$. The usual arguments from calculus imply that it is differentiable at θ^o, provided the partial derivatives exist and are continuous within some neighbourhood of θ^o.

What kind of a measure family is the first order approximation of a differentiable experiment ? Note first that if $\dot{\mathscr{E}}_{\theta^o} = (P_{\theta^o}, \dot{P}_{\theta^o}, \dot{P}_{\theta^o,1}, \ldots, \dot{P}_{\theta^o,m})$ is the first order approximation of $\mathscr{E}$ at θ^o then $P_{\theta^o} >> \dot{\mathscr{E}}_{\theta^o}$ and each measure $\dot{P}_{\theta^o,i}$, $i = 1,\ldots,m$, has total mass zero. All general properties of first order approximations may be deduced from these two properties. In fact any measure family $(\sigma, \sigma_1, \ldots, \sigma_m)$ such that the measures $\sigma_1, \ldots, \sigma_m$ all have total mass zero and such that σ is a probability measure dominating $\sigma_1, \ldots, \sigma_m$ is of the form $\dot{\mathscr{E}}_{\theta^o}$, where $\mathscr{E} = (P_\theta : \theta \in \Theta)$ and $P_\theta = \text{constant} \cdot |\sigma + \sum_i (\theta_i - \theta^o_i)\sigma_i|$ when $\theta \in \Theta$.

It is easily checked that if $\mathscr{E}$ is differentiable at $\theta = \theta^o$ and if $\mathscr{E}$ is at least as informative as $\mathscr{F}$ then $\mathscr{F}$ is also differentiable at $\theta = \theta^o$. In fact, if $\mathscr{E} = (P_\theta : \theta \in \Theta)$ is 0-deficient w.r. to $\mathscr{F} = (Q_\theta : \theta \in \Theta)$ then $Q_\theta \equiv P_\theta M$ for a transition M and consequently

$\dot{Q}_{\theta^o,i} = \dot{P}_{\theta^o,i} M$, $i = 1,\ldots,m$. Thus $\dot{\mathscr{F}}_{\theta^o}$ is obtained from $\dot{\mathscr{E}}_{\theta^o}$ by the transition M.

Likewise, finite products of experiments which are differentiable at θ^o are themselves differentiable at θ^o. More precisely if $\mathscr{E} = (P_\theta : \theta \in \Theta)$ and $\mathscr{F} = (Q_\theta : \theta \in \Theta)$ are differentiable at θ^o and if $\mathscr{G} = \mathscr{E} \times \mathscr{F}$ then the first order approximation $\mathscr{G}_{\theta^o}$ of $\mathscr{G}$ is the measure family

$$(P_{\theta^o} \times Q_{\theta^o}, P_{\theta^o} \times \dot{Q}_{\theta^o,i} + \dot{P}_{\theta^o,i} \times Q_{\theta^o} ; i = 1,\ldots,m).$$

As we will make clearer later on, the local properties of the differentiable experiment $\mathscr{E} = (\chi, \mathscr{A}, P_\theta : \theta \in \Theta)$ at θ^o are determined by the distribution $F(\cdot|\theta^o, \mathscr{E})$ of the random vector $(d\dot{P}_{\theta^o,i}/dP_{\theta^o} ; i=1,\ldots,m)$ under P_{θ^o}.

Clearly, $\int x F(dx|\theta^o, \mathscr{E}) = 0$ and any probability distribution on R^m having expectation zero is of the form $F(\cdot|\theta^o, \mathscr{E})$ for some differentiable experiment $\mathscr{E}$.

We shall here only point out that if $\mathscr{E}$ and $\mathscr{F}$ are both differentiable at θ^o and if $\mathscr{G} = \mathscr{E} \times \mathscr{F}$ then $F(\cdot|\theta^o, \mathscr{G}) = F(\cdot|\theta^o, \mathscr{E}) * F(\cdot|\theta^o, \mathscr{F})$, where $*$ denotes convolution.

We will see later on how $F(\cdot|\theta^o, \mathscr{E})$ and $F(\cdot|\theta^o, \mathscr{F})$ must be related in order to ensure that $\mathscr{E}$ is locally at least as informative as $\mathscr{F}$ at θ^o.

Consider now two experiments, $\mathscr{E} = (P_\theta : \theta \in \Theta)$ and $\mathscr{F} = (Q_\theta : \theta \in \Theta)$, which both are differentiable at $\theta = \theta^o$. Equip R^m with the L-norm $\| \ \|$: $x \to \sum_i |x_i|$. For each $\varepsilon > 0$, denote the restrictions of $\mathscr{E}$ and $\mathscr{F}$ to the ε-ball $N(\theta^o, \varepsilon) = \{\theta : \|\theta - \theta^o\| \leq \varepsilon\}$, respectively, by $\mathscr{E}_\varepsilon$ and $\mathscr{F}_\varepsilon$.

The problem of local comparison of $\mathscr{E}$ and $\mathscr{F}$ within small neighbourhoods of θ^o may now be discussed in terms of the behaviour of the deficiency of $\delta(\mathscr{E}_\varepsilon, \mathscr{F}_\varepsilon)$ for small values of ε. If $\mathscr{M}_i$ denotes the totally non informative experiment then continuity implies that the deficiencies $\delta(\mathscr{M}_i, \mathscr{F}_\varepsilon)$ and $\delta(\mathscr{M}_i, \mathscr{E}_\varepsilon)$ both tend to zero as $\varepsilon \to 0$. Thus $\delta_k(\mathscr{E}_\varepsilon, \mathscr{F}_\varepsilon) \to 0$ as $\varepsilon \to 0$, for each $k = 1,2,\ldots,\infty$.

Next let us determine the rate of the convergence of $\delta_k(\mathcal{E}_\varepsilon, \mathcal{F}_\varepsilon)$ as $\varepsilon \to 0$. If $\mathcal{E} = (P_\theta : \theta \in \Theta)$ is differentiable in θ^o then we may expand P_θ as $P_\theta = P_{\theta^o} + \Sigma(\theta - \theta^o)_i \dot{P}_{\theta^o, i} + \tau(\mathcal{E}, \theta^o, \theta)$, where the measure $\tau(\mathcal{E}, \theta^o, \theta)$ is defined by this expansion. The differentiability assumption implies then that $\|\tau(\mathcal{E}, \theta^o, \theta)\| / \|\theta - \theta^o\| \to 0$ as $\theta \to \theta^o$.

In the following we will find it convenient to utilize the symbol o in the usual way, i.e. o will denote any real valued function on $]0, \infty[$ such that $o(t)/t \to 0$ as $t \to 0$.

If $\mathcal{E} = (P_\theta : \theta \in \Theta)$ and $\mathcal{F} = (Q_\theta : \theta \in \Theta)$ are both differentiable at θ^o, then the smallest (it exists) constant η such that $\dot{\mathcal{E}}_{\theta^o}$ is $(0, \eta, \ldots, \eta)$ deficient w.r. to $\dot{\mathcal{F}}_{\theta^o}$ for k-point sets will be called <u>the local deficiency at θ^o of $\mathcal{E}$ w.r. to $\mathcal{F}$ for k-decision problems</u>. The local deficiency at θ^o of $\mathcal{E}$ w.r. to $\mathcal{F}$ for k-decision problems will be denoted as $\dot{\delta}_{k,\theta^o}(\mathcal{E}, \mathcal{F})$. Here, as elsewhere we may omit the qualification "for k-decision problems" and the subscript k if $k = \infty$.

The local deficiency determines the rate of convergence of $\delta_k(\mathcal{E}_\varepsilon, \mathcal{F}_\varepsilon)/\varepsilon$ as $\varepsilon \to 0$ by

<u>Theorem 3.1</u>: Asymptotic behaviour of deficiencies within small neighbourhoods

With the notation introduced above for differentiable experiments $\mathcal{E} = (P_\theta : \theta \in \Theta)$ and $\mathcal{F} = (Q_\theta : \theta \in \Theta)$ we have

$$\delta_k(\mathcal{E}_\varepsilon, \mathcal{F}_\varepsilon) \leq \varepsilon \dot{\delta}_{k,\theta^o}(\mathcal{E}, \mathcal{F}) + o(\varepsilon),$$

where $o(\varepsilon) = \sup\{\|\tau(\mathcal{E}, \theta^o, \theta)\| + \|\tau(\mathcal{F}, \theta^o, \theta)\| : \|\theta - \theta^o\| \leq \varepsilon\}$. Furthermore,

$$\delta_k(\mathcal{E}_\varepsilon, \mathcal{F}_\varepsilon)/\varepsilon \to \dot{\delta}_{k,\theta^o}(\mathcal{E}, \mathcal{F}) \quad \text{as} \quad \varepsilon \to 0.$$

<u>Remark 1</u>. The local deficiency may, by the randomization criterion, be expressed as

$$\dot{\delta}_{\theta^o}(\mathcal{E}, \mathcal{F}) = \min\{\max_i \|\dot{P}_{\theta^o, i} M - \dot{Q}_{\theta^o, i}\| : P_{\theta^o} M = Q_{\theta^o}\},$$

where M varies within the set of transitions (randomizations) from $L(\mathcal{E})$ to the L-space $\Gamma(\mathcal{E})^*$ of bounded additive set functions on the sample space of $\mathscr{F}$.

We may (see Remark 1 after Theorem 2.2) limit our attention to transitions from $L_1(P_{\theta^0})$ to $L_1(Q_{\theta^0})$.

Remark 2. The proof implies that the statements of the theorem remain true, if, for each $\varepsilon > 0$, $\mathscr{E}_\varepsilon$ and $\mathscr{F}_\varepsilon$ are replaced by the restrictions of the experiments $\mathscr{E}$ and $\mathscr{F}$ to the sub set of $N(\theta^0,\varepsilon)$ consisting of the $2m$ points (vertices) $\theta^0 \pm (0,\ldots,\overset{(i)}{\varepsilon},\ldots,0)$, $i = 1,\ldots,m$.

Proof: Consider first the case $k = \infty$. If $P_{\theta^0}M = Q_{\theta^0}$ for a transition M and if $\|\theta-\theta^0\| \leq \varepsilon$ then

$$\|P_\theta M-Q_\theta\| = \|(P_\theta-P_{\theta^0})M-(Q_\theta-Q_{\theta^0})\|$$

$$= \|\sum_i(\theta-\theta^0)_i(\dot{P}_{\theta^0,i}M-\dot{Q}_{\theta^0,i})+\tau(\mathscr{E},\theta^0,\theta)M-\tau(\mathscr{F},\theta^0,\theta)\|$$

$$\leq \sum_i|(\theta-\theta^0)_i|\max_i\|\dot{P}_{\theta^0,i}M-\dot{Q}_{\theta^0,i}\| + \|\tau(\mathscr{E},\theta^0,\theta)\| + \|\tau(\mathscr{F},\theta^0,\theta)\|$$

$$\leq \varepsilon\max_i\|\dot{P}_{\theta^0,i}M-\dot{Q}_{\theta^0,i}\| +o(\varepsilon),$$

so that $\delta(\mathscr{E}_\varepsilon,\mathscr{F}_\varepsilon) \leq \varepsilon\dot{\delta}_{\theta^0}(\mathscr{E},\mathscr{F})+o(\varepsilon)$. It follows that $\limsup\ \delta(\mathscr{E}_\varepsilon,\mathscr{F}_\varepsilon)/\varepsilon \leq \dot{\delta}_{\theta^0}(\mathscr{E},\mathscr{F})$ as $\varepsilon \to 0$.

Consider on the other hand any number $c > \liminf_{\varepsilon\to 0} \delta(\mathscr{E}_\varepsilon,\mathscr{F}_\varepsilon)/\varepsilon$. Then $\delta(\mathscr{E}_\varepsilon,\mathscr{F}_\varepsilon) < c\varepsilon$ for all ε belonging to a sequence $\varepsilon_1,\varepsilon_2,\ldots$ which decreases to zero. Assume that ε belongs to this subsequence. The randomization criterion (Theorem 2.2) yields then a transition M_ε from $L(\mathscr{E})$ to $\Gamma(\mathscr{F})^*$ so that $\|P_\theta M_\varepsilon-Q_\theta\| < c\varepsilon$ when $\theta \in N(\theta^0,\varepsilon)$. We may, by compactness, assume that M_ε converges pointwise on $L(\mathscr{E}) \times \Gamma(\mathscr{F})$ to a transition M from $L(\mathscr{E})$ to $\Gamma(\mathscr{F})^*$. Expanding around θ^0 we find that

$$\|P_{\theta^0}M_\varepsilon-Q_{\theta^0}+\sum_i(\theta_i-\theta^0_i)(\dot{P}_{\theta^0,i}M_\varepsilon-\dot{Q}_{\theta^0,i})+\tau(\mathscr{E},\theta^0,\theta)M_\varepsilon-\tau(\mathscr{F},\theta^0,\theta)\| < c\varepsilon$$

when $\|\theta-\theta^0\| \leq \varepsilon$. In particular $\|P_{\theta^0}M_\varepsilon-Q_{\theta^0}\| < c\varepsilon$. $\varepsilon \to 0$ yields then $P_{\theta^0}M = Q_{\theta^0}$.

Putting $\theta' = \theta^{o} + (0,\ldots,\overset{(i)}{\varepsilon},\ldots,0)$ and $\theta'' = \theta^{o} - (0,\ldots,\overset{(i)}{\varepsilon},\ldots,0)$ we find that

$$\|\varepsilon(\dot{P}_{\theta^{o},i} M_{\varepsilon} - \dot{Q}_{\theta^{o},i}) + P_{\theta^{o}} M_{\varepsilon} - Q_{\theta^{o}} + \tau(\mathscr{E},\theta^{o},\theta') M_{\varepsilon} - \tau(\mathscr{F},\theta^{o},\theta')\| < c\varepsilon$$

and

$$\|\varepsilon(\dot{P}_{\theta^{o},i} M_{\varepsilon} - \dot{Q}_{\theta^{o},i}) - (P_{\theta^{o}} M_{\varepsilon} - Q_{\theta^{o}}) - (\tau(\mathscr{E},\theta^{o},\theta'') M_{\varepsilon} - \tau(\mathscr{F},\theta^{o},\theta''))\| < c\varepsilon.$$

Hence $2\varepsilon\|\dot{P}_{\theta^{o},i} M_{\varepsilon} - \dot{Q}_{\theta^{o},i}\| + o(\varepsilon) < 2c\varepsilon$. Dividing through by ε and letting $\varepsilon \to 0$ we find that $\|\dot{P}_{\theta^{o},i} M - \dot{Q}_{\theta^{o},i}\| \leq c$. Thus $c \geq \dot{\delta}_{\theta^{o}}(\mathscr{E},\mathscr{F})$. It follows that $\liminf \delta(\mathscr{E}_{\varepsilon},\mathscr{F}_{\varepsilon})/\varepsilon \geq \dot{\delta}_{\theta^{o}}(\mathscr{E},\mathscr{F})$ as $\varepsilon \downarrow 0$ so that $\delta(\mathscr{E}_{\varepsilon},\mathscr{F}_{\varepsilon})/\varepsilon \to \dot{\delta}_{\theta^{o}}(\mathscr{E},\mathscr{F})$ as $\varepsilon \to 0$. This completes the proof when $k = \infty$.

The case of a finite k may be reduced to the case $k = \infty$ as follows: Assume that $k < \infty$ and let $(\mathscr{Y},\mathscr{B})$ denote the sample space of $\mathscr{F}$. Let $\mathscr{B}_0$ be a sub algebra of $\mathscr{B}$ containing at most 2^k sets. The "k=∞" part of the theorem implies that $\delta(\mathscr{E}_{\varepsilon},\mathscr{F}_{\varepsilon}|\mathscr{B}_0) \leq \varepsilon\dot{\delta}_{\theta^{o}}(\mathscr{E},\mathscr{F}|\mathscr{B}_0)$ $+o(\varepsilon)$, where $o(\varepsilon)$ is defined as in the theorem and, consequently, does not depend on $\mathscr{B}_0$. Taking the supremum for all such algebras $\mathscr{B}_0$ we find, by Corollary 6 in Torgersen (1970), that

$\delta_k(\mathscr{E}_{\varepsilon},\mathscr{F}_{\varepsilon}) \leq \varepsilon \sup \dot{\delta}_{\theta^{o}}(\mathscr{E},\mathscr{F}|\mathscr{B}_0) + o(\varepsilon)$. The sumpremum on the right hand side is, by essentially the same argument that was used in the proof of this corollary, precisely the "dotted"

deficiency $\dot{\delta}_{k,\theta^{o}}(\mathscr{E},\mathscr{F})$. Hence $\delta_k(\mathscr{E}_{\varepsilon},\mathscr{F}_{\varepsilon}) \leq \varepsilon\dot{\delta}_{k,\theta^{o}}(\mathscr{E},\mathscr{F}) + o(\varepsilon)$. If follows that $\limsup_{\varepsilon\to 0} \delta_k(\mathscr{E}_{\varepsilon},\mathscr{F}_{\varepsilon})/\varepsilon \leq \dot{\delta}_{k,\theta^{o}}(\mathscr{E},\mathscr{F})$.

On the other hand, $\liminf_{\varepsilon\to 0} \delta_k(\mathscr{E}_{\varepsilon},\mathscr{F}_{\varepsilon})/\varepsilon \geq \liminf_{\varepsilon\to 0} \delta(\mathscr{E}_{\varepsilon},\mathscr{F}_{\varepsilon}|\mathscr{B}_0)/\varepsilon =$ $\dot{\delta}_{\theta^{o}}(\mathscr{E},\mathscr{F}|\mathscr{B}_0)$ for any sub algebra $\mathscr{B}_0$ of $\mathscr{B}$ containing at most 2^k sets.

Taking the sumpremum for all such algebras $\mathscr{B}_0$ we see that $\liminf_{\varepsilon\to 0} \delta_k(\mathscr{E}_{\varepsilon},\mathscr{F}_{\varepsilon})/\varepsilon \geq \dot{\delta}_{k,\theta^{o}}(\mathscr{E},\mathscr{F})$.

Define the <u>local deficiency distance at θ^{o} between $\mathscr{E}$ and $\mathscr{F}$</u> for k-decision problems as the largest of the numbers $\dot{\delta}_{k,\theta^o}(\mathscr{E},\mathscr{F})$ and $\dot{\delta}_{k,\theta^o}(\mathscr{F},\mathscr{E})$. This number will be denoted as $\dot{\Delta}_{k,\theta^o}(\mathscr{E},\mathscr{F})$. Here the subscript k and the qualification "for k-decision problems" may be omitted when $k=\infty$. It is easily checked that $\dot{\Delta}_{k,\theta^o}$ is a pseudodistance for experiments. The theory yields readily

<u>Corollary 3.2</u>: Asymptotic behaviour of deficiency distances within small neighbourhoods

With the notation introduced above, we have $\Delta_k(\mathscr{E}_\varepsilon,\mathscr{F}_\varepsilon)\le\varepsilon\dot{\Delta}_{k,\theta^o}(\mathscr{E},\mathscr{F})+o(\varepsilon)$, where $o(\varepsilon)$ is defined as in the theorem. Furthermore "=" holds for some function $o(\varepsilon)$.

Related to the local deficiencies are the local orderings and the local equivalences of experiments. Thus we shall say that <u>$\mathscr{E}$ is locally at least as informative as $\mathscr{F}$ for k-decision problems at $\theta=\theta^o$</u> if $\dot{\delta}_{k,\theta^o}(\mathscr{E},\mathscr{F})=0$. This defines, for $k=1,2,\ldots,\infty$, a partial ordering $\underset{k,\theta^o}{\geq}$ of differentiable experiments. If $\mathscr{E}\underset{k,\theta^o}{\geq}\mathscr{F}$ and $\mathscr{F}\underset{k,\theta^o}{\geq}\mathscr{E}$, i.e. if $\dot{\Delta}_{k,\theta^o}(\mathscr{E},\mathscr{F})=0$, then we shall say that <u>$\mathscr{E}$ and $\mathscr{F}$ are locally equivalent for k-decision problems at $\theta=\theta^o$</u>. This defines for each $k=1,2,\ldots,\infty$ a local equivalence $\underset{k,\theta^o}{\sim}$ for differentiable experiments. It turns out, however, that the non trivial equivalences $\underset{k,\theta^o}{\sim}$, $k=2,3,\ldots,\infty$, are all the same. More generally, it may be shown that all the pseudodistances $\dot{\Delta}_{k,\theta^o}$, $k=2,3,\ldots,\infty$, define the same notion of convergence.

What is the statistical significance of local deficiencies and related notions ? Some insight may be gained from the following characterization in terms of performance functions.

<u>Proposition 3.3</u>: Local comparison of performance functions

Let $\mathscr{E}=(P_\theta:\theta\in\Theta)$ and $\mathscr{F}=(Q_\theta:\theta\in\Theta)$ be both differentiable at $\theta=\theta^o$.

Let $(T,\mathscr{S})$ be a decision space and consider a decision rule σ in $\mathscr{F}$.

Then there is a decision rule ρ in $\mathscr{E}$ such that

$$\limsup_{\theta\to\theta^o} \| P_\theta\rho - Q_\theta\sigma \| / \| \theta-\theta^o \| \le \dot{\delta}_{\theta^o}(\mathscr{E},\mathscr{F}).$$

Furthermore, $(T,\mathscr{S})$ and σ may be chosen so that

$$\limsup_{\theta\to\theta^o} \| P_\theta\rho - Q_\theta\sigma \| / \| \theta-\theta^o \| \ge \dot{\delta}_{\theta^o}(\mathscr{E},\mathscr{F})$$

for all decision rules ρ in $\mathscr{E}$.

<u>Remark</u>. If $(T,\mathscr{S})$ is a k-decision space then the first inequality may be sharpened by replacing $\dot{\delta}_{\theta^o}(\mathscr{E},\mathscr{F})$ by the usually smaller number $\delta_{k,\theta^o}(\mathscr{E},\mathscr{F})$. Furthermore, if $(T,\mathscr{S})$ is a k-decision space then σ in $\mathscr{F}$ may be chosen so that

$$\limsup_{\theta\to\theta^o} \| P_\theta\rho - Q_\theta\sigma \| / \| \theta-\theta^o \| \ge \dot{\delta}_{k,\theta^o}(\mathscr{E},\mathscr{F})$$

for all decision rules ρ in $\mathscr{E}$.

This may be seen by applying the proposition to the restrictions of $\mathscr{F}$ to algebras of events containing at most 2^k events.

<u>Proof</u>: First note that for any transition M, we have $\limsup_{\theta\to\theta^o} \| P_\theta M - Q_\theta \| / \| \theta-\theta^o \| = \max_i \| \dot{P}_{\theta^o,i} M - \dot{Q}_{\theta^o,i} \|$ or $= \infty$ as $P_{\theta^o} M = Q_{\theta^o}$ or $P_{\theta^o} M \neq Q_{\theta^o}$. The first statement of the proposition follows now by putting $\rho = M\sigma$, where $P_{\theta^o} M = Q_{\theta^o}$ and $\max_i \| \dot{P}_{\theta^o,i} M - \dot{Q}_{\theta^o,i} \| = \dot{\delta}_{\theta^o}(\mathscr{E},\mathscr{F})$. The second statement follows by observing that we may let $(T,\mathscr{S})$ be the sample space of $\mathscr{F}$ and then choose σ as the identity map.

If ψ is a sub linear function on $R \times R^m$ and if $\mathscr{E} = (P_\theta : \theta \in \Theta)$ is differentiable at $\theta = \theta^o$ then $\int \psi(dP_{\theta^o}, d\dot{P}_{\theta^o,1}, \ldots, d\dot{P}_{\theta^o,m}) = \int \psi(1, x_1, \ldots, x_m) F(dx|\theta^o, \mathscr{E})$. It follows readily that $\mathscr{E} = (P_\theta : \theta \in \Theta)$ is locally at least as informative as $\mathscr{F}$ at $\theta = \theta^o$ if and only if $\int \phi dF(\cdot|\theta^o, \mathscr{E}) \ge \int \phi dF(\cdot|\theta^o, \mathscr{F})$ for all convex functions ϕ on R^m. As we will see in the next section, this in turn is equivalent to the condition that $F(\cdot|\theta^o, \mathscr{E}) = DF(\cdot|\theta^o, \mathscr{F})$ for a Markov kernel (randomization) D from R^m to R^m such that $\int y D(dx|y) = y$ for all points y

in R^m. A Markov kernel having the latter property is called a <u>dilation</u>.

<u>The Fisher information matrix</u> $I(\theta^o, \mathcal{E})$ is the covariance matrix of $F(\cdot|\theta^o, \mathcal{E})$-provided of course that $F(\cdot|\theta^o, \mathcal{E})$ possesses finite second order moments. It follows that if $\mathcal{E}$ is locally at least as informative as $\mathcal{F}$ at θ^o and if the Fisher information matrix of $\mathcal{E}$ at θ^o exists then the Fisher information matrix of $\mathcal{F}$ at θ^o also exists and then the difference matrix $I(\theta^o, \mathcal{E}) - I(\theta^o, \mathcal{F})$ is non negative definite. This proves the local, and hence the "global", monotonicity of the Fisher information matrix.

<u>Example 3.4</u>: Local orderings of linear normal models

For each $p \times n_A$ matrix A let $\mathcal{E}_A$ denote the linear normal experiment $(N(A'\beta, I_A) : \beta \in R^p)$, I_A being the $n_A \times n_A$ unit matrix. The Fisher information matrix of $\mathcal{E}_A$ is the $p \times p$ matrix AA'. If B is another matrix with p rows and if $\mathcal{E}_A$ is locally at least as informative as $\mathcal{E}_B$ then, by the remarks above, $AA' \geq BB'$. The ordering "$\geq$" for matrices which is used in this example is the ordering which declares that $M \geq N$ if $M-N$ is non negative definite. If $AA' \geq BB'$ then $\mathcal{E}_A \sim \mathcal{E}_B \times \mathcal{E}_M$, where M is the non negative definite square root of $AA'-BB'$. It follows that the local orderings as well as the global orderings of linear normal models with known variances coincide with the usual ordering of Fisher information matrices.

When we turn to the case of unknown variances the matters become a bit more involved. Let $\mathcal{F}_A$ denote the experiment $(N(A'\beta, \sigma^2 I_A) : \beta \in R^p, \sigma > 0)$, where A and I_A are as above. The $(p+1) \times (p+1)$ Fisher information matrix of $\mathcal{F}_A$ w.r. to the unknown parameters $\beta_1, \ldots, \beta_p$ and σ is

$$I(\beta, \sigma, \mathcal{F}_A) = \begin{pmatrix} AA'/\sigma^2, & 0 \\ 0, & 2n_A/\sigma^2 \end{pmatrix}.$$

By Hansen and Torgersen (1974) the experiment $\mathcal{F}_A$ is at least as informative as $\mathcal{F}_B$ if and only if $AA' \geq BB'$ and $n_A \geq n_B + \text{rank}(AA'-BB')$. In fact (Torgersen (1984)) this is equivalent to the condition that $\mathcal{F}_A \sim \mathcal{F}_B \times \mathcal{F}_C$, i.e. $AA' = BB'+CC'$ and $n_A = n_B + n_C$, for some $p \times n_C$

matrix C.

The above mentioned result of Hansen and Torgersen (1974) was extended by Lehmann (1983) to the case of multivariate regression. In that case $\mathscr{F}_A$ is, for each $n_A \times p$ matrix A', realized by observing a random $n_A \times q$ matrix X such that $EX = A'\beta$ for an unknown $p \times q$ matrix β, while the rows of X are independent and multinormally distributed with the same unknown non singular covariance matrix σ^2. (Actually Lehmann assumes that A is in a "reduced" form where rank $A=p$.) If we compare the distribution of the minimal sufficient statistics we see again that $\mathscr{F}_A \sim \mathscr{F}_B \times \mathscr{F}_C$ when $AA' = BB'+CC'$ and $n_A = n_B+n_C$. On the other hand if we consider e.g. the first column of β and restrict β and σ^2 so that all the other entries of β are 0, while σ^2 is a diagonal matrix so that all the diagonal elements are known except the first one, then we are back in the univariate case with $q = 1$. It follows that the above mentioned criteria extends directly to the multivariate case.

Returning to the univariate case we see that the Fisher information matrix of $\mathscr{F}_A$ majorizes the Fisher information matrix of $\mathscr{F}_B$ if and only if $AA' \geq BB'$ and $n_A \geq n_B$. This amounts to the condition that the two restrictions we may obtain from $\mathscr{F}_A$ by assuming that exactly one of the quantities β and σ are known are at least as informative as the corresponding restrictions of $\mathscr{F}_B$. It follows that the local orderings, the global ordering as well as the ordering by the Fisher information matrices are all stronger than the ordering of Fisher information matrices for known σ. Assuming that $AA' \geq BB'$ we may distinguish the three cases "$n_A<n_B$", "$n_A \geq n_B$ and $n_A<n_B+\text{rank}(AA'-BB')$" and "$n_A \geq n_B+\text{rank}(AA'-BB')$".

In the first case $\mathscr{F}_A$ and $\mathscr{F}_B$ are not comparable for any of the mentioned orderings. In the last case $\mathscr{F}_B$ is as we noted above, a factor of $\mathscr{F}_A$ so that $\mathscr{F}_A$ majorizes $\mathscr{F}_B$ for the global ordering and thus also for the weaker orderings. In the second case we know that $\mathscr{F}_A$ does not majorize $\mathscr{F}_B$ globally while it does majorize $\mathscr{F}_B$ for the ordering of Fisher information matrices.

It remains therefore to consider the problem of local ordering for the case when $n_B+\text{rank}(AA'-BB') > n_A \geq n_B$ only.

As a particular case consider univariate normally distributed variables X and Y such that X is distributed as $N(\xi,\sigma^2)$ while Y is distributed as $N(\xi,\sigma^2/2)$. Assume that ξ and $\sigma > 0$ are both unknown. Replacing X with $\sqrt{2}X$ we see that we are in this situation with $n_A = n_B = 1$, $AA' = 2$ and $BB' = 1$. We shall see below, in spite of the fact that the experiments defined by X and Y are not comparable that X is locally more informative than Y at any point (ξ,σ).

In fact we shall now show that <u>the local ordering of these experiments coincides with</u> the usual ordering of the Fisher information matrices. Thus let us assume that $AA' \geq BB'$ and that $n_A \geq n_B$. Using the dilation criterion mentioned above we shall prove our claim by showing that $\dot{\delta}_{\beta^o,\sigma^o}(\mathscr{F}_A,\mathscr{F}_B) = 0$ for all $\beta^o \in R^p$ and all $\sigma^o > 0$.

As we will see, the consequence that the local orderings do not depend on where localization takes place follows from the arguments described below. However, this fact follows also by general considerations on invariance under groups acting transitively and "smoothly" on the parameter set.

Differentiating the log likelihoods we find that $F(\cdot|\sigma,\beta,\mathscr{F}_A) = ((AX)'/\sigma,(\sum_1^{n_A} X_i^2-n_A)/\sigma)$, where X is distributed as $N(0,I_A)$. Similarly $F(\cdot|\sigma,\beta,\mathscr{F}_B) = ((BY)'/\sigma,(\sum_1^{n_B} Y_i^2-n_B)/\sigma)$, where Y is distributed as $N(0,I_B)$. Thus we must show that $E\phi((AX)'/\sigma,(\sum_1^{n_A} X_i^2-n_A)/\sigma)$ is greater or equal than the analogous expression for $\mathscr{F}_B$ when ϕ is convex on R^{p+1} and, say, a maximum of a finite number of linear functions. Replacing ϕ with $\phi(\cdot/\sigma)$ we may without loss of generality assume that $\beta = 0$ (in R^p) and $\sigma = 1$. Proceeding as in Hansen and Torgersen (1974) we consider first the case where AA' is the $p \times p$ identity matrix while $BB' = \Lambda$ is a $p \times p$ diagonal matrix. Let λ_i denote the (i,i)th element of Λ. For convenience of formulation we will assume that the diagonal elements of Λ are ordered in decreasing order. If $s = \text{rank } B = \text{rank } \Lambda$ then $\lambda_i > 0$ or $=0$ as $i \leq s$ or $i > s$.

As the rows of A are orthonormal we may add $n_A - p$ rows to A so that the extended $n_A \times n_A$ matrix $\tilde{A}$ is orthonormal. Putting $\tilde{X} = \tilde{A}X$ we see that $AX = (\tilde{X}_1,\ldots,\tilde{X}_p)'$ and that $\Sigma X_i^2 = \Sigma \tilde{X}_i^2$. Thus, since $\tilde{X}$ is also distributed as $N(0,I_A)$, we find that

$E\phi((AX)',\Sigma X_i^2-n_A) = E\phi(X_1,\ldots,X_p,X_1^2+\ldots+X_{n_A}^2-n_A)$.

Likewise the rows of B are orthogonal. The first s rows became orthonormal after having been divided by, respectively, $\sqrt{\lambda_1},\sqrt{\lambda_2},\ldots,\sqrt{\lambda_s}$. The p-s remaining rows are all the $1\times n_B$ zero matrix. Extend the described orthonormal system of row matrices to a $n_B\times n_B$ orthonormal matrix U and put $\tilde{Y}=UY$. Then $\tilde{Y}$ is distributed as Y and $BY = (\sqrt{\lambda_1}\tilde{Y}_1,\ldots,\sqrt{\lambda_s}\tilde{Y}_s,\ldots,0)'$, while $\Sigma Y_i^2 = \Sigma\tilde{Y}_i^2$. Hence $E\phi((BY)',\Sigma Y_i^2-n_B) = E\phi(\sqrt{\lambda_1}Y_1,\ldots,\sqrt{\lambda_s}Y_s,0,\ldots,0,\Sigma Y_i^2-n_B)$. Our task is therefore to show that $E\phi(X_1,\ldots,X_p,\sum_1^{n_A}X_i^2-n_A) \geq E\phi(\sqrt{\lambda_1}X_1,\ldots,\sqrt{\lambda_p}X_p,\sum_1^{n_B}X_i^2-n_B)$ when $X_1,X_2,\ldots$ are independent $N(0,1)$ variables. Assume first that $n_B\geq p$. Then, by Jensen's inequality, $E\phi(X_1,\ldots,X_p,\sum_1^{n_A}X_i^2-n_A) = EE(\phi(\ldots)|X_1,\ldots,X_{n_B}) \geq E\phi(X_1,\ldots,X_p,\sum_1^{n_B}X_i^2-n_B)$. If, on the other hand, $n_B<p$ then, since $n_B\geq s$, the same argument implies that $E\phi(\sqrt{\lambda_1}X_1,\ldots,\sqrt{\lambda_p}X_p,\sum_1^{n_B}X_i^2-n_B) \leq E\phi(\sqrt{\lambda_1}X_1,\ldots,\sqrt{\lambda_p}X_p,\sum_1^{p}X_i^2-p)$ and that $E\phi(X_1,\ldots,X_p,\sum_1^{p}X_i^2-p) \leq E\phi(X_1,\ldots,X_p,\sum_1^{n_A}X_i^2-n_A)$. Putting $t=\max(p,n_B)$ we see that we will be through in both cases if we can show that $E\phi(X_1,\ldots,X_p,\sum_1^{t}X_i^2-t) \geq E\phi(\sqrt{\lambda_1}X_1,\ldots,\sqrt{\lambda_p}X_p,\sum_1^{t}X_i^2-t)$. Let the variables $\xi_1=\pm1,\ \xi_2=\pm1,\ldots,\xi_p=\pm1$ be independent and independent of $(X_1,\ldots,X_t)$. Assume that $\Pr(\xi_i=1) = (\sqrt{\lambda_i}+1)/2$. (This is feasible since $\lambda_i\in[0,1]$.) Then $E\xi_i=\sqrt{\lambda_i}$. By symmetry and Jensen's inequality we obtain

$$E\phi(X_1,\ldots,X_p,\sum_1^{t}X_i^2)=E\phi(\xi_1X_1,\ldots,\xi_pX_p,\sum_1^{t}X_i^2) \geq \phi(\sqrt{\lambda_1}X_1,\ldots,\sqrt{\lambda_p}X_p,\sum_1^{t}X_i^2).$$

This establishes the desired inequality when AA' is the $p\times p$ unit matrix and BB' is a diagonal matrix. If, more generally, rank $A=p$ then there is a $p\times p$ matrix F such that $FAA'F'$ is the $p\times p$ unit matrix while $FBB'F'$ is a diagonal matrix (with the diagonal elements being in decreasing order). Then $AX=F^{-1}FAX$ and $BY=F^{-1}FBY$ and we are back to the previous case with ϕ replaced by $\tilde{\phi}$, where $\tilde{\phi}(x_1,\ldots,x_p,z) \equiv \phi((F^{-1}x)',z)$.

Finally, if rank A $= r < p$ then we may choose a basis $v_{\cdot 1}, v_{\cdot 2}, \ldots, v_{\cdot r}$ for the column space of A. Any vector in the column space of B belongs to this space since $AA' \geq BB'$. If $a_{\cdot j}$ is the j-th column in A and $b_{\cdot j}$ is the j-th column in B then we may write $a_{\cdot j} = \sum_i s_{ij} v_{\cdot i}$ and $b_{\cdot j} = \sum t_{ij} v_{\cdot i}$. Putting $S = \{s_{ij}\}$ and $T = \{t_{ij}\}$ we obtain $A = VS$ and $B = VT$, where V is the $p \times r$ matrix $(v_{\cdot 1}, \ldots, v_{\cdot r})$. Then S and T have, respectively, dimensions $r \times n_A$ and $r \times n_B$. Furthermore, rank S=r and $SS' \geq TT'$. If $\phi_V(y_1, \ldots, y_r, z) \equiv \phi(Vy, z)$ then $\phi((AX)', \Sigma X_i^2 - n_A) = \phi_V((SX)', \Sigma X_i^2 - n_A)$ and $\phi((BY)', \Sigma Y_i^2 - n_B) = \phi_V((TY)', \Sigma Y_i^2 - n_B)$. Thus we may apply the previous arguments to $\mathscr{F}_S$ and $\mathscr{F}_T$.

It would be interesting to know if there are general and manageable expressions for the local deficiencies between linear normal models. The reader is referred to LeCam (1975) and Swensen (1980) for information on the "global" deficiencies in this case.

The statistical significance of local deficiencies is particularly transparent in the one dimensional case, i.e. $m = 1$. In this case, as explained in Section 2, the deficiencies $\dot{\delta}_{k,\theta^o}(\mathscr{E}, \mathscr{F})$, $k = 2,3,\ldots,$ are all the same. They may then be expressed in terms of powers of most powerful tests or in terms of slopes of power functions of locally most powerful tests.

These results may be expressed as follows. Let for each (ordered) pair (P_1, P_2) of probability measures on the same measurable space the power of the most powerful level α-test for testing "P_1" against "P_2" be denoted by $\beta(\alpha | P_1, P_2)$. Then

$$\dot{\delta}_{\theta^o}(\mathscr{E}, \mathscr{F}) = 2 \lim_{\varepsilon \to 0} \sup_\alpha [\beta(\alpha | Q_{\theta^o}, Q_{\theta^o + \varepsilon}) - \beta(\alpha | P_{\theta^o}, P_{\theta^o + \varepsilon})] / \varepsilon .$$

In terms of locally most powerful tests - or rather slope maximizing tests - we have

$$\dot{\delta}_{\theta^o}(\mathscr{E}, \mathscr{F}) = \sup_\alpha [\tau(\alpha | \mathscr{F}) - \tau(\alpha | \mathscr{E})]^+ ,$$

where for a differentiable experiment $\mathscr{E} = (P_\theta : \theta \in \Theta)$ and a number $\alpha \in [0,1]$ the quantity $\tau(\alpha | \mathscr{E})$ is the maximal slope at θ^o for power functions of level α tests for testing "$\theta = \theta^o$" against "$\theta > \theta^o$". The function $\tau(\cdot|\mathscr{E})$ is, and may be any, continuous concave function on $[0,1]$ vanishing at $\alpha = 0$ and $\alpha = 1$.

Actually, $\mathscr{E}_{\theta^o} \sim (\lambda, \tau)$, where λ is the uniform distribution on $[0,1]$ and τ is the measure on $[0,1]$ posessing $\tau(\cdot|\mathscr{E})$ as distribution function.

In particular, $\mathscr{E}$ is locally at least as informative as $\mathscr{F}$ at $\theta = \theta^o$ if and only if $\tau(\alpha|\mathscr{F}) \leq \tau(\alpha|\mathscr{E})$ for all numbers $\alpha \in [0,1]$. Thus the function $\tau(\cdot|\mathscr{E})$ characterizes $\mathscr{E}$ up to local equivalence at θ^o. The local deficiency distance $\dot{\Delta}_{\theta^o}$ becomes then just the sup norm distance for such functions.

If G is a probability distribution on the real line then the corresponding translation family is differentiable at θ^o, and hence for all θ, if and only if G has an absolutely continuous density g such that $\int_{-\infty}^{\infty} |g'(x)|dx < \infty$. Any experiment $\mathscr{E} = (P_\theta : \theta \in \Theta)$ which is differentiable at $\theta = \theta^o$ and which is not locally equivalent to a totally non informative experiment (i.e. $\dot{P}_{\theta^o} \neq 0$) is locally equivalent to a differentiable translation experiment $(G(\cdot\ ,-\theta) : \theta \in \Theta)$. In fact we may require that G is strongly unimodal and then G is determined up to a translation.

It follows that, from the point of view of local comparison which we have described here, all differentiable exeriments which are not locally non informative are equivalent to an essentially unique strongly unimodal translation experiment. This construction is as follows. Let $\mathscr{E} = (P_\theta : \theta \in \Theta)$ be differentiable at θ^o and assume that $\dot{P}_{\theta^o} \neq 0$. Then the set of solutions of the differential equation $G' = \tau(1-G|\mathscr{E})$ is the set of translates of a strongly unimodal distribution function G. If $\mathscr{G} = (G(\cdot - \theta) : \theta \in \Theta)$ is the translation experiment determined by G then it follows that $\tau(\cdot|\mathscr{E}) = \tau(\cdot|\mathscr{G})$ so that $\dot{\Delta}_{\theta^o}(\mathscr{E}, \mathscr{G}) = 0$.

If two strongly unimodal translation experiments are locally equivalent, then they are "globally" equivalent as well. In general, however, translation experiments may be locally equivalent without being "globally" equivalent. On the other hand, whatever the dimension m, $\mathscr{E}$ is locally at least as informative as $\mathscr{F}$ at θ^o, whenever $\mathscr{E}$ is "globally" at least as informative as $\mathscr{F}$. In fact any transition which carries $\mathscr{E}$

onto $\mathscr{F}$ also carries $\dot{\mathscr{E}}_{\theta^o}$ onto $\dot{\mathscr{F}}_{\theta^o}$. It should be kept in mind that the theory for local comparison which have been outlined here is only concerned with first order approximations. Certainly, there are cases of interest where the first order approximations are quite inadequate. Thus any experiment $\mathscr{E} = (P_\theta : \theta \in \Theta)$ such that $[\partial P_\theta / \partial \theta_i]_{\theta^o} = 0$; $i = 1, \ldots, m$ is locally equivalent (for first order approximations) to a totally non informative experiment. If, however, we want to test "$\theta = \theta^o$" against close alternatives then such experiments may behave quite differently. Using the theory of Section 2 we may then proceed and consider higher order approximations.

Along with the notion of local information considered here goes a notion of local sufficiency. Thus if $\mathscr{E} = (\chi, \mathscr{A}, P_\theta : \theta \in \Theta)$ is differentiable at $\theta = \theta^o$ and if $\mathscr{F}$ is the restriction $\mathscr{E}|\mathscr{B}$ of $\mathscr{E}$ to some sub σ-algebra $\mathscr{B}$ of $\mathscr{A}$ then $\dot{\Delta}_{\theta^o}(\mathscr{E}, \mathscr{F}) = 0$ if any only if $d\dot{P}_{\theta^o, i} / dP_{\theta^o}$ may be specified $\mathscr{B}$ measurable for all $i = 1, \ldots, m$. Thus these Radon-Nikodym derivatives describe the minimal locally sufficient sub σ-algebra of $\mathscr{A}$.

4. Comparison of measures. Dilations

Orderings of experiments and of measures are often expressed in terms of dilations. However, the usual notion of a dilation as a kernel which dilates is too narrow and should for our purposes be replaced by some notion of (almost) density preserving kernels. The typical situation is as follows.

Assume that we are given two measure families $\mathscr{E} = (\chi, \mathscr{A}, \mu_\theta : \theta \in \Theta)$ and $\mathscr{F} = (\mathscr{Y}, \mathscr{B}, \nu_\theta : \theta \in \Theta)$ such that the measurable spaces $(\chi, \mathscr{A})$ and $(\mathscr{Y}, \mathscr{B})$ are both Euclidean.

Assume also that we are given a point $\theta^o \in \Theta$ such that the measures μ_{θ^o} and ν_{θ^o} are non negative and dominate, respectively, $\mathscr{E}$ and $\mathscr{F}$. Deficiency for deficiency functions vanishing at θ^o may then be characterized by

Theorem 4.1: Almost density preserving kernels

Assume that the requirements described above are satisfied and that $\varepsilon_{\theta^0} = 0$. For each $\theta \in \Theta$, let f_θ and g_θ be, respectively, the densities of μ_θ and ν_θ w.r. to μ_{θ^0} and ν_{θ^0}.

Then $\mathscr{E}$ is ε-deficient w.r. to $\mathscr{F}$ if and only if $\mu_{\theta^0} = D\nu_{\theta^0}$ for a Markov kernel D from $(\mathscr{Y}, \mathscr{B})$ to $(\chi, \mathscr{A})$ such that

$$\int \left| \int f_\theta(x) D(dx|y) - g_\theta(y) \right| \nu_{\theta^0}(dy) \leq \varepsilon_\theta ; \theta \in \Theta .$$

Proof: It follows from Theorem 2.2 that $\mathscr{E}$ is ε-deficient w.r. to $\mathscr{F}$ if and only if there is a Markov kernel M form $(\chi, \mathscr{A})$ to $(\mathscr{Y}, \mathscr{B})$ such that $\|\mu_\theta M - \nu_\theta\| \leq \varepsilon_\theta$; $\theta \in \Theta$. If M has this property, then $\varepsilon_{\theta^0} = 0$, since $\mu_{\theta^0} M = \nu_{\theta^0}$. It follows that there is a joint distribution on $(\chi \times \mathscr{Y}, \mathscr{A} \times \mathscr{B})$ with marginals μ_{θ^0} and ν_{θ^0} so that the conditional distribution of the second coordinate given $x \in \chi$ is $M(\cdot|x)$. Let D be a regular distribution of the first coordinate (in χ) given the second coordinate (in $\mathscr{Y}$). Then $\mu_{\theta^0} \times M = D \times \nu_{\theta^0}$. It is easily checked that $y \to \int f_\theta(x) D(dx|y)$ is a density of $\mu_\theta M$ w.r. to ν_{θ^0} so that $\varepsilon_\theta \geq \|\mu_\theta M - \nu_\theta\| = \int \left| \int f_\theta(x) D(dx|y) - g_\theta(y) \right| \nu_{\theta^0}(dy)$.

Assume conversely that there exists a randomization D mapping ν_{θ^0} onto μ_{θ^0} and having this property. Let M be a regular conditional distribution of the second coordinate (in $\mathscr{Y}$) given the first coordinate (in χ) for the distribution $D \times \nu_{\theta^0}$ on $\mathscr{A} \times \mathscr{B}$. Thus $D \times \nu_{\theta^0} = \mu_{\theta^0} \times M$. We find again that $y \to \int f_\theta(x) D(dx|y)$ is a density of $\mu_\theta M$ w.r. to ν_{θ^0} and thus $\|\mu_\theta M - \nu_\theta\| = \int \left| \int f_\theta(x) D(dx|y) - g_\theta(y) \right| \nu_{\theta^0}(dy) \leq \varepsilon_\theta$; $\theta \in \Theta$.

The assumption that the kernels should map ν_{θ^0} into μ_{θ^0} (or μ_{θ^0} into ν_{θ^0}) is a generalized form of the condition that a Markov matrix should be doubly Markov (stochastic). We obtain the latter condition if we impose the condition that χ and $\mathscr{Y}$ are the same finite sets and the measures μ_{θ^0} and ν_{θ^0} both are the uniform distribution on this set.

Let us express these facts in terms of matrices in the slightly more general situation where μ_{θ^0} is a non negative distribution μ on $\{1,\ldots,m\}$ and ν_{θ^0} is a non negative distribution ν on $\{1,\ldots,n\}$.

Let us also, for simplicity, assume that the measure families are finite. Thus we assume that we are given a finite family $\mathscr{E} = (\mu, \mu_1, \ldots, \mu_r)$ of measures on $\{1,\dots,m\}$ and a finite family $\mathscr{F} = (\nu,\nu_1,\dots,\nu_r)$ of measures on $\{1,\dots,n\}$.

Identify $\mathscr{E}$ with (μ, A), where μ is the row matrix $(\mu(1),\dots,\mu(m))$ and A is the $r \times m$ matrix whose (k,i)th element is the number $\mu_k(i)$, $k = 1,\dots,r$, $i = 1,\dots,m$.

Identify $\mathscr{F}$ with (ν, B), where ν is the row matrix $(\nu(1),\dots,\nu(n))$ and B is the $r \times n$ matrix whose (k,j)th element is $\nu_k(j)$, $k=1,\dots,r$, $j=1,\dots,n$.

Thus the entries of the row matrices μ and ν are non negative and we shall assume that they are positive. Otherwise there are no restrictions on the (real valued) entries of the matrices μ, A, ν and B.

The theorem yields the following comparison rules for matrices.

Corollary 4.2: Informational inequalities for matrices

Let $\varepsilon_1,\dots,\varepsilon_r$ be non negative numbers and let A and B be matrices with real valued entries and with, respectively, dimensions $r \times m$ and $r \times n$.

Let also μ and ν be row matrices with positive entries and of dimensions m and n, respectively. Put $\tilde{A}_{k,i} = A_{k,i}\mu_i$ and $\tilde{B}_{k,j} = B_{k,j}\nu_j$; $k = 1,\dots,r$, $i = 1,\dots,m$, $j = 1,\dots,n$.

Then the following conditions are equivalent:

(i) $\mu M = \nu$ for a $m \times n$ Markov matrix M such that

$$\sum_j |(\tilde{A}M)_{k,j} - \tilde{B}_{k,j}| \leq \varepsilon_k, \quad k = 1,\dots,r;$$

(ii) $\mu = \nu D$ for a $n \times m$ Markov matrix D such that

$$\sum_j |(AD)_{k,j} - B_{k,j}|\nu_j \leq \varepsilon_k, \quad k = 1,\dots,r;$$

(iii) $$\sum_i \psi(\mu_i, A_{1,i}, A_{2,i},\dots,A_{r,i}) \geq \sum_j \psi(\nu_j, B_{1,j}, B_{2,j},\dots,B_{r,j}) - \sum_{k=1}^{r} \varepsilon_k[\psi(0,0,\dots,\overset{(k)}{-1},\dots,\overset{(r)}{0}) \vee \psi(\overset{(1)}{0},0,\dots,\overset{(k)}{1},\dots,\overset{(r)}{0})]$$ for each sub linear function ψ on $R \times R^r$.

Remark. We leave it to the reader to simplify these statements when μ and ν are uniform. If $\varepsilon_1 = \varepsilon_2 = \ldots = \varepsilon_k = 0$, then (iii) reduces to

(iii') $\sum_i \phi(A_{1i}/\mu_i, \ldots, A_{ri}/\mu_i)\mu_i \geq \sum_j \phi(B_{1j}/\nu_j, \ldots, B_{rj}/\nu_j)\nu_j$ for each convex function ϕ on R^r.

Consider a non negative measure T on R^m and a Markov kernel D from R^m to R^m. If $\varepsilon = (\varepsilon_1, \ldots, \varepsilon_m)$, where $\varepsilon_1, \ldots, \varepsilon_m$ are non negative numbers then we shall say that D is a (T,ε) dilation if $\int |(\int x_i D(dx|y) - y_i| T(dy) \leq \varepsilon_i$, $i = 1, \ldots, m$. A Markov kernel D from R^m to R^m such that $\int x_i D(dx|y) \underset{y}{\equiv} y_i$, $i = 1, \ldots, m$, is called a dilation. Clearly, a dilation is a $(0,T)$ dilation and, conversely, any $(0,T)$ dilation is T equivalent to a dilation.

Now we give a more general corollary to Theorem 4.1 which includes some of the standard results on dilations in R^m as special cases.

Corollary 4.3: ε-dilations in R^m

Let S and T be non negative measures on R^m such that the projections are integrable. Let the measure S_i and the measure T_i, for each $i = 1, \ldots, m$, be the measure having the projection on the i-th coordinate space as a density w.r. to the measure S and the measure T, respectively. Then the following conditions are equivalent for non negative numbers $\varepsilon_1, \ldots, \varepsilon_k$:

(i) $SM = T$ for a Markov kernel M form R^m to R^m such that

$\|S_i M - T_i\| \leq \varepsilon_i$, $i = 1, \ldots, m$;

(ii) $S = DT$ for a (T,ε) dilation D from R^m to R^m;

(iii) $\int \psi(1,x)S(dx) \geq \int \psi(1,x)T(dx) -$

$$\sum_{i=1}^m \varepsilon_i [\psi(\overset{(0)}{0}, \overset{(1)}{0}, \ldots, \overset{(i)}{-1}, \ldots, \overset{(m)}{0}) \vee \psi(\overset{(0)}{0}, \overset{(1)}{0}, \ldots, \overset{(i)}{1}, \ldots, \overset{(m)}{0})]$$

each sub linear function ψ on $R \times R^m$.

Remark. If $\varepsilon_1 = \ldots = \varepsilon_k = 0$ then (ii) becomes

(ii') $S = DT$ for a dilation D.

In this case (iii) may be written as

(iii') $\int \phi(x) S(dx) \geq \int \phi(x) T(dx)$ for all convex functions ϕ on R^m.

Proof: Apply Theorem 4.1. and the sub linear function criterion ((iii) in Section 2) to the measure families $(S, S_1, \ldots, S_m)$ and $(T, T_1, \ldots, T_m)$.

Let us phrase this corollary in terms of standard measures of measure families.

If $\mathcal{E} = (\chi, \mathcal{A}, \mu_\theta : \theta \in \Theta)$ is a measure family with finite parameter set Θ then the measure induced from $\mu = \sum_\theta |\mu_\theta|$ by the map $x \to ([d\mu_\theta/d\mu]_x; \theta \in \Theta)$ from χ to R^Θ is its standard measure. Thus the corollary may be phrased as follows.

Corollary 4.4: Ordering in terms of standard measures

Let S and T denote, respectively, the standard measures of the measure families $\mathcal{E} = (\mu_1, \ldots, \mu_m)$ and $\mathcal{F} = (\nu_1, \ldots, \nu_m)$.

Let, for each $i = 1, \ldots, m$, the measure S_i (the measure T_i) be the measure on R^m having the projection on the i-th coordinate space as a density w.r. to the measure S (the measure T).

Then the measure family $(\sum_i |\mu_i|, \mu_1, \ldots, \mu_m)$ is $(0, \varepsilon_1, \ldots, \varepsilon_m)$ deficient w.r. to $(\sum_i |\nu_i|, \nu_1, \ldots, \nu_m)$ if and only if the (equivalent) conditions (i), (ii) and (iii) of the previous corollary are satisfied.

In particular if the measures $\mu_1, \ldots, \mu_m$ and $\nu_1, \ldots, \nu_m$ are all non negative then $\mathcal{E}$ is 0-deficient w.r. to $\mathcal{F}$ if and only if $S = DT$ for a dilation D.

Another application is to differentiable experiments. The result below provides in particular the criterion for "being locally at least as informative as" which we commented on and utilized in the previous section.

Corollary 4.5: Local deficiencies

Assume that the parameter set Θ is a sub set of R^m and that the point θ^o is an interior point of Θ.

Let the experiments $\mathcal{E} = (P_\theta : \theta \in \Theta)$ and $\mathcal{F} = (Q_\theta ; \theta \in \Theta)$ both be differentiable at θ^o and let $\varepsilon_1, \ldots, \varepsilon_m$ be non negative numbers.

Using the notations of Section 3, put $S = F(\cdot|\theta^o, \mathcal{E})$ and $T = F(\cdot|, \theta^o, \mathcal{F})$.

Then each of (the equivalent) conditions (i)-(iii) of Corollary 4.3 is equivalent to the condition that $\dot{\mathcal{E}}_{\theta^o} = (P_{\theta^o}, \dot{P}_{\theta^o,1}, \ldots, \dot{P}_{\theta^o,m})$ is

$(0, \varepsilon_1, \ldots, \varepsilon_m)$ deficient w.r. to $\dot{\mathcal{F}}_{\theta^o} = (Q_{\theta^o}, \dot{Q}_{\theta^o,1}, \ldots, \dot{Q}_{\theta^o,m})$.

In particular $\mathcal{E}$ is locally at least as informative as $\mathcal{F}$ at θ^o if and only if $\int \phi(x) S(dx) \geq \int \phi(x) T(dx)$ for each convex function ϕ on R^m.

The dilation results for standard measures may be viewed as a particular case of dilation criteria for comparison of measures. A fairly general situation is described in

Theorem 4.6: Orderings of measures

Let μ and ν be non negative finite measures on a measurable space $(\chi, \mathcal{A})$.

Consider also a convex set H of $\mu + \nu$ integrable functions containing 0 and having the property that $h_1 \vee h_2 \in H$ when $h_1, h_2 \in H$.

Then $\int h d\mu \geq \int h d\nu$ for all $h \in H$ if and only if there is a transition M from $L_1(\mu)$ to $L_1(\nu)$ such that

$$(h\nu)M \geq h\nu, \quad h \in H.$$

If, in addition, H contains the constants and if $(\chi, \mathcal{A})$ is Euclidean then this condition is also equivalent to the condition that $\mu = D\nu$ for a Markov kernel D such that $\int h(x) D(dx|y) \geq h(y)$ for ν almost all y for all $h \in H$.

Remark 1. If μ is a measure and if $\int h d\mu$ exists then $h\mu$ denotes the measure which assigns to each measurable set A the mass $\int_A h d\mu$.

Remark 2. If H is countable or if H is "sufficiently separable" then the exceptional sets in the last statement may be chosen as the empty set.

Proof: Consider the measure families $\mathscr{E} = (h\mu : h \in H)$ and $\mathscr{F} = (h\nu : h \in H)$. Let $h_1, \ldots, h_r \in H$ and let ψ be a sub linear function on R^r which is both monotonically increasing and a maximum of a finite set of linear functionals on R^r. Thus $\psi(x) \equiv \bigvee_{t=1}^{n} \sum_{i=1}^{r} a_{ti} x_i$ for non negative constants a_{ti}. If $a_1, \ldots, a_r$ are non negative constants then, since $0 \in H$ and H is convex, $\frac{1}{N} \sum_{i=1}^{r} a_i h_i \in H$ when $N \geq \sum_{i=1}^{r} a_i$. It follows that $\frac{1}{N}\psi(h_1, \ldots, h_r) \in H$ when N is sufficiently large. Thus $\int \psi(h_1, \ldots, h_r) d\mu \geq \int \psi(h_1, \ldots, h_r) d\nu$. Consider so the maximum ψ of an arbitrary finite family of linear functionals on R^r. The map $x \to \psi(x) + \sum_{i=1}^{r} x_i \psi(0, \ldots, \overset{(i)}{-1}, \ldots, 0)$ is then monotonically increasing on R^r. Thus $\int \psi(h_1, \ldots, h_r) d\mu \geq \int \psi(h_1, \ldots, h_r) d\nu - \sum_{i=1}^{r} \varepsilon(h_i) \psi(0, \ldots, \overset{(i)}{-1}, \ldots, 0)$, where $\varepsilon(h) = \int h d\mu - \int h d\nu$, $h \in H$. It follows that $\mathscr{E}$ is ε-deficient w.r. to $\mathscr{F}$. Hence, by the randomization criterion, there is a transition M from $L_1(\mu)$ to $L_1(\nu)$ such that $\| (h\mu)M - (h\nu) \| \leq \varepsilon(h)$, $h \in H$. Now $\| (h\mu)M - (h\nu) \| = [(h\mu)M - h\nu](1) + 2\|[(h\mu)M - (h\nu)]^-\| = \varepsilon(h) + 2\|[(h\mu)M - (h\nu)]^-\|$. Thus $\|[(h\mu)M - (h\nu)]^-\| = 0$ when $h \in H$ so that $(h\mu)M \geq h\nu$, $h \in H$. Conversely, these inequalities imply that $\int h d\mu = (h\mu)(1) = ((h\mu)M)(1) \geq (h\mu)(1) = \int h d\nu$ when $h \in H$.

Assume that this condition is satisfied, that H contains the constants and that the measurable space $(\chi, \mathscr{A})$ is Euclidean. Then $(\pm\mu)M \geq \pm\nu$ so that $\mu M = \nu$. Let D be a regular conditional distribution of the first coordinate given the second for the distribution $\mu \times M$. Thus $\mu \times M = \nu \times D$. If $A \in \mathscr{A}$ and $h \in H$ then $\int_A [\int h(x) D(dx|y)] \nu(dy) = \int h(x) I_A(y) (D \times \nu)(d(x,y)) = \int h(x) I_A(y) (\mu \times M)(d(x,y)) = \int M(A|x) h(x) \mu(dx) = ((h\mu)M)(A) \geq (h\nu)(A) = \int_A h d\nu$. Hence $\mu = D\nu$ and $\int h(x) D(dx|\cdot) \geq h$ a.e. ν.

This in turn implies, by essentially the same computations, that $(h\mu)M \geq h\nu$, where the Markov kernel M satisfies $\mu \times M = D \times \nu$.

Consider the particular case where H consists of all functions on R^m which are maxima of a finite set of (monotonically increasing) linear functionals. If $\int h d\mu \geq \int h d\nu$ for all $h \in H$ then, by Theorem 4.6 and Remark 2, $\mu = D\nu$ for a Markov kernel D such that $\int x D(dx|y) \geq y$ for all $y \in R^m$. Conversely the existence of such a D implies, by Jensen's inequality that $\int h d\mu \geq \int h d\nu$, $h \in H$. If, more generally, $S_1, S_2, \ldots$ is a sequence of probability distributions on R^m such that the sequence $\int h dS_1, \int h dS_2, \ldots$ is monotonically increasing for each $h \in H$ then $S_{n+1} = D_n S_n$, where $\int x D_{n+1}(dx|y) \geq y$, $y \in R^m$. Let the joint distribution of $(X_1, X_2, \ldots)$ be determined by the requirements that $(X_1) = S_1$ and that $D_n(\cdot|X_n)$, for each n, is a conditional distribution of X_{n+1} given X_n. Then the Markov process $X_1, X_2, \ldots$ is a sub martingale such that X_n, for each n, has distribution S_n. This and Jensen's inequality prove the following result of Strassen (1965).

Corollary 4.7. Sub martingales with prescribed marginal distributions

Assume that the probability distributions $S_1, S_2, \ldots$ on R^m all possess finite expectations. Then there is a (sub) martingale $X_1, X_2, \ldots$ with, respectively, marginal distributions $S_1, S_2, \ldots$ if and only if $\int h dS_1 \leq \int h dS_2 \leq \ldots$ whenever the function h is convex (and monotonically increasing) on R^m.

Another interesting situation occurs when we assume that the functions in H are monotonically increasing w.r. to some partial ordering. In the latter case we obtain, see e.g. Karlin (1983), the joint distribution characterization of stochastic orderings of distributions on sets with prescribed partial orderings. As more general situations will be treated in the next section we shall not write this out here.

If we just require in Theorem 4.6 that $\int dh\mu \geq \int h d\nu$ for the functions h in H which are non negative, i.e. for the functions in $H_+ = \{h: h \in H, h \geq 0\}$ then the first part of the theorem applies with H replaced by H_+. The argument used in the proof of the last part of the theorem is, however, no longer valid since it required that $\int (-1) d\mu \geq \int (-1) d\nu$.

For this situation there is a usefull result of Fischer and Holbrook (1980). As the general underlying idea is quite simple we shall expose it here as

Proposition 4.8: Existence of minorizing functionals

Let γ be a real valued convex functional on a convex sub set H of a linear space.

Let $\mathscr{P}$ be a sub convex class of real valued concave functionals on H.

Assume that $\mathscr{P}$ is compact for the weakest topology which makes $P(h)$ lower semicontinuous in $P \in \mathscr{P}$ for each $h \in H$.

Then there is a P in $\mathscr{P}$ such that $\gamma \geq P$ on H if and only if $\gamma(h) \geq \inf\{P(h): P \in \mathscr{P}\}$ for all $h \in H$.

Remarks. We may, for intuition, think about H as a convex set of real valued functions and of each $P \in \mathscr{P}$ as an expectation.

$\mathscr{P}$ is called sub convex if there to each pair $(P_0, P_1) \in \mathscr{P} \times \mathscr{P}$ and to each $t \in]0,1[$ corresponds a $P_t \in \mathscr{P}$ so that $P_t \leq (1-t)P_0 + tP_1$. The compactness condition amounts to require that to each net $\{P_\alpha\}$ in $\mathscr{P}$ there corresponds a $P \in \mathscr{P}$ such that $\lim_\alpha \sup P_\alpha(h) \geq P(h)$, $h \in H$.

Proof: The "only if" part is trivial and the "if" part follows by applying minimax theory, e.g. Theorem 4.2 in Sion (1958), to the concave-convex payoff function $(P,h) \to \gamma(h) - P(h)$.

Applying this proposition to the situation of Corollary 4.3 with $\varepsilon_1 = \ldots = \varepsilon_m = 0$ we obtain

Corollary 4.9: Sub Markov kernels and non negative convex functions

Let S and T be non negative measures on R^m such that the projections are integrable. For each $i = 1,\ldots,m$, let S_i and T_i be the measures having the projection on the i-th coordinate space as a density w.r.t the measure S and the measure T, respectively. Then the following conditions are equivalent:

(i) $SM = T$ for a sub Markov kernel M from R^m to R^m so that $S_i M = T_i$, $i = 1,\ldots,m$;

(ii) $S \geq DT$ for a dilation D from R^m to R^m;

(iii) $\int \phi dS \geq \int \phi dT$ for each non negative convex function ϕ on R^m;

(iv) there is a non negative measure U on R^m such that $\int \phi dS \geq \int \phi d(T+U)$ for all convex functions ϕ on R^m. If so then the measure U may always be chosen as a one point mass distribution.

Remark. If S and T are required to possess compact supports then the equivalence of (iii) and (iv) follows from Fischer and Holbrook (1980).

Proof: Suppose (i) is satisfied. Then $\|S\| \geq \|SM\| = \|T\|$. If $\|S\| = \|T\|$ then M may be chosen as a proper Markov kernel and then we know from the remark after Corollary 4.3 that $\int \phi dS \geq \int \phi dT$ for all convex functions ϕ on R^m. If $\|S\| > \|T\|$ then we may put $\xi_i = [\ \|S_i\| - \|T_i\|\ ]/[\ \|S\| - \|T\|]$, $i = 1,\ldots,m$. Put $\xi = (\xi_1,\ldots,\xi_m)$ and define a Markov kernel $\tilde{M}$ by putting $\tilde{M}(B|x) = M(B|x)+[1-M(R^m|x)]I_B(\xi)$. Then $S\tilde{M} = T+U$ where U is the one point distribution which assigns mass $\|S\| - \|T\|$ to ξ. It is then a matter of checking that $S_i\tilde{M} = (T+U)_i = T_i+U_i$ where, for each measure κ, κ_i denotes the measure having $x \to x_i$ as density w.r. to κ. Thus the remark after Corollary 4.3 applies again and we get $\int \phi dS \geq \int \phi d(T+U) = \int \phi dT + \int \phi dU \geq \int \phi dT$ when ϕ is non negative and convex on R^m. Hence (i) ⇒ (iv) and clearly (iv) ⇒ (iii). Assume so that (ii) is satisfied. If ϕ is non negative and convex then $\int \phi dS \geq \int \phi d(DT) \geq \int \phi dT$, where the last "$\geq$" follows from the remark after Corollary 4.3. Thus (ii) ⇒ (iii).

Assume next that (iii) is satisfied. Put $\rho(x) = \bigvee_i |x_i|$ when $x = (x_1,\ldots,x_m) \in R^m$. Then $\|S\| = \int 1 dS \geq \int 1 dT = \|T\|$. If $\|S\| = \|T\|$ and if ϕ is convex and bounded from below by the constant b then $\int \phi dS = \int (\phi - b) dS + b\|S\| \geq \int (\phi - b) dT + b\|T\| = \int \phi dT$. This implies readily (iv) with $U = 0$. Put $\Gamma = [\ \|S\| - \|T\|]^{-1}(S-T)$ when $\|S\| > \|T\|$. Then $\Gamma(\phi) \geq 0$ for all non negative convex functions ϕ on R^m and $\Gamma(1) = 1$. Let H_N, $N = 1,2,\ldots$, denote the set of convex functions h such that $h \leq \rho$, while $h(x) \geq 0$ when $\rho(x) \geq N$. Take $\mathscr{P}_N$ as the set of probability distributions on $\{x:\rho(x) \leq N\}$. If $h \in H_N$ then

$\Gamma(h) \geq \inf_x h(x) = \inf\{P(h): P \in \mathscr{P}_N\}$. It follows, since $\mathscr{P}_N$ is tight, that there is a $P_N \in \mathscr{P}_N$ so that $\Gamma(h) \geq P_N(h)$, $h \in H_N$. In particular $\Gamma(\rho) \geq P_N(\rho)$. It follows that $(P_1,P_2,\ldots,)$ is tight. Let the probability distribution P be a weak limit point for this sequence. If h is convex and if $h \in H_N$ then $h \in H_{N_0}$ when $N \geq N_0$. Hence $\Gamma(h) \geq \limsup_N P_N(h) \geq P(h)$. In particular $\Gamma(h \vee (\rho - N_0)) \geq P(h \vee (\rho - N_0))$ when h is convex and $h \leq \rho$. $N_0 \to \infty$ yields $\Gamma(h) \geq P(h)$. It follows that $\Gamma(\phi) \geq P(\phi)$ when ϕ is convex and $\phi \leq k\rho$ for some positive constant k. Jensen's inequality tells us that $P(\phi) \geq \phi(\int xP(dx))$ when ϕ is convex. It follows that we may assume that P is a one point distribution. Altogether this shows that(iv) holds with U being the one point distribution which assigns mass $\|S\| - \|T\|$ to the point $\int xP(dx)$. Thus (iii) $\Rightarrow$ (iv). If (iv) holds then, by the remark after Corollary 4.3, there is a Markov kernel M so that $SM = T+U$ and $S_iM = T_i+U_i$, $i = 1,\ldots,m$. Furthermore $T = (T+U)\hat{M}$ where $\hat{M}$ is the sub Markov kernel $(x,B) \to I_B(x)g(x)$ where g is a version of $dT/d(T+U)$ such that $0 \leq g \leq 1$. Then $SM\hat{M} = T$ and $S_iM\hat{M} = T_i$ for the sub Markov kernel $M\hat{M}$. Hence (iv) $\Rightarrow$ (i). Still assuming (iv) and applying the same remark we find that $S = D(T+U)$ for a dilation D. Thus $S \geq DT$ so that (iv) $\Rightarrow$ (ii). The proof is now completed by combining the established implications.

Returning to matrices again we obtain

<u>Corollary 4.10</u>: The matrix case

Let μ and ν be row matrices of, respectively, dimensions $1 \times m$ and $1 \times n$. We shall assume that the entries $\mu_1,\ldots,\mu_m$ of μ and $\nu_1,\ldots,\nu_n$ of ν are all positive.

Let A and B be matrices with real entries and with dimensions $r \times m$ and $r \times n$, respectively. Then the following statements are equivalent:

(i) $\mu M = \nu$ for a $m \times n$ sub Markov matrix M such that $\sum_i A_{ki}\mu_i M_{i,j} = B_{k,j}\nu_j$, $k = 1,\ldots,r$, $j = 1,\ldots,n$;

(ii) $\mu \geq \nu D$ for a $n \times m$ Markov matrix D such that $AD = B$;

(iii) $\sum_i \phi(A_{1,i}/\mu_i,\ldots,A_{r,i}/\mu_i)\mu_i \geq \sum_j \phi(B_{1,j}/\nu_j,\ldots,B_{r,j}/\nu_j)\nu_j$ for each non negative convex function ϕ on R^r.

Proof. Apply the previous corollary to S and T where S assigns mass μ_i to each point $(A_{1,i},\ldots,A_{r,i})$ while T assigns mass ν_j to each point $(B_{1,j},\ldots,B_{r,j})$.

Consider again the situation treated in Theorem 4.6 under the additional assumptions that the functions in H are bounded and that the constant functions are in H.

If we weaken the requirement that $\mu(h) \geq \nu(h)$ for all $h \in H$ to the requirement that $\mu(h) \geq \nu(h)$ for all non negative functions $h \in H$ then there is a non negative and additive set function κ on $\mathcal{A}$ such that $\mu(h) \geq \nu(h) + \kappa(h)$ for all $h \in H$. This may be seen as follows. The assumption that $1 \in H$ implies that $\|\mu\| \geq \|\nu\|$. If $\|\mu\| = \|\nu\|$ and if c is a lower bound for h then $(1/2)(h-c) \in H_+$ so that $(1/2)\mu(h-c) \geq (1/2)\nu(h-c)$ yielding $\mu(h) \geq \nu(h)$. Thus we may put $\kappa = 0$ in this case. If $\|\mu\| > \|\nu\|$ then we may apply Proposition 4.7 to the functional $\Gamma = [\|\mu\| - \|\nu\|]^{-1}(\mu-\nu)$ yielding a finitely additive probability set function P such that $\Gamma(h) \geq P(h)$ for all $h \in H$. We may then put $\kappa = [\|\mu\| - \|\nu\|]P$.

If we were able to show that κ might be chosen countably additive then Proposition 4.6 would become applicable with ν being replaced by $\nu+\kappa$. A particular case where this works is, by the Riesz representation theorem, when $\mathcal{A}$ is the class of Borel sub sets of a compact Hausdorff space χ and the functions in H are continuous. This yields finally

Corollary 4.11: Comparison for integrals of non negative functions

Let μ and ν be non negative finite measures on the Borel class of a compact metric space χ.

Let H be a convex set of continuous functions on χ such that $h_1 \vee h_2 \in H$ when $h_1 \in H$ and $h_2 \in H$. We shall also assume that H contains the constant functions. Then the following conditions are equivalent:

(i) $\int h d\mu \geq \int h d\nu$ when $0 \leq h \in H$;

(ii) there is a non negative measure κ on $\mathcal{A}$ so that $\int h d\mu \geq \int h d\nu + \int h d\kappa$ for all $h \in H$;

(iii) there is a sub Markov kernel M such that $(h\mu)M \geq h\nu$, $h \in H$;

(iv) $\mu \geq D\nu$ for a Markov kernel D such that $\int h(x)D(dx|y) \geq h(y)$ for ν almost all y and for all $h \in H$.

Remark: If additional assumptions on linear and topological structures are satisfied then the equivalence of (i) and (ii) follows from Fischer and Holbrook (1980).

5. Application to a problem on probability distributions with given marginals

We shall here apply the theory presented in Section 2 to derive a result of Strassen in his 1965 paper on probability distribution with given marginals.

The result is related to the problem of determining the set of possible probabilities $Pr((X,Y) \in S)$ for a specified set S and for prescribed marginal distributions P and Q for X and Y. By convexity this set is clearly an interval and it is also fairly clear that it is closed when we permit joint distributions which are not necessarily countably additive. Thus it suffices, at least under this permission, to find the end points of this interval. As the problem of minimizing $Pr((X,Y) \in S)$ is equivalent to the problem of maximizing $Pr((X,Y) \notin S)$ it should then also suffice to determine the right end point of this interval.

We shall here, as it is actually done in Strassen's paper, consider the problem of maximizing $Pr((X,Y) \in S)$ when S is closed and the sample spaces of X and Y are both complete separable metric spaces. The right end point of the above mentioned interval may then by described as follows.

Let B be any event in the sample space of Y. Then $Pr((X,Y) \in S)$ may be decomposed as

$$Pr((X,Y) \in S \text{ and } Y \notin B) + Pr((X,Y) \in S \text{ and } Y \in B).$$

Hence

$$Pr((X,Y) \in S) \leq Pr(Y \notin B) + Pr((X,y) \in S \text{ for some } y \in B),$$

where the number on the right hand side is determined by P and Q.

Let us agree to use for each subset B of the sample space $\mathcal{Y}$ of Y the notation $B^{(S)}$ for the set of points x in the sample space χ of X such that $(x,y) \in S$ for some y in B. The upper bound derived above may then be written as

$$\inf_B [Q(B^c)+P(B^{(S)})]$$

and it follows from Strassen's paper (under the above mentioned regularity conditions) that this upper bound is achieved.

Usually it is not required to consider all sets B. If e.g. X and Y take their values in the same partially ordered set $(\chi,\leq)$ and if $S=\{(x,y):x\geq y\}$ then it suffices to consider monotonically increasing sets B (B is called <u>monotonically increasing (decreasing)</u> if its indicator function is monotonically increasing (decreasing)). Thus if χ is a sub set of the real line with the usual ordering then we obtain the greatest lower bound of all numbers $1-G(x)+F(x)$; $x\in\chi$, where F and G are, respectively, the distribution functions of X and Y.

In general there is a joint distribution such that $\Pr((X,Y)\in S)\geq 1-\varepsilon$ if and only if $Q(B)\leq P(B^{(S)})+\varepsilon$ for each (measurable) sub set B of χ.

If ρ is a real valued function on $\chi\times\mathcal{Y}$ and if $S=\{(x,y):\rho(x,y)\leq\varepsilon\}$ then this amounts to the requirement that $Q(B)\leq P(\{x:\rho(x,y)\leq\varepsilon$ for some $y\in B\})+\varepsilon$ when B is a (measurable) subset of $\mathcal{Y}$. This yields as it was shown in Strassen (1965), the joint distribution characterization of the Prohorov distance for a given distance ρ on χ.

There is a rephrasing of this result which is obtained by relaxing the condition that the distribution of Y should by exactly Q while at the same time strengthening the requirement on $\Pr((X,Y)\in S)$ by requiring that $\Pr((X,Y)\in S)=1$. Using this rephrasing we shall now see how this type of results may be deduced from the basic principles for comparing measure families.

The essential idea used in the proof may be grasped by restricting attention to finite sets χ and $\mathcal{Y}$. The reader might then see that the arguments using measure families may be replaced by simple arguments using well known results on support functions. The point here, however, is to provide another example on the relationship between the principles for comparison of measure families on the one hand and existence problems for probability distributions with given marginals on the other hand.

Except for the rephrasing, condition (iii) below, the following result is a particular case of a theorem in Strassen's paper.

Proposition 5.1: Probabilities for a specified set assigned by joint distributions possessing prescribed marginals

Let $(\chi, \mathscr{A}, P)$ and $(\mathscr{Y}, \mathscr{B}, Q)$ be probability spaces and let S be a sub set of $\chi \times \mathscr{Y}$. Assume that χ and $\mathscr{Y}$ are compact metric spaces with, respectively, Borel classes $\mathscr{A}$ and $\mathscr{B}$. Assume also that S is a closed sub set of $\chi \times \mathscr{Y}$. Then the following two conditions are equivalent for a number $\varepsilon \in [0,1]$:

(i) $Q(B) \leq P(B^{(S)}) + \varepsilon$ for all Borel sets B;

(ii) there is a joint distribution R on $\mathscr{A} \times \mathscr{B}$ with marginals P and Q such that $R(S) \geq 1-\varepsilon$.

If, furthermore, the projection of S into χ is onto then these conditions are equivalent to:

(iii) there is a joint distribution R on $\mathscr{A} \times \mathscr{B}$ with marginals P and μ such that $R(S) = 1$ and $\|\mu - Q\| \leq 2\varepsilon$.

Remark 1. The set $B^{(S)}$ is the projection on χ of the measurable set $(\chi \times B) \cap S$ and is consequently analytic and therefore completion measurable.

Remark 2. It is apparent from this proof that several of the assumptions may be weakened. Thus we shall only make use of (i) for closed sets B.

If S is defined by a partial ordering then it suffices to consider monotonically increasing sets B in (i) and then $B^{(S)} = B$. In this case these conditions are equivalent to the condition that $Q(h) \leq P(h) + \varepsilon \| h \|$ when the function h is bounded, monotonically increasing and measurable.

Proof. We know already that (ii) $\Rightarrow$ (i). Assume now that (iii) holds. Then there is, by the joint distribution characterization of the statistical distance (see e.g. Torgersen (1970)), a joint distribution U on $\mathcal{B}\times\mathcal{B}$ with marginals μ and Q such that $U(\{(y,y):y\in\mathcal{Y}\})\geq 1-\varepsilon$. Define a joint distribution V for variables X,Z and Y such that (X,Z) has distribution R while the conditional distribution of Y given (X,Z) is given by a regular conditional distribution of Y given Z in the situation where U is the joint distribution of Y and Z. Then (ii) holds if R is replaced by the joint distribution of (X,Y). Thus in any case (and we did not use compactness) (iii) $\Rightarrow$ (ii) $\Rightarrow$ (i). Assume now that (i) holds and that the projection of S into χ is onto. Put for any bounded measurable function h on $\mathcal{Y}$ and $x\in\chi$ $\hat{h}(x)=\sup\{h(y):y\in S_x\}$. If t is a real number then clearly $[h>t]^{(S)}=[\hat{h}>t]$. Thus, by (i), $Q(h)=\int_0^{\|h\|}Q(h>t)\,dt\leq\int_0^{\|h\|}[P(\hat{h}>t)+\varepsilon]\,dt=P(\hat{h})+\varepsilon\|h\|$ when h is non negative. It follows from the expression for $[\hat{h}\geq t]$ above and Remark 1 that $\hat{h}$ is completion measurable. Using that $\hat{c}(x)\equiv c$ for a constant function c we find, since $\widehat{h-c}=\hat{h}-c$, that $Q(h)\leq P(\hat{h})+2\varepsilon\|h\|$ for any bounded measurable function h. Here we used the assumption that the projection of S into χ was onto. Let h vary through the set $C(\mathcal{Y})$ of continuous functions on $\mathcal{Y}$. Then the linear functional Q is majorized by the sum of the two sub linear functionals $h\to P(\hat{h})$ and $h\to 2\varepsilon\|h\|$. It follows that Q may be decomposed as $Q=\mu+\nu$, where μ and ν are linear functionals (measures) such that $\mu(h)\leq P(\hat{h})$, $h\in C(\mathcal{Y})$ while $\nu(h)\leq 2\varepsilon\|h\|$. If $h\leq 0$ then $\hat{h}\leq 0$ and thus $\mu(h)\leq 0$. If $h=c$ is a constant then $\mu(c)\leq P(\hat{c})=$ $=P(c)=c$. It follows that μ is a probability distribution and that ν is a measure whose total variation is at most 2ε.

Consider the measure families $\mathcal{E}=(\hat{h}P:h\in C(\mathcal{Y}))$ and $\mathcal{F}=(h\mu:h\in C(\mathcal{Y}))$. Put $\eta_h=P(\hat{h})-\mu(h)$ when $h\in C(\mathcal{Y})$.

Let $h_1,\ldots,h_r\in C(\mathcal{Y})$ and let $h=\psi(h_1,\ldots,h_r)$ where ψ is a sub linear monotonically increasing function on R^r. Put $\hat{h}=\psi(\hat{h}_1,\ldots,\hat{h}_r)$. Then $\int\psi(h_1,\ldots,h_r)\,d\mu=\int h\,d\mu\leq\int\hat{h}\,dP\leq\int\psi(\hat{h}_1,\ldots,\hat{h}_r)\,dP$. If ψ is any sub linear function on R^r then $x\to\psi(x)+\sum_i x_i\psi(0,\ldots,-1,\ldots,0)$ is monotonically increasing. It follows then from the sub linear function criterion (iii) of Section 2 that $\mathcal{E}$ is η deficient w.r. to $\mathcal{F}$. Thus there is a Markov kernel M from $(\chi,\mathcal{A})$ to $(\mathcal{Y},\mathcal{B})$ such that $\|(\hat{h}P)M-h\mu\|\leq\eta_h$, $h\in C(\mathcal{Y})$. Thus $P(\hat{h})-\mu(h)+2\|[(\hat{h}P)M-h\mu]^-\|\leq\eta_h$. Now $h=1$ yields $PM\geq\mu$ so that $PM=\mu$. Let D be a regular conditional distribution of "x" given "y" for $P \times M$. Then $P \times M = D \times \mu$. If $h \in C(\mathscr{Y})$ and $B \in \mathscr{B}$ then $\int_B [\int \hat{h}(x) D(dx|y)]\mu(dy) = \int \hat{h}(x) I_B(y)(D\times M)(d(x,y)) = \int \hat{h}(x) I_B(y)(P \times M)(d(x,y)) = \int M(B|x)\hat{h}(x)(dx) = ((\hat{h}P)M)(B) \geq (h\mu)(B) = \int_B h(y)\mu(dy)$. Thus $\int \hat{h}(x) D(dx|y) \geq h(y)$ for μ almost all y when $h \in C(\mathscr{Y})$. Since $C(\mathscr{Y})$ is separable we may modify D so that $\int \hat{h}(x) D(dx|y) \geq h(y)$ for all $y \in \mathscr{Y}$ and all $h \in C(\mathscr{Y})$. This inequality extends, since $\hat{h}_n \uparrow \hat{h}$ when $h_n \uparrow h$, to all lower semicontinuous functions h on $\mathscr{Y}$. Let $y_0 \in \mathscr{Y}$ and let, for $n = 1,2,\ldots$, the open ball around y_0 with radius $1/n$ be denoted as B_n. Applying the inequality to the indicator function h_n of B_n we find that $D(B_n^{(S)}|y_0) = \int \hat{h}_n(x) D(dx|y) \geq h_n(y_0) = 1$. Furthermore, since $B_1^{(S)} \cap B_2^{(S)} \cap \ldots = \{x:(x,y_0) \in S\}$, we find that $D(\{x:(x,y_0) \in S\}|y_0) = 1$ for all points $y_0 \in \mathscr{Y}$. Thus $(P \times M)(S) = (D \times \mu)(S) = \int D(\{x:(x,y) \in S\}|y)\mu(dy) = 1$.

It follows that condition (iii) is satisfied with $R = P \times M$.

Altogether this shows that conditions (i), (ii) and (iii) are equivalent when the projection of S into χ is onto.

It remains to show that (i) $\Rightarrow$ (ii) in the general case. Assume then that (i) holds. Let the metric d which metricizes $\mathscr{Y}$ be bounded by 1. Add a point a to $\mathscr{Y}$ and extend d by putting $d(a,a) = 0$ while $d(a,y) = d(y,a) = 1$ for all points $y \in \mathscr{Y}$. Put $\mathscr{Y}_1 = \mathscr{Y} \cup \{a\}$ and $S_1 = S \cup \{(x,a): x \in \chi\}$. Then (i) holds if $\mathscr{Y}$ and S are replaced by, respectively, $\mathscr{Y}_1$ and S_1 and if Q is replaced by its extension $\tilde{Q}$ to $\mathscr{Y}_1$. Thus, by what we have proved, there is a joint distribution $\tilde{R}$ on $\chi \times \mathscr{Y}_1$ with marginals P and $\tilde{Q}$ such that $\tilde{R}(S_1) \geq 1-\varepsilon$. Then, since $\tilde{Q}(\mathscr{Y}) = 1$, $\tilde{R}$ is supported by $\chi \times \mathscr{Y}$ and thus (ii) holds with R being the restriction of $\tilde{R}$ to $\chi \times \mathscr{Y}$.

Using the fact that a complete separable metric space is homeomorphic to a G_δ sub set of compact metric space we obtain the theorem of Strassen mentioned above. This theorem states that (i) and (ii) are equivalent for probability spaces $(\chi, \mathscr{A}, P)$ and $(\mathscr{Y}, \mathscr{B}, Q)$ and sub sets S of $\chi \times \mathscr{Y}$ such that χ and $\mathscr{Y}$ are complete separable metric spaces and S is closed.

6. Majorization

The concepts of majorization defined in Marshall and Olkin (1979) generalizes as follows.

Let $\mathcal{E} = (\chi, \mathcal{A}, \mu_\theta : \theta \in \Theta)$ and $\mathcal{F} = (\mathcal{Y}, \mathcal{B}, \nu_\theta : \theta \in \Theta)$ be two measure families. We say that $\mathcal{E}$ weakly supermajorizes $\mathcal{F}$ if $\mathcal{E}$ is $\nu.(\mathcal{Y}) - \mu.(\chi)$ deficient w.r. to $\mathcal{F}$. We say that $\mathcal{E}$ weakly sub majorizes $\mathcal{F}$ if $\mathcal{E}$ is $\mu.(\chi) - \nu.(\mathcal{Y})$ deficient w.r. to $\mathcal{F}$ and that $\mathcal{E}$ majorizes $\mathcal{F}$ if $\mathcal{E}$ is 0-deficient w.r. to $\mathcal{F}$.

It is then easily checked that the two kinds of weak majorization as well as majorization are partial orderings for measure families.

The basic properties of weak supermajorization are collected in

Proposition 6.1: Weak supermajorization

The following properties are equivalent for measure families $\mathcal{E} = (\chi, \mathcal{A}, \mu_\theta : \theta \in \Theta)$ and $\mathcal{F} = (\mathcal{Y}, \mathcal{B}, \nu_\theta : \theta \in \Theta)$:

(i) $\mathcal{E}$ weakly supermajorizes $\mathcal{F}$;

(ii) $\mu_\theta M \leq \nu_\theta$ for all θ for some transition M from $L(\mathcal{E})$ to $L(\mathcal{F})$;

(iii) if $\theta_1, \ldots, \theta_r \in \Theta$ and ψ is a sublinear and monotonically decreasing function on R^r then $\int \psi(d\mu_{\theta_1}, \ldots, d\mu_{\theta_r}) \geq \int \psi(d\nu_{\theta_1}, \ldots, d\nu_{\theta_r})$.

Remark 1. If $\mu_\theta M \leq \nu_\theta$ for a transition M and if $\mu_\theta(\chi) = \nu_\theta(\mathcal{Y})$ then actually $\mu_\theta M = \nu_\theta$. It follows that these conditions imply that the inequality in (iii) holds for a sub linear function ψ provided ψ is monotonically decreasing in x_i when $\mu_{\theta_i}(\chi) \neq \nu_{\theta_i}(\mathcal{Y})$.

Remark 2. If $\theta_0 \in \Theta$ is such that $\mu_{\theta_0}, \nu_{\theta_0} \geq 0$, $\|\mu_{\theta_0}\| = \|\nu_{\theta_0}\|$, $\mu_{\theta_0} >> \mathcal{E}$, $\nu_{\theta_0} >> \mathcal{F}$ and the sample spaces are Euclidean then we may, as we saw in Section 4, express weak super majorization in terms of dilations. In that case (iii) may be replaced by

(iii') $\int \phi(d\mu_{\theta_1}/d\mu_{\theta_0}, \ldots, d\mu_{\theta_r}/d\mu_{\theta_0})\, d\mu_{\theta_0}$

$\geq \int \phi(d\nu_{\theta_1}/d\nu_{\theta_0}, \dots, d\nu_{\theta_r}/d\nu_{\theta_0})d\nu_{\theta_0}$ when ϕ is monotonicaly decreasing and convex on R^r.

Remark 3. If $\Theta = \{\theta_0, \theta_1\}$, $\mu_{\theta_0} \geq 0$, $\nu_{\theta_0} \geq 0$ and $\|\mu_{\theta_0}\| = \|\nu_{\theta_0}\|$ then it suffices, as explained in Section 2, to consider comparison for 2-points sets (testing problems) and we find that (iii) is equivalent to

(iii") $\mu_{\theta_1}(\chi) \leq \nu_{\theta_1}(\mathcal{Y})$ and $\int [c d\mu_{\theta_0} - d\mu_{\theta_1}]^+ \geq \int [c d\nu_{\theta_0} - d\nu_{\theta_1}]^+$; $c \in R$.

Considering the support function

$$(c_0, c_1) \to \begin{cases} \int [c_0 d\mu_{\theta_0} + c_1 d\mu_{\theta_1}]^+ & \text{when } c_1 \leq 0 \\ \infty & \text{when } c_1 > 0 \end{cases}$$

we see that (iii) now states

(iii"') $(\mu_{\theta_1}(\chi) \leq \nu_{\theta_1}(\mathcal{Y})$ and $\min\{\mu_{\theta_1}(\delta) : \mu_{\theta_0}(\delta) = \alpha\}$

$\leq \min\{\nu_{\theta_1}(\phi) : \nu_{\theta_0}(\phi) = \alpha\}$; $\alpha \geq 0$, where δ (ϕ) runs through the set of test functions in $\mathcal{E}(\mathcal{F})$.

The minima in (iii"') may be obtained in the same way as maximum power is obtained by the Neyman-Pearson lemma.

Putting $\alpha = \mu_{\theta_0}(\chi)$ we see that the condition which appears first in (iii") and in (iii"') is superfluous when $\mu_{\theta_0} >> \mu_{\theta_1}$.

Proof: Assume that $\mathcal{E}$ is $\nu.(\mathcal{Y}) - \mu.(\chi)$ deficient w.r. to $\mathcal{F}$. Then there is a transition M such that $\|\mu_\theta M - \nu_\theta\| \leq \nu(\mathcal{Y}) - \mu_\theta(\chi)$ for all $\theta \in \Theta$. This may, by the identity $\|\lambda\| = \int 1 d\lambda + 2\|\lambda^-\|$, be written $\nu_\theta \geq \mu_\theta M : \theta \in \Theta$. This in turn implies for any monotonically decreasing sub linear function ψ on R^r that $\int \psi(d\mu_{\theta_1}, \dots, d\mu_{\theta_r}) \geq \int \psi(d(\mu_{\theta_1}M), \dots, d(\mu_{\theta_r}M)) \geq \int \psi(d\nu_{\theta_1}, \dots, d\nu_{\theta_r})$. Altogether we have shown that (i) $\Rightarrow$ (ii) $\Rightarrow$ (iii). If (iii) holds and ψ is any sub linear function on R^r then, putting $e^i = (0, \dots, \overset{(i)}{i}, \dots, 0)$, we see that $x \to \psi(x) - \int x_i \psi(e^i)$ is monotonically decreasing so that

$$\int\psi(d\mu_{\theta_1},\ldots,d\mu_{\theta_r}) \geq \int\psi(d\nu_{\theta_1},\ldots,d\nu_{\theta_r}) - \sum_{i=1}^{r} \psi(e^i)(\nu_{\theta_i}(\mathscr{Y})-\mu_{\theta_i}(\chi)).$$

Hence (i) holds.

We omit the proof of the quite similar case of weak submajorization.

Proposition 6.2: Weak sub majorization

The following conditions are equivalent for measure families $\mathscr{E} = (\chi, \mathscr{A}, \mu_\theta : \theta\in\Theta)$ and $\mathscr{F} = (\mathscr{Y}, \mathscr{B}, \nu_\theta : \theta \in \Theta)$:

(i) $\mathscr{E}$ weakly sub majorizes $\mathscr{F}$;

(ii) $\mu_\theta M \geq \nu_\theta$ for all θ for some transition M from $L(\mathscr{E})$ to $L(\mathscr{F})$;

(iii) $\int\psi(d\mu_{\theta_1},\ldots,d\mu_{\theta_r}) \geq \int\psi(d\nu_{\theta_1},\ldots,d\nu_{\theta_r})$ when $\theta_1,\ldots,\theta_r \in \Theta$ and ψ is a sub linear and monotonically increasing function on R^r.

Remarks: We may here make similar remarks as those we made after the previous proposition. Thus Remarks 1 and 2 apply provided we substitute "increasing" for "decreasing". If $\Theta = \{\theta_0,\theta_1\}$, $\mu_{\theta_0}, \nu_{\theta_0} \geq 0$, $\|\mu_{\theta_0}\| = \|\nu_{\theta_0}\|$ then (iii) may be replaced by

"$\mu_{\theta_1}(\chi) \geq \nu_{\theta_1}(\mathscr{Y})$ and $\int[dc\mu_{\theta_0}+d\mu_{\theta_1}]^+ \geq \int[cd\nu_{\theta_0}+d\nu_{\theta_1}]^+$, $c \in R$",

or by

"$\mu_{\theta_1}(\chi) \geq \nu_{\theta_1}(\mathscr{Y})$ and $\max\{\mu_{\theta_1}(\delta):\mu_{\theta_0}(\delta) = \alpha\} \geq \max\{\nu_{\theta_1}(\phi):\nu_{\theta_0}(\phi)=\alpha\}$, $\alpha > 0$",

where $\delta(\phi)$ runs through the set of test functions in $\mathscr{E}(\mathscr{F})$.

The condition $\mu_{\theta_1}(\chi) \geq \nu_{\theta_1}(\mathscr{Y})$ is superfluous when $\mu_{\theta_0} >> \mu_{\theta_1}$.

Combining the two kinds of weak majorization we obtain

Theorem 6.3: Majorization criteria for measure families

The following properties are equivalent for measure families

$\mathscr{E} = (\mu_\theta : \theta \in \Theta)$ and $\mathscr{F} = (\nu_\theta : \theta \in \Theta)$:

(i) $\mathscr{E}$ majorizes $\mathscr{F}$;
(ii) $\mathscr{E}$ weakly supermajorizes $\mathscr{F}$ and weakly submajorizes $\mathscr{F}$;
(iii) $\mu_\theta M \underset{\theta}{\equiv} \nu_\theta$ for some transition M from $L(\mathscr{E})$ to $L(\mathscr{F})$;
(iv) if $\theta_1, \ldots, \theta_r \in \Theta$ and ψ is sublinear on R^r then

$$\int \psi(d\mu_{\theta_1}, \ldots, d\mu_{\theta_r}) \geq \int \psi(d\nu_{\theta_1}, \ldots, d\nu_{\theta_r}).$$

Remark 1. If $\theta_0 \in \Theta$ is such that $\mu_{\theta_0} \geq 0$, $\nu_{\theta_0} \geq 0$, $\| \mu_{\theta_0} \| = \| \nu_{\theta_0} \|$, $\mu_{\theta_0} >> \mathscr{E}$ and $\nu_{\theta_0} >> \mathscr{F}$ then we may express majorization in terms of dilations, provided the sample spaces are Euclidean. In that case (iv) may be expressed as

(iv') $\int \phi(d\mu_{\theta_1}/d\mu_{\theta_0}, \ldots, d\mu_{\theta_r}/d\mu_{\theta_0})/d\mu_{\theta_0}$

$\geq \int \phi(d\nu_{\theta_1}/d\nu_{\theta_0}, \ldots, d\nu_{\theta_r}/d\nu_{\theta_0}) d\nu_{\theta_0}$ when ϕ is convex on R^r.

If in addition μ_{θ_0} and ν_{θ_0} are probability measures then this is a particular case of Corollary 4.6.

Remark 2. If $\Theta = \{\theta_0, \theta_1\}$, $\mu_{\theta_0} \geq 0$, $\nu_{\theta_0} \geq 0$, $\| \mu_{\theta_0} \| = \| \nu_{\theta_0} \|$ and $c_1 \neq 0$ is a fixed number, then (iv) is equivalent to

(iv") $\mu_{\theta_1}(\chi) = \nu_{\theta_1}(\mathscr{Y})$ and $\int [c_0 d\mu_{\theta_0} + c_1 d\mu_{\theta_1}]^+ \geq \int [c_0 d\nu_{\theta_0} + c_1 d\nu_{\theta_1}]^+$

when $c_0 \in R$.

The function $(c_0, c_1) \to \int [c_0 d\mu_{\theta_0} + c_1 d\mu_{\theta_1}]^+$ is the support function of the closed convex range of $(\mu_{\theta_0}, \mu_{\theta_1})$. Thus conditions (i)-(iv) are in this case also equivalent to the condition that $\mu_{\theta_1}(\chi) = \nu_{\theta_1}(\mathscr{Y})$ and that the convex hull of the range of $(\mu_{\theta_0}, \mu_{\theta_1})$ contains the range $(\nu_{\theta_0}, \nu_{\theta_1})$. Furthermore the latter condition implies the first one when $\mu_{\theta_0} >> \mu_{\theta_1}$.

If $\mu_{\theta_1}(\chi) = \nu_{\theta_1}(\mathscr{Y})$ then $(\mu_{\theta_0}, \mu_{\theta_1})$ majorizes $(\nu_{\theta_0}, \nu_{\theta_1})$ if and only if $(\mu_{\theta_o}, \mu_{\theta_1})$ super (sub) majorizes $(\nu_{\theta_o}, \nu_{\theta_1})$.

If $\Theta = \{\theta_o, \theta_1\}$, $\chi = \mathscr{Y} = \{1,\ldots,n\}$ and $\mu_{\theta_o}(i) = \nu_{\theta_o}(i) = 1$, $i = 1,\ldots,n$, then we obtain the equivalence of conditions (i)-(iv) in Example 1.5.

7. References

[1] R.R.Bahadur, 1954. Sufficiency and statistical decision functions. Ann.Math.Statist. 25, 423-462.

[2] R.R.Bahadur, 1955. A characterization of sufficiency. Ann.Math. Statist. 26, 286-293.

[3] D.Blackwell, 1951. Comparison of experiments. Proc. Second Berkeley Sympos. Math. Statist. Probab. 93-102.

[4] D.Blackwell, 1953. Equivalent comparisons of experiments. Ann.Math. Statist. 24, 265-272.

[5] C.Boll, 1955. Comparison of experiments in the infinite case. Ph.D. thesis. Standford University.

[6] G.Dahl, 1983. Pseudo experiments and majorization. Thesis. Univ. of Oslo. Statistical research report, 1984.

[7] N.Dunford and J.T.Schwartz, 1958. Linear operators. Part I, Interscience Publishers, New York.

[8] P.Fischer and J.A.R.Holbrook, 1980. Balayage defined by the non negative convex functions. Proc.Amer.Soc. 79, 445.

[9] O.H.Hansen and E.N.Torgersen, 1974. Comparison of linear normal experiments. Ann.Statist. 2, 367-373.

[10] G.H.Hardy, J.E.Littlewood and G.Polya, 1934. Inequalities. Cambridge University Press.

[11] H.Heyer, 1973. Mathematische Theorie statistischer Experimente. Springer Verlag.

[12] S.Kakutani, 1941. Concrete representation of abstract (L)-spaces and the mean ergodic theorem. Ann.of Math. (2) 42, 523-537.

[13] S.Karlin, 1983. Comparison of measures, multivariate majorization, and applications to statistics. In: Karlin, Amemiya, Goodman, Studies in econometric time series and multivariate statistics, Academic Press.

[14] I.L.Kelley, 1955. General topology. D. van Nostrand. Princeton, New Jersey.

[15] L.LeCam, 1964. Sufficiency and approximate sufficiency. Ann.Math. Statist. 35, 1419-1455.

[16] L.LeCam, 1974. Notes on asymptotic metods in statistical decision theory. Centre de Recherches Math.Univ. de Montréal.

[17] L.LeCam, 1975. Distances between experiments. In: J.N.Srivastava, Ed., A survey of Statistical Design and Linear Models. North--Holland, Amsterdam, pp. 383-396.

[18] E.L.Lehmann, 1983. Comparison of experiments for some multivariate normal situations. In: Karlin, Amemiya, Goodman. Studies in econometric time series and multivariate statistics, Academic Press.

[19] A.W.Marshall, and I,Olkin, 1979. Inequalities: Theory of majorization and its applications. Academic Press.

[20] P.W.Millar, 1983. The minimax principle in asymptotic statistical theory. Lecture Notes in Mathematics No 976, 76-265. Springer--Verlag.

[21] M.Sion, 1958. On general minimax theorems. Pacific. J.Math. 8, 171-176.

[22] V.Strassen, 1965. The existence of probability measures with given marginals. Ann.Math. Statist. 36, 423-439.

[23] A.R.Swensen, 1980. Deficiencies in linear normal experiments. Ann.Statist. 8, 1142-1155.

[24] E.Torgersen, E.1969. On ε-comparison of experiments. Mimeographed notes. Univ. of California, Berkeley.

[25] E.Torgersen, 1970. Comparison of experiments when the parameter space is finite. Z.Wahrscheinlichkeitstheorie Verw.Geb. 16, 219-249.

[26] E.Torgersen, 1972a. Local comparison of experiments when the parameter set is one dimensional. Statist. res. report no.4 1972, Univ. of Oslo.

[27] E.Torgersen, 1972b. Local comparison of experiments. Statist. res. report no.5, Univ. of Oslo.

[28] E.Torgersen, 1972c. Comparison of translation experiments. Ann. Math. Stat. 43, 1383-1399.

[29] E.Torgersen, 1976. Comparison of statistical experiments. Scand. Journ. of Statistics. 3, 186-208.

[30] E.Torgersen, 1982. Comparison of some statistical experiments associated with sampling plans. Probability and Mathematical Statistics 3, 1-17.

[31] E.Torgersen, 1984. Orderings of linear models. Journal of Statistical Planning and Inference 9, 1-17.

[32] A.Wald, 1950. Statistical decision functions. New York, Wiley.

CHARACTERIZATION OF LINEAR ADMISSIBLE ESTIMATORS IN GAUSS-MARKOV MODEL UNDER NORMALITY

S.Zontek
Institute of Mathematics
Polish Academy of Sciences
Wrocław, Poland

1. Introduction

Throughout the paper Y will stand for a random n-vector normally distributed with expectation $\mu = X\beta$ and covariance matrix $\sigma^2 V$. Here X is an $n \times p$ known matrix, V is an $n \times n$ positive definite (p.d., for short) known matrix, and $\beta \in \mathbb{R}^p$ and $\sigma^2 \geq 0$ are the unknown parameters.

It is desired to estimate $C'\mu$, where C is an $n \times m$ matrix, while C' stands for the transpose matrix of C. Let t be any estimator of $C'\mu$ based on Y and let

$$R_C(t;\mu,\sigma^2 V) = E(t-C'\mu)'(t-C'\mu)$$

be the risk function. We restrict attention to estimators t for which the above defined risk is finite for each $\beta \in \mathbb{R}^p$ and for each $\sigma^2 \geq 0$. If t is a linear estimator, say $t = L'Y$, where L is an $n \times m$ matrix, then

$$R_C(L'Y;\ \mu,\sigma^2 V) = \mathrm{tr}[\sigma^2 L'VL+(L-C)'\mu\mu'(L-C)],$$

where $\mathrm{tr}A$ stands for the trace of the square matrix A.

The relations "as good as" and "better than" are understood in the usual way. An estimator t is said to be admissible for $C'\mu$, if no other estimator is better than t. A linear estimator $L'Y$ is said to be admissible for $C'\mu$ among the class of linear estimators, if no other linear estimator is better than $L'Y$.

For a given matrix A, the symbols $r(A)$, $\mathcal{R}(A)$, $\mathcal{N}(A)$ and A^+ will denote, respectively, the rank, the range, the null space and the Moore-Penrose generalized inverse of A. Moreover, for given square matrices A and B, we write $A \leq B$ and $A < B$ when the matrix $B-A$ is nonnegative definite and positive definite, respectively.

As well known Stein (1956) has shown that in case when $X=V=I_n$ and when σ^2 is known, the random vector Y is admissible for μ iff $n \leq 2$. Here I_n stands for the unit $n \times n$ matrix. Using this result Cohen (1966) has given a full characterization of admissible linear estimators of μ within the model considered by Stein. He proved that (i) $L'Y$ is admissible for μ iff

$$L'Y \text{ is admissible for } \mu \text{ among linear estimators} \tag{1.1}$$

and

$$\text{at most two eigenvalues of } L \text{ are equal to one} \tag{1.2}$$

and that (ii) for (1.1) to hold it is necessary and sufficient that

$$\begin{aligned} &L \text{ be symmetric},\\ &L L \leq L. \end{aligned} \tag{1.3}$$

It is worth while to notice that the assumption of normality of Y and the assumption that σ^2 is known are not needed to show the equivalence of (1.1) and (1.3). Only the existence of the first two moments is required.

In a well known paper Rao (1976) has extended the latter result (ii) to any arbitrary matrix X, any p.d. matrix V and any parametric function of the form $C'\mu$. Under the assumption that the first two moments of Y exist he has shown that $L'Y$ is admissible for $C'\mu$ among the class of all linear estimators iff

$$\begin{aligned} &L'VC \text{ is symmetric},\\ &L'VL \leq L'VC,\\ &\mathcal{R}(VL) \subset \mathcal{R}(X). \end{aligned} \tag{1.4}$$

For the considerations to follow it should be noted that if we drop the assumption that σ^2 is known the above mentioned result of Stein is not more true (see Section 2). It remains true, as shown by James and Stein, when it is assumed in addition that there is also observed a random variable distributed as $\hat{\sigma}^2\chi_k^2$, $k \geq 1$, which is independent upon Y.

The purpose of this paper is to extend the former result (i) to more general linear models than the one considered by Cohen. More precisely, the purpose of this paper is to establish a condition (see Theorem 2.2) which combined with Rao's condition (1.4) would form necessary and sufficient conditions for $L'Y$ to be admissible for $C'\mu$. In view of the James and Stein's paper (1961) it is natural to focus attention on the following class of linear models: Y is normally distributed with expectation $X\beta$ and covariance $\sigma^2 V$ with V being p.d. and with no condition on X when σ^2 is a known positive constant (model A, say) and with the condition $r(X) < n$ when σ^2 is not known (model B).

2. Preliminaries

For the above specified models A and B with $V=I_n$ the theorems of Stein and of James and Stein may be modified as follows.

Theorem 2.1: Within model (A) as well as within model (B) with $V=I_n$ an estimator $C'Y$ is admissible for $C'\mu$ iff $C = XX^+C$ and $r(C) \leq 2$.

Proof: We shall establish this result first under model (A). If $C \neq XX^+C$, then $C'Y$ is inadmissible for $C'\mu$ because $C'XX^+Y$ is better.

Put $r=r(C)$. There exists a nonsingular matrix W such that

$$W'C'CW = \begin{bmatrix} I_r & O \\ O & O \end{bmatrix}.$$

Clearly, $C'Y$ is admissible for $C'\mu$ iff $W'C'Y$ is admissible for $W'C'\mu$ by the Shinozaki's lemma (1975).

We have

$$R_{CW}(W'C'Y;\ \mu,\sigma^2 I_n) = R_{I_r}(Z;[I_r,O]W'C'\mu,\ \sigma^2 I_r),$$

where $Z = [I_r,O]W'C'Y$. Since $\mathscr{R}([I_r,O]W'C'X) = \mathscr{R}([I_r,O]W'C') = \mathbb{R}^r$ the assertion follows from the above mentioned Stein's result.

Under model (B) the proof follows from the above mentioned result due to James and Stein by observing that $Q = Y'(I-XX^+)Y$ is distributed as $\sigma^2\chi^2_{n-r(X)}$, since by assumption $r(X) < n$, and that Q and $C'Y$ are independent provided $C=XX^+C$.

To characterize the admissible linear estimators for $C'\mu$ we will need a characterization of admissible linear estimators for μ within both models (A) and (B).

Theorem 2.2: Within model (A) as well as within model (B) with $V=I_n$ an estimator $L'Y$ is admissible for μ iff conditions (1.1) and (1.2) hold.

Theorem 2.2 gives a slight exstension of Cohen's theorem which is concerned with the case when $X=I_n$, only. The proof of the theorem is similar to that of Cohen's theorem except that it is based on Theorem 2.1. We omit the details.

It is interesting to note that if Y has an expectation which ranges over $\mathbb{R}^n$ and a covariance matrix of the form $\sigma^2 I_n$, where $\sigma^2 \geq 0$, then Y is admissible for μ. In fact, suppose that there exists an estimator $t = t(Y)$ such that

$$R_{I_n}(t;\mu,\sigma^2 I_n) - R_{I_n}(Y;\mu,\sigma^2 I_n) \geq 0 \tag{2.1}$$

with strict inequality for at least one $\sigma^2 \geq 0$ and $\mu \in \mathbb{R}^n$. For $\sigma^2=0$ the left hand side of (2.1) is equal to $(t(\mu)-\mu)'(t(\mu)-\mu)$ so that $t(x)=x$ for every $x \in \mathbb{R}^n$, which shows the assertion.

One can also show easily using the technique developed by Cohen (1966) that if we assume in addition that Y is normally distributed, then each linear estimator of μ which is admissible among the class of linear estimators is admissible.

3. Main results

In this section we restrict attention to models (A) and (B), only, and we establish a full characterizations of the admissible linear estimators for $C'\mu$ within these models.

Lemma 3.1: An estimator $L'Y$ is admissible for $C'\mu$ iff there exists a matrix H such that $L=HC$ and such that $H'Y$ is admissible for μ.

Proof: Suppose that $L=HC$ and that $H'Y$ is admissible for μ. From the Shinozaki's lemma if follows straightforwardly that $L'Y$ is admissible for $C'\mu$.

Now suppose that $L'Y$ is admissible for $C'\mu$. Without loss of generality we may suppose that $V=I_n$. Really, let $Z=V^{-1/2}Y$. Then $L_*'Z$ is admissible for $C_*'V^{-1/2}X$ within the model with $\mathrm{cov}Z=\sigma^2 I_n$, where $L_*=V^{1/2}L$ and $C_*=V^{1/2}C$.

From the above mentioned Rao's theorem it follows that

$$\begin{aligned} &\text{(i)} \quad L'C = C'L,\\ &\text{(ii)} \quad L'L \leq L'C,\\ &\text{(iii)} \quad \mathscr{R}(L) \subset \mathscr{R}(X). \end{aligned} \tag{3.1}$$

Now let $H=L(C'L)^{+}L'$. As noted in Klonecki and Zontek (1983), the estimator $H'Y$ is admissible for μ among linear estimators. From $\mathscr{R}(L') = \mathscr{R}(L'L) \subset \mathscr{R}(L'C) \subset \mathscr{R}(L')$ by (3.1) (iii) and from (3.1)(ii) it follows that $C'L(C'L)^{+}$ is the orthogonal projection on $\mathscr{R}(L')$. Hence $L=HC$.

To prove that $H'Y$ is admissible for μ it is sufficient, in view of Theorem 2.2, to show that H has at most two eigenvalues equal to one.

Suppose, to the contrary, that H has three or more eigenvalues equal to one. Let P be the orthogonal projection on the subspace spanned by the eigenvectors corresponding to the eigenvalue one .Then $q=r(P) \geq 3$ and $\mathscr{R}(P) \subset \mathscr{R}(X)$ by (3.1) (i). Moreover

$$r(C'P) = q. \tag{3.2}$$

To prove this it is sufficient to show that $\mathscr{N}(CP) = \mathscr{N}(P)$. If $Px \neq 0$, then $Px \notin \mathscr{N}(L')$ and consequently $C'Px = C'L(C'L)^{+}L'Px \neq 0$. Thus (3.2) holds.

The risk of $L'Y$ may be written as

$$R_C(L'Y;\mu,\sigma^2 I_n) = R_{PC}(C'PY;\ \mu,\sigma^2 I_n)+R_{(I-P)C}(L'(I-P)Y;\ \mu,\sigma^2 I_n).$$

Now an estimator $C'PY$ is inadmissible for $C'P\mu$ by Theorem 2.1 which implies that $L'Y$ is inadmissible for $C'\mu$. This contradiction terminates the proof.

Remark: Zontek (1983) has defined a class of general linear models for which Lemma 3.1 is valid provided we restrict attention to linear estimators only. This class includes the Gauss-Markov model too.

Theorem 3.2: An estimator $L'Y$ is admissible for $C'\mu$ iff

(i) $L'Y$ is admissible for $C'\mu$ among linear estimators,
(ii) at most two eigenvalues of $VL(C'VL)^+L'$ are equal to one.

Proof: Proving Lemma 3.1 we established that $L_*(C_*'L_*)^+L_*'Z$ is admissible for $V^{-1/2}\mu$, where $Z=V^{-1/2}Y$, $L_*=V^{1/2}L$ and $C_*=V^{1/2}C$. Hence by the Shinozaki's lemma $VL(C'VL)^+L'Y$ is admissible for μ. The proof terminates the observation that the sets of eigenvalues of $L_*(C_*'L_*)^+L_*'$ and of $VL(C'VL)^+L'$ coincide.

The next two corollaries characterize admissible linear estimators for $C'\beta$ in the case when X is of full rank.

Corollary 3.3: The following three conditions are equivalent:

(i) $L'Y$ is admissible for $C'\beta$,
(ii) there exists a matrix H_1 such that $L=H_1C$ and such that $H_1'Y$ is admissible for β,
(iii) there exists a matrix H_2 such that $L=H_2X(X'X)^{-1}C$ and such that $H_2'Y$ is admissible for $X\beta$.

Proof. The equivalence of (i) and (iii) follows from Lemma 3.1. The implications (iii) $\Rightarrow$ (ii) and (ii) $\Rightarrow$ (i) follow from the Shinozaki's lemma.

Corollary 3.4: An estimator $L'Y$ is admissible for $C'\beta$ iff

(i) $L'Y$ is admissible for $C'\beta$ among linear estimators,
(ii) at most two eigenvalues of $VL(C'(X'X)^{-1}X'VL)^+L$ are equal to one.

Acknowledgement. I am greatly indebted to Professor W.Klonecki for his suggestions while formulating the final version of this paper.

References

[1] A.Cohen (1966). All admissible linear estimates of the mean vector, Ann.Math.Statist. 37 485-463.

[2] W.James and C.Stein (1961). Estimation with quadratic loss. Proc.Fourth Berkeley Symp.Math.Statist. Prob. 1 311-319.

[3] W,Klonecki and S.Zontek (1983). On the structure of admissible linear estimators. Preprint No 270, Institute of Math.,Polish Academy of Sciences.

[4] C.R.Rao (1976). Estimation of parameters in a linear model. Ann.Statist. 4 1023-1037.

[5] N.Shinozaki (1975). A study of generalized inverse of matrix and estimation with quadratic loss. Ph.D. thesis, Keino University, Japan.

[6] C.Stein (1956). Inadmissibility of the usual estimator for the mean of a multivariate distribution. Proc. Third Berkeley Symp. Math.Statist.Prob. 1 197-206.

[7] S.Zontek (1983). On characterization of linear admissible estimators, an extension of a result due to C.R.Rao. Preprint No 271, Institute of Math.,Polish Academy of Sciences.

ACKNOWLEDGEMENT

List of referees

1. J. K. Baksalary
2. T. Bednarski
3. W. Bernacki
4. B. Ceranka
5. L.C.A. Corsten
6. H. Drygas
7. U. Gajewska
8. S. Gnot
9. Z. Hanusz
10. J. Jurečkowá
11. S. Kageyama
12. R. Kala
13. C.G. Khatri
14. A. Kozek
15. M. Kozłowska
16. L.R. LaMotte
17. H. Läuter
18. A. Markiewicz
19. T. Mathew
20. A. Molińska
21. E. Neuwirth
22. E. Niedokos
23. W. Oktaba
24. R. Pincus
25. P. Pordzik
26. C. Stępniak
27. D.M. Titterington
28. D. von Rosen
29. E.R. Williams
30. R. Zmyślony
31. S. Zontek